W9-ADA-464

THEORY AND INTERPRETATION OF FLUORESCENCE AND PHOSPHORESCENCE

Theory and Interpretation of Fluorescence and Phosphorescence

RALPH S. BECKER

University of Houston

WILEY INTERSCIENCE

A Division of John Wiley & Sons

New York · London · Sydney · Toronto

10 9 8 7 6 5 4 3 2 1

Library of Congress Catalog Card Number: 79-76049
SBN 471 06126 3

Printed in the United States of America

To the future generation. . . .
Mark, Sherryl, Janet, and Scott

Preface

There has long been a need for a comprehensive consideration of both theoretical and experimental aspects of fluorescence and phosphorescence processes. It has been twenty years since the classic treatise of Pringsheim. Further, there have been particularly rapid expanding efforts regarding both the theoretical and experimental aspects of emission spectroscopy over this period. These, in turn, have had a substantial impact in the area of photochemistry. The central purpose of this book is to make available to those interested in molecular spectroscopy and photochemistry the current thinking in theory and the maximum in data. In addition, the purpose is to point out areas in which problems still exist and thereby hopefully to motivate additional research efforts.

I feel it is important to make several comments regarding certain aspects of the book. Every attempt has been made to include almost all subjects of interest to those needing knowledge of spectroscopy. Rigorous treatment to the best of my ability to interpret has been maintained. However, at certain times, approximations are used to give a more qualitative meaning and physical feeling for those not requiring a rigorous approach. The book is broadly organized into chapters titled by spectroscopic phenomenology and in some cases by the class of compounds. For the sake of completeness and easy reference, the discussions of fluorescence and phosphorescence for a class of compounds such as aromatic hydrocarbons, molecular complexes, and the like have been combined whenever possible. Many of my professional colleagues have sent preprints of material to be published. In this way the book could be current despite the time gap from finishing the manuscript to publishing. I want to express my appreciation for their help.

It is important to point out that the information packing density in some areas is high and quick perusal will result in the reader's missing key points. Also, particular effort has been made in the tables to provide large amounts of different types of datum on each compound as well as a footnote for each entry. My colleagues have reminded me of the frustration of finding widely varying data for a particular entry with no explanation. Also, every attempt has been made to include error limits when such information was originally available.

Different levels of molecular orbital theory and the area of absorption spectroscopy are deemed important because these topics provide much of the necessary language and foundation for further considerations. Also, it is inevitable that the concepts of the absorption and emission processes are interwoven. A group-theory section is included to clarify the language but also to show the power of group theory in its use in electronic spectroscopy. One of the areas that has received considerable emphasis in the literature is the nature of the radiationless processes. Consequently, the theories relating to internal conversion and vibronic perturbations have been explored.

Because of the particular growth in interest in biomolecules, several sections deal with such molecules. In some instances, molecules not of direct interest in biology are included in order to draw comparison of properties. In addition, the physical chemistry of lowest excited states has been examined for both bio- and nonbiomolecules. The significance of dimer and polymers in relation to exciton theory is considered. Finally, aspects relative to energy transfer, and photochemistry are presented.

Several sections, particularly those included in Chapter 9, represent considerations of some complexity. I particularly wish to thank Dr. Martin Gouterman and Dr. Andrew Albrecht for their comments on these sections. I also wish to thank several graduate students, including Joe Steelhammer, Norris Tyer, Robert Bost, David Balke and Al Hugh, for their suggestions in improving some aspects of several chapters. Finally, my wife Jan deserves recognition for patience with a head and mind ofttimes resembling a ball of tangled twine.

Ralph S. Becker

Houston, Texas
December 20, 1968

Contents

List of Tables

THEORY AND INTERPRETATION OF FLUORESCENCE AND PHOSPHORESCENCE

Chapter 1 Introduction

Fluorescence and phosphorescence are processes in which radiation is emitted by molecules or atoms that have been excited by the absorption of radiation. The initial stage of the molecule before excitation is usually the singlet ground electronic state S_0. As a result of the emission from electronic excited states, the molecule returns to the ground electronic state, although frequently to a vibrationally excited form of the ground electronic state (Figure 1). Any state notation for the molecule is obtained by first taking the product of the appropriate quantum numbers over all the electrons in the orbitals as in atoms. In the case of a diatomic molecule the state potential energy curve is only a function of internuclear distance (Chapter 3-C). In a polyatomic molecule the situation is much more complicated but a similar picture is qualitatively adequate in the sense that it represents a cross section of the potential diagram.

State multiplicities (see Chapter 2 for more details) are important since the nature of the emission processes depends on them. If the states from which the emission originates and terminates have the same multiplicity, the emission is called *fluorescence*. This process most commonly occurs between the first excited singlet S_1 and ground state singlet (Figure 1). If the states from which the emission originates and terminates differ in spin ($\Delta S \geq 1$), however, then the emission is known as *phosphorescence* (Figure 1). The states of principal concern to us relative to phosphorescence will be the lowest excited triplet state, T_1, and ground state singlet. Because highly probable transitions can occur only when $\Delta S = 0$, phosphorescence emissions have relatively long lifetimes (10^{-3} to 10 sec) while fluorescences have relatively short lifetimes (10^{-7} to 10^{-10} sec). More details concerning the general nature of these emissions will be given in succeeding chapters, particularly 7 and 8. It should be re-emphasized that the rigorous differentiation between fluorescence and

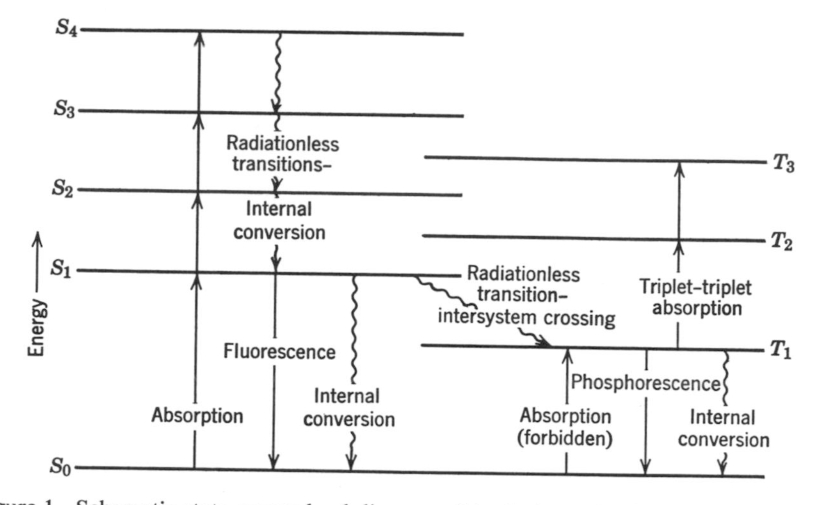

Figure 1 Schematic state energy level diagram: S is singlet and T is triplet. The S_0 state is the ground state and the subscript numbers identify individual states.

phosphorescence is the nature of the states between which the emission occurs (see Figure 1 and Chapter 7). For fluorescence the multiplicities of the states are the same, most commonly from the first excited singlet state S_1 to the ground singlet state. For phosphorescence the multiplicities of the states are different, most commonly from the lowest triplet state T_1 to the singlet ground state (Figure 1).

In addition to the radiative processes discussed above there are radiationless processes that are possible (Figure 1). These are the result of conversion of electronic energy into vibrational energy. These can occur between excited states of the same or different multiplicity (Figure 1). When a radiationless process occurs between the first excited singlet and lowest excited triplet, it is known as *intersystem crossing* (Figure 1). In all other cases the process is usually referred to as *internal conversion* (Figure 1).

Excitation to any excited singlet state above the first is normally followed by internal conversion such that fluorescence or intersystem crossing occurs only from the first excited singlet state. Thus, individually fluorescence and phosphorescence will normally be the same in all respects, such as wavelength, shape, lifetime, no matter which excited singlet state is initially occupied. The foregoing occur because the internal conversion process is some 10^3 (or greater) more likely than is emission directly from any excited singlet state above the first excited singlet. Internal conversion from the first excited singlet state can occur, thus competing with and thereby quenching fluorescence and intersystem crossing (therefore, also phosphorescence). However, internal conversion from the first excited singlet is less likely than among the higher excited singlets and the complete quenching of fluorescence and phosphorescence is extremely rare.

The following is a sequential outline of the processes of concern, see Figure 1, with approximate lifetimes where appropriate:

$$S_0 \longrightarrow S_n \qquad \text{(absorption)}$$

$$S_n \xrightarrow{10^{-11}\text{–}10^{-14}\ \text{sec}} S_1 \qquad \text{(internal conversion)}$$

$$S_1 \begin{cases} \xrightarrow{10^{-7}\text{–}10^{-9}\ \text{sec}} S_0 + h\nu & \text{(fluorescence)} \\ \xrightarrow{10^{-8}\ \text{sec}} T_n & \text{(intersystem crossing)} \\ \xrightarrow[10^{-5}\text{–}10^{-7}\ \text{sec}]{} S_0 & \text{(internal conversion)} \end{cases}$$

$$T_1 \begin{cases} \xrightarrow{10\text{–}10^{-3}\ \text{sec}} S_0 + h\nu & \text{(phosphorescence)} \\ \xrightarrow[10\text{–}10^{-3}\text{sec}]{} S_0 & \text{(internal conversion)} \end{cases}$$

More details of all these processes—such as the factors determining the lifetimes, the nature of the couplings involved, the triplet state to which intersystem crossing occurs, and many others—are given in succeeding chapters, particularly 7 to 16.

Chapter 2 Molecular Orbital Theories

It is of particular importance to point out that theoretical and experimental approaches are complementary. It is not possible to expect theoretical approaches to predict absolute quantities for those molecules of most general interest. Rather, theory is most useful because one can make important qualitative and relative conclusions. Theory in some forms, such as semi-empirical, permits certain absolute values to be calculated with a good accuracy. Therefore, theory can be varied such that its application can furnish a general broad understanding or more fine detail. Our discussion will be limited principally to the former purpose. The primary molecules to be considered will be those containing multiple bonds and their substituted derivatives.

Certain general considerations are important no matter what theoretical approach may be taken—that is, at least within the context of molecular orbital theory. No attempt will be made to discuss and compare this approach with that of valence bond theory. The wave function and energy for any electron of interest is associated with the term "orbital" in which only one electron resides; that is, we consider all wave functions as one-electron wave functions. The orbitals of interest are polycentric and are called *molecular orbitals.* Those of most concern to us are called pi (π) and are generally formed from combination of the appropriate p atomic orbitals. These particular restrictions are employed since for the molecules of interest to us, the spectra result from electronic transitions involving π electrons, $\pi^* \leftarrow \pi$.

Each orbital has quantum numbers associated with it. The electrons are added to orbitals in order of increasing energy and according to the Pauli principle. A state for a molecule is determined by the orbital electron configuration. An electron in an atom or molecule behaves as if it generates

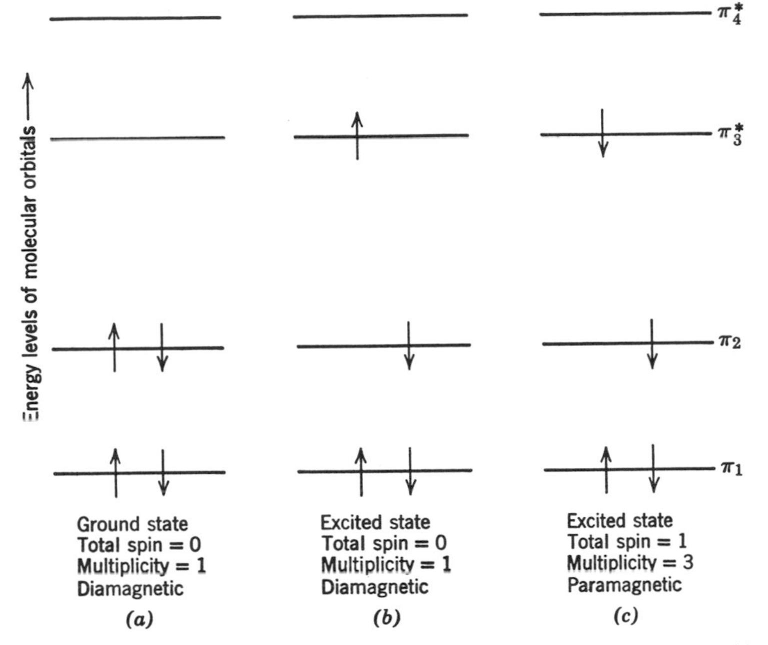

Figure 2 Electron orbitals for butadiene (diagrammatic). In the normal state (*a*) the two lowest orbitals are each occupied by two electrons of opposite spin, yielding a singlet state. In the excited state (*b*) one of the electrons has been raised to the next higher level without change in spin to yield an excited singlet state. In state (*c*) excitation is accompanied by a change in spin of one of the electrons to yield an excited triplet state. The energy of the state derived from the orbitals in (*c*) is less than that derived from the orbitals in (*b*).

two types of magnetic momenta and mechanical angular momenta. These arise from its motion about the nucleus, and from the "spin" of the electron. As an illustrative example consider a molecule such as butadiene, whose spectral features are approximately determined by four π electrons, one from each carbon atom. According to molecular orbital theory a molecule has as many π-molecular orbitals as it does π electrons. Two electrons of opposite spin can be assigned to each orbital. In the case of butadiene, the two lowest molecular orbitals contain four electrons and the ground state of the molecule can be shown to be a singlet; that is, the resultant spin S of the ground state is zero, and the multiplicity of the state, $2S + 1$, is one (Figure 2). An excited state is derived by removing one of the electrons from the uppermost filled orbital (bonding π) of the ground state to a vacant orbital (antibonding π^*) of higher energy, and if the spin of the excited electron is conserved in this process, the total spin of the excited state is zero, and the multiplicity is again singlet (Figure 2). If, however, the spin of the excited electron is no longer opposed to that of the odd electron remaining in the second orbital, the total spin is unity. The multiplicity is three and the excited state is a triplet. The magnetic interaction between the spin and orbital motion of an electron in an atom is very small for light atoms, and if several electrons are present, the individual spins couple to form a resultant spin, S. Further, the individual orbital momenta couple to form a resultant orbital angular momentum, L, and these two vectors couple to form a resultant total angular momentum, J. Only a discrete number of relative orientations of the spin and orbital momenta is allowed; for example, if $L = 1$ and $S = 1$, three orientations are possible, giving rise to three energy levels. The states corresponding to these values of L and S are therefore triplets, the individual members of which are generally only slightly separated from each other. Thus the helium atom and the atoms of the second group of elements in the periodic table have, in addition to the singlet levels corresponding to opposed spins of the two optical electrons, sets of triplet levels, the separation of which is very small in helium and beryllium, and increases rapidly with increasing atomic number. The triplet levels associated with a given orbital angular momentum are lower in energy than the corresponding singlet levels. Moreover, when the interaction between the orbital motion and the spin is small, there will be little perturbation in an electronic transition which could cause a change in spin. In the lighter members of the group of atoms with two optical electrons, therefore, there will be a very low probability of singlet$\leftrightarrow$ triplet transitions. In atoms of higher atomic number the interaction between the spin and orbital momenta becomes relatively large, so that during an electronic transition the perturbation can change the spin. Thus spectral lines involving a change in multiplicity become relatively intense in heavy atoms. For example, the well-known ultraviolet mercury line of wavelength 2537 Å is associated in absorption with the transition from the $^1S_0$ ground level to the middle level $^3P_1$ of the first excited ^3P state.

The case is similar in simple molecules and in molecules with conjugated chains. For every excited singlet state there is a corresponding triplet state of lower energy. If all the atoms in the molecule are light—for example hydrogen, carbon, nitrogen, and oxgen—transitions between the ground singlet and the excited triplet states will be very improbable and the corresponding absorption will be extremely weak. On the other hand if the molecule contains heavy atoms such as bromine, iodine, or heavy metals, the singlet $\leftrightarrow$ triplet transitions can be expected to become stronger.

The ground state of most stable molecules containing an even number of electrons is diamagnetic because of the arrangement of the electrons into pairs with opposed spin magnetic moments. Thus the ground state for most molecules is singlet (e.g. O_2 is an exception). A molecule excited to the triplet state is paramagnetic by virtue of the total spin magnetic moment of the two unpaired electrons.

If the potential energy of the system is time independent, the sum of the potential V and kinetic energy T is identical with what has been known in classical mechanics as the "Hamiltonian." Thus the conservation of energy equation can be written as

$$H = E \tag{1}$$

where

$$H = \tfrac{1}{2}m(\rho_x^2 + \rho_y^2 + \rho_z^2) + V(x, y, z) \tag{2}$$

and ρ is the symbol for momentum. The quantum mechanical expression results from substitution $(h/2\pi i)/(d/dn)$ wherever ρ_n appears. Thus

$$H = -\frac{h^2}{8\pi^2 m}\left(\frac{d^2}{dx^2}\frac{+d^2}{dy^2}\frac{+d^2}{dz^2}\right) + V(x, y, z) \tag{3}$$

However, d^2/dx^2, etc., are operators† and as such must operate on something. This something is taken to be the wave function appropriate to the molecular orbital of concern. Therefore, the elegantly simple (though deceptively so) expression

$$H\psi = E\psi \tag{4}$$

arises. This expression is the essential basis for all quantum mechanical calculations.

The electron energies can be considered to consist of two main parts—that arising from no interaction with the other electrons and that arising from interactions with the other electrons in the molecule. Consequently, the Hamiltonian operator has the form

$$H = \sum_i^p H_i + \sum_{m,n} \frac{e^2}{r_{m,n}} \tag{5}$$

where H_i is the one-electron part and $e^2/r_{m,n}$ is the two-electron part in which $r_{m,n}$ is the distance between the m and nth electrons. In some approximations the second or electron repulsion part is neglected. Although this is obviously incorrect, it is found that for many molecules certain qualitatively and relatively meaningful data can result. More concerning these points will be given later.

Although certain other considerations are usually common to the various theoretical approaches, it would be more meaningful to discuss them within the context of one or more of the approximations.

A. LINEAR COMBINATION OF ATOMIC ORBITALS—MOLECULAR ORBITALS (LCAO-MO)

In this case the molecular orbital wave function is a one-electron wave function and thus the Hamiltonian is also for a single election. The energy is

† Operators can be considered to be disembodied instructions: a definition given to me many years ago by a good friend, Ian Ross.

therefore determined only by the coordinates of the one electron. The molecular orbital function ψ_i is expressed as a linear combination of atomic orbital functions ϕ_n. It is not necessary to give an explicit form for the atomic orbitals,

$$\psi_i = C_{i1}\phi_1 + C_{i2}\phi_2 + C_{i3}\phi_3 + \cdots = \sum_n c_{in}\phi_n \tag{6}$$

where the i index represents a particular molecular orbital and the second index, n, represents the atomic orbital belonging to that particular atom of the molecule. From the above function, we obtain

$$H\psi_i = \sum_n c_{in}H\phi_n = \varepsilon_i \sum_n c_{in}\phi_n \tag{7}$$

Using the variational procedure which permits calculation of the best possible molecular orbital energy, ε_i,

$$\varepsilon_i = \frac{\int \sum_n c_{in}\phi_n H \sum_n c_{in}\phi_n \, d\tau}{\int \left(\sum_n c_{in}\phi_n\right)^2 d\tau} \tag{8}$$

The energy ε_i is minimized with respect to the coefficients c_{in}. Expansion of (8) results in several integrals of the form,

$$\begin{aligned} \alpha_{jj} &= \int \phi_j H \phi_j \, d\tau \\ \beta_{jk} &= \int \phi_j H \phi_k \, d\tau \\ S_{jk} &= \int \phi_j \phi_k \, d\tau \\ S_{jj} &= \int \phi_j \phi_j \, d\tau = 1 \end{aligned} \tag{9}$$

The integrals β_{jk}, over the atomic wave functions (j and k) and the Hamiltonian operator are known as *resonance integrals*. These are, individually, sums of contributions the most important of which will be those where ϕ_j and ϕ_k are both not small. The two functions will be simultaneously the largest in the region of space near the two cores j and k. The integrals α_{jj} are referred to as *coulombic integrals*. Each has the physical significance of being the energy of an electron whose motion is described by ϕ_j and which is subject to a potential V. The function ϕ_j has a notable value only in the vicinity of the jth core and in this part of the molecule the potential V becomes principally that of the jth core. The energy then may be regarded to be that required to remove an electron from the jth atom. The S_{jk} integrals are called the *overlap integrals*. Finally, the S_{jj} integrals are normally equal to one and when they are, the atomic orbitals are normalized. We will always consider this latter condition to be true. Further, expansion of equation (8) and minimization with respect to the coefficients result in a set of secular equations

$$c_{ij}(\alpha_{jj} - \varepsilon_i) + \sum_{\substack{k \\ j \neq k}} (\beta_{jk} - \varepsilon_i S_{jk})c_{ik} = 0 \tag{10}$$

The number of equations is equal to the number of π electrons in the molecule. This set of equations will have nontrivial solutions (where all $c_j \neq 0$) if the determinant is equal to zero.

$$\left| (\alpha - \varepsilon) + \sum_{\substack{k \\ j \neq k}} (\beta_{jk} - \varepsilon S_{jk}) \right| = 0 \tag{11}$$

The eigenvalues (ε_i) are the molecular orbital energies and the eigenvectors (ψ_i) the molecular orbitals.

1. Hückel Approximation

In this approximation it is assumed that in addition to the approximations just given,

All α_{jj} are equal
All β_{jk} are equal ($j \neq k$)
All $\beta_{jk} = 0$ when j and k are not neighbors
All $S_{jk} = 0$

For the simplest case of ethylene a 2×2 determinant results

$$\begin{vmatrix} \alpha_{11} - \varepsilon & \beta_{12} \\ \beta_{21} & \alpha_{22} - \varepsilon \end{vmatrix} = 0 \tag{12}$$

and

$$\begin{aligned} \varepsilon_1 &= \alpha + 1\beta \qquad \psi_1 = \phi_1 + \phi_2 \\ \varepsilon_2 &= \alpha - 1\beta \qquad \psi_2 = \phi_1 - \phi_2 \end{aligned} \tag{13}$$

The lower energy orbital(s) is associated with a plus (+) sign. The two electrons are put in ε_1 and the lowest allowed electronic excitation results from $\psi_1^2 \rightarrow \psi_1\psi_2$ with $\Delta\varepsilon = 2\beta$.

The general solution for a linear polyene of A carbon atoms each with a p electron is

$$\varepsilon_n^\pi = \alpha + 2\beta \cos \frac{n\pi}{A + 1} \tag{14}$$

where $n = 1, 2, 3 \ldots A$. It can be seen that as the number of π electrons increases, the coefficient of β (see 14) decreases. Therefore $\Delta\varepsilon$ between the

Table 1 Wavelength and Ionization Potential of First Transition of Some Hydrocarbons

Compound	Band Maximum (Å)	Ionization Potential
Ethylene	1614 (onset ~1800)	10.52
Butadiene	2170	9.07
Hexatriene	2510	8.47 (calc)
Octatetraene	3040	
1-chloroethylene	~2150 (onset)	10.0
1-methyl ethylene	1730	9.73
Benzene	2600	9.24
Naphthalene	3120	8.26, 8.12
Anthracene	3780	7.55, 7.66
Naphthacene	4700	7.15
Phenanthrene	3530	8.03, 8.06

highest filled and lowest empty orbitals decreases. Consequently the transition energy decreases and the lowest electronic absorption band moves toward the visible spectral region (see Table 1). Substituted polyenes can be treated within the same framework.

A similar treatment can be applied to aromatic hydrocarbons as well as their substituted derivatives. For linear polyene systems and cyclic aromatic hydrocarbons it is possible to classify the members as alternant, odd and even, and nonalternant. This classification is important in terms of certain possible general solutions for the energy of the molecular orbitals as well as their relative energy location.

Alternant hydrocarbons are those that are planar and contain no odd membered rings. Further, the carbon atoms are divided into two groups, starred (*) and unstarred and starred atoms only have unstarred neighbors and vice versa.

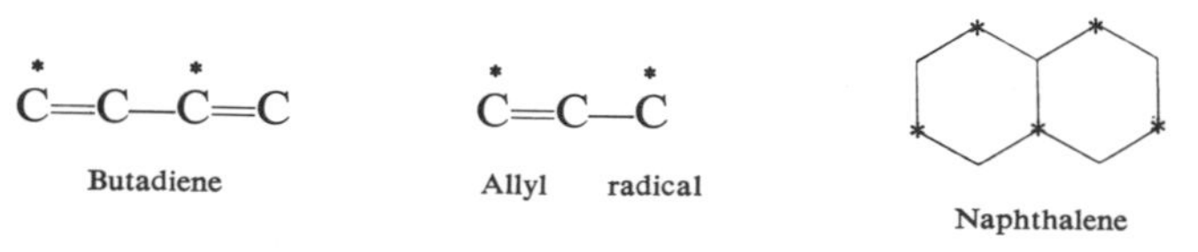

Butadiene Allyl radical Naphthalene

In even alternant hydrocarbons the number of starred and unstarred atoms is equal (e.g., butadiene and naphthalene), whereas in odd alternants the number of starred atoms is commonly one greater than unstarred. In non-alternant hydrocarbons either starred or unstarred atoms will be adjacent (cyclopentadiene, for example).

A general solution for the energy of even alternant polyenes was given in

(14). The general solution for an even alternant cyclic hydrocarbon is

$$\varepsilon_n^\pi = \alpha + 2\beta \cos \frac{2n\pi}{A} \tag{15}$$

where A is the number of atoms and $n = 0, \pm 1, \pm 2, \pm 3, \ldots, A - 1$. Of further important consequence is the relative distribution of the resulting energy levels. In even alternant hydrocarbons the roots of the secular determinant occur in pairs with opposite sign, $\pm x$; thus, the energy levels occur likewise, $\varepsilon_n^\pi = \alpha \pm x\beta$:

$$\varepsilon_n^\pi = \pm\varepsilon_1, \pm\varepsilon_2, \pm\varepsilon_3 \cdots \tag{16}$$

In the case of odd alternant hydrocarbons, where the number of starred atoms is one greater than unstarred (as allyl radical), the roots and thus the energy levels occur in $\pm$ pairs with the additional root equal to zero. The latter signifies that the energy is equal to α and therefore is a nonbonding orbital.

The above generalizations immediately permit qualitative understanding of several important spectroscopic parameters. Figure 3 will aid in the discussion. The pairing properties of the molecular orbital energy levels and the trend in wavelength of the first transition of the polyenes (Table 1) are clearly seen based on the decreasing energy spacing between the highest filled and lowest empty π orbitals. Further, similar pairing properties and trends can be seen for the aromatic hydrocarbons. The case of allyl radical points up the existence of the pairing property and the nonbonding orbital (Figure 3). Bonding orbitals lie below and antibonding orbitals above α. Certain precautions are necessary in the case of certain aromatic hydrocarbons because of added features that modify the levels responsible for the experimental spectra versus those determined solely from Hückel calculations. This will be discussed in more detail later (Section 3). Nevertheless the qualitative considerations are valid.

In addition there are other spectroscopic parameters whose qualitative trend can be predicted based on the preceding discussion. The first ionization potential (IP) is the energy required to remove an electron from the highest filled orbital to infinity. In general,

$$\varepsilon_i^\pi = -\text{IP}_i \tag{17}$$

and this is known as *Koopmans' theorem.* The first electron affinity (EA) would be the energy released when an electron is brought from infinity and added to the lowest empty orbital. Both the general solutions and Figure 3 make it clear why there is a progressive decrease in ionization potential and increase in electron affinity with an increase in the number of atoms in

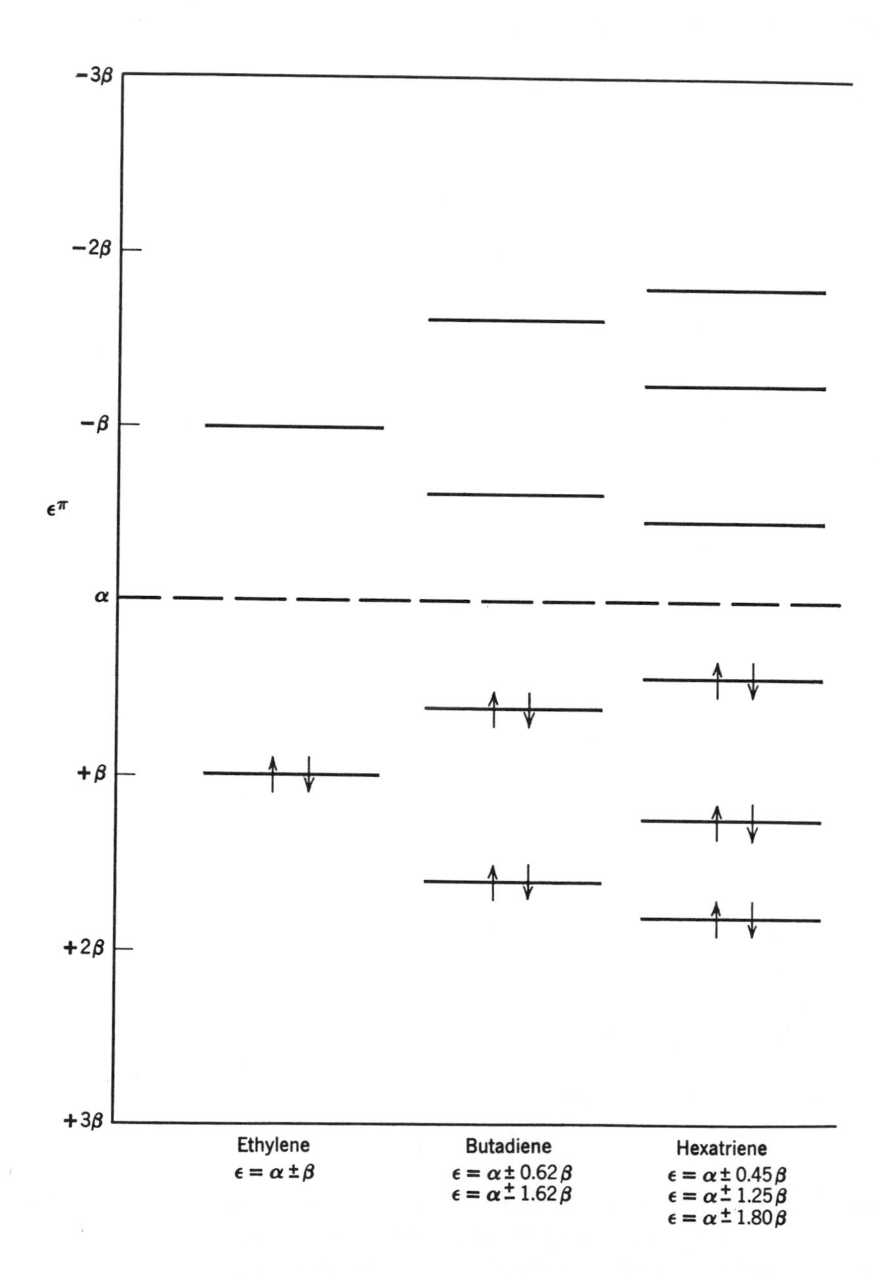

Figure 3 π-energy levels and electron configurations for some hydrocarbons.

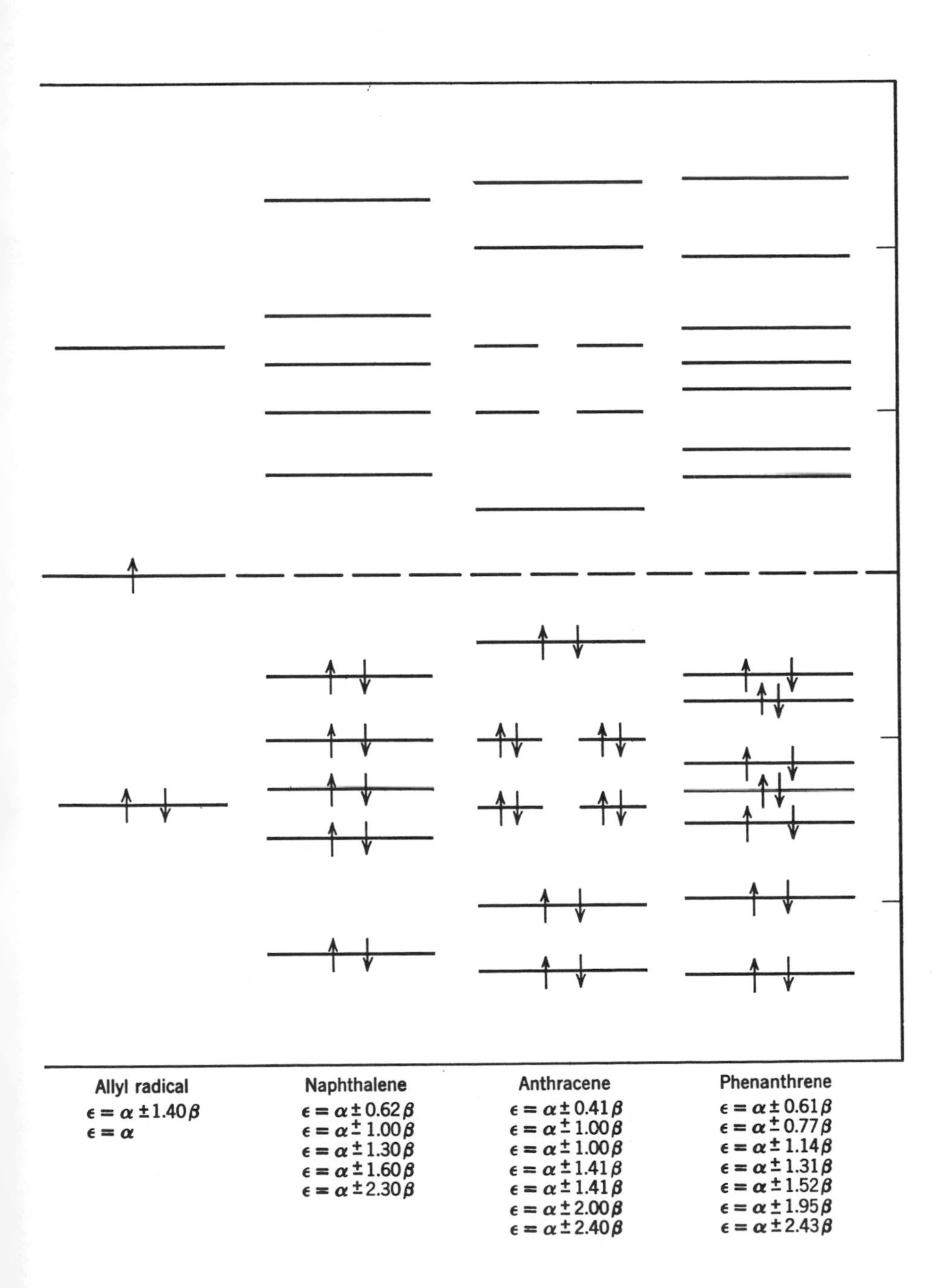
Allyl radical
$\epsilon = \alpha \pm 1.40\beta$
$\epsilon = \alpha$
Naphthalene
$\epsilon = \alpha \pm 0.62\beta$
$\epsilon = \alpha \pm 1.00\beta$
$\epsilon = \alpha \pm 1.30\beta$
$\epsilon = \alpha \pm 1.60\beta$
$\epsilon = \alpha \pm 2.30\beta$
Anthracene
$\epsilon = \alpha \pm 0.41\beta$
$\epsilon = \alpha \pm 1.00\beta$
$\epsilon = \alpha \pm 1.00\beta$
$\epsilon = \alpha \pm 1.41\beta$
$\epsilon = \alpha \pm 1.41\beta$
$\epsilon = \alpha \pm 2.00\beta$
$\epsilon = \alpha \pm 2.40\beta$
Phenanthrene
$\epsilon = \alpha \pm 0.61\beta$
$\epsilon = \alpha \pm 0.77\beta$
$\epsilon = \alpha \pm 1.14\beta$
$\epsilon = \alpha \pm 1.31\beta$
$\epsilon = \alpha \pm 1.52\beta$
$\epsilon = \alpha \pm 1.95\beta$
$\epsilon = \alpha \pm 2.43\beta$

alternant polyenes. The same is true for the linearly condensed aromatic hydrocarbons, also see Table 1.

It may well be worthwhile to point out that any "bends" resulting from angular benzo condensation in aromatic hydrocarbons increase the energy difference between the orbitals responsible for the lowest energy transition; for example, the first transition in anthracene is at longer wavelength than that of phenanthrene. The general theoretical approach outlined previously gives a completely satisfactory qualitative account of this phenomenon (see Figure 3). Similar considerations apply to the ionization potential and electron affinity (see Figure 3); that is, anthracene has a lower ionization potential than does phenanthrene, also see Table 1.

Further, it is possible to calculate bond orders, charge densities, free valence, etc., by using this theory. We will not pursue this since it is not of direct interest to us.

2. Free-Electron Approximation (FEMO)

The simplest approach to this theory is that of considering a bound electron in a one-dimensional box or square well. Further, the potential within the well is constant and for convenience we will consider it zero. The potential at the edges of the well is infinite. Inside the well the wave equation is

$$\left(\frac{-h^2}{8\pi^2 m}\right)\left(\frac{d^2\psi}{dx^2}\right) = E\psi \tag{18}$$

which has the solution (the eigenfunction),

$$\psi_n(x) = B \sin \frac{n\pi x}{L} \tag{19}$$

where B is a constant, x is the distance along the bottom of the well, L is length of the well, and n is a quantum number with values $1, 2, 3, \ldots$. Outside the well ψ is equal to zero and, furthermore, in order for ψ to be continuous ψ must go to 0 at $x = 0$ and $x = L$. The eigenvalues associated have the values,

$$\varepsilon_n = \frac{n^2 h^2}{8mL^2} \tag{20}$$

The evolution of quantized states results naturally from the boundary conditions, which does not happen in classical mechanics.

Thus for linear polyenes there are a series of molecular orbitals determined by ψ_n with associated energies, ε_n (see Figure 4). The number of nodal points (other than at the boundaries) in ψ is equal to $n - 1$ and the energy increases as n^2 and decreases as L^2,

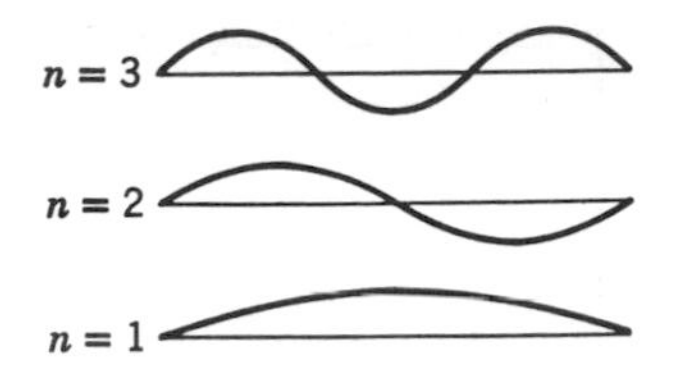

Of particular interest again is the energy of the lowest energy transition. If N represents the number of π electrons (then N is equal to the number of p atomic orbitals or atoms), the highest filled orbital $n_f = N/2$ and the lowest empty orbital $n_g = N/2 + 1$. The energy of the transition will be

$$\Delta\varepsilon\,(\text{cm}^{-1}) = \frac{h^2}{8mL^2}(N+1)$$

or (21)

$$\lambda(\text{Å}) = \frac{8mc}{h}\,\frac{L^2}{N+1}$$

where L is the length of the chain (dependent on N). For large N the theory predicts the wavelength of the first transition should be linearly related to $1/N$.

In the case of cyclic hydrocarbons of the even alternant catacondensed type a slightly altered result evolves. A catacondensed hydrocarbon is one in which no ring carbon is common to more than two rings; for example, pyrene is *not* catacondensed (it is a pericondensed hydrocarbon). Platt [1]

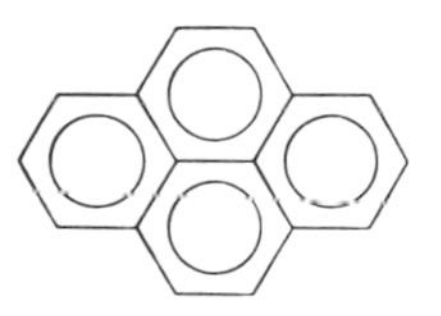

approximated the perimeter as a circle in which the electrons traveled under a constant potential. This results in a plane rotator problem with resulting molecular orbitals having

$$\varepsilon\,(\text{cm}^{-1}) = \frac{q^2h^2}{2mL^2} \tag{22}$$

where $q = 0, 1, 2, 3, \ldots$ is a quantum number, and L is the length of the perimeter or circle. All levels but $q = 0$ are twofold degenerate. The highest filled level is identified as f with progressively lower lying filled orbitals being e, d, c, etc. The lowest empty orbital is denoted as g with successively higher lying orbitals denoted as h, i, etc. (Figure 4).

This nomenclature system is also applicable to the linear polyenes. Figure 4 gives examples for a polyene and an aromatic hydrocarbon. Further, in the case of polyenes the state notations depend on Δn

$\Delta n = 1$	$g \leftarrow f$	${}^1B \leftarrow {}^1A$
$\Delta n = 2$	$g \leftarrow e$	${}^1C \leftarrow {}^1A$
	$h \leftarrow f$	${}^1C_2 \leftarrow {}^1A$
$\Delta n = 3$		${}^1D_{1,2,3} \leftarrow {}^1A_1$

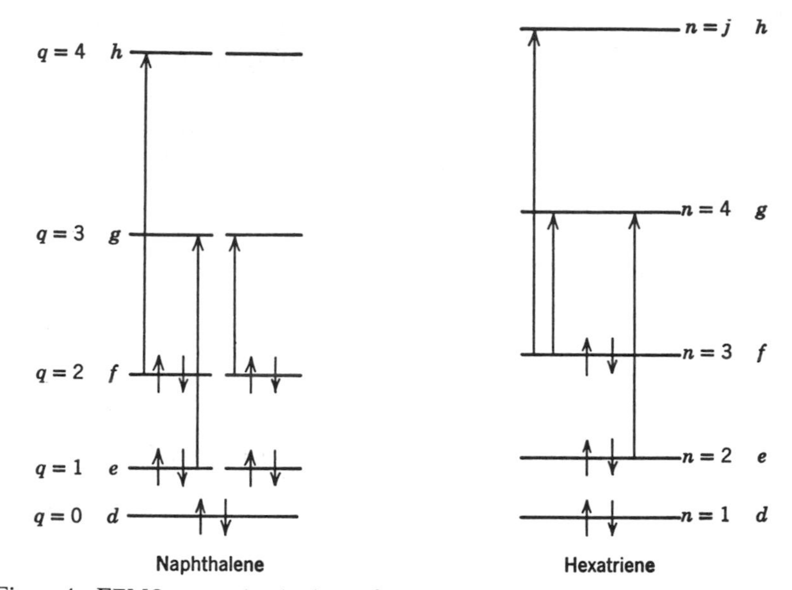

Figure 4 FEMO energy level scheme for a polyene and an aromatic hydrocarbon.

In the case of the catacondensed hydrocarbons there is a further problem relative to the orbital degeneracies. This degeneracy is removed for the real molecules because of a nonconstant and periodically varying potential field resulting from the crosslinking and existence of atomic centers in the perimeter or circle. The degeneracy is not removed only in those cases where the molecular symmetry is unusually high such as benzene, triphenylene, and coronene. The states are denoted as follows:

$$\begin{array}{lll} q = 1 & g \leftarrow f & L_a, L_b, B_a, B_b \\ q = 2 & h \leftarrow f & C_a, C_b, M_a, M_b \\ & g \leftarrow e & C_a, C_b, K_a, K_b \end{array}$$

Further, the L states refer to the lower-lying excited states and the subscripts a and b denote short and long axis polarization, respectively. Generally the B_b state is assigned on the basis of intensity; that is, it is the one to which the transition probability is greatest. Nonetheless, some confusion regarding assignments is still present. More concerning these considerations will be given later.

3. Effect of Electron Repulsion and Configuration Interaction

No attempt will be made to go into all of the details pertinent to the subject [1a]. Certain results will be used because of the particular significance they have for our considerations. The π-transition energy between the ground state and a singlet monoexcited configuration, $k \leftarrow j$, or the $k \leftarrow j$ singlet state is

$$\Delta E = \varepsilon_k - \varepsilon_j - J_{j,k} + 2K_{j,k}$$

where ε_k and ε_j are commonly the energies of the jth and kth Hartree-Fock self-consistent field molecular orbitals (although Hückel orbital energies could be used), $J_{j,k}$ and $K_{j,k}$ are the molecular coulomb and exchange integrals, respectively, between the jth and kth molecular orbitals. J and K represent the interaction between the two π electrons. The J integral is

$$J_{j,k} = \int \Psi_j(1)\Psi_k(2) \frac{e^2}{r_{12}} \Psi_j(1)\Psi_k(2)\, d\tau_1\, d\tau_2$$

and Ψ_j and Ψ_k are the LCAO-MO orbital functions (usually antisymmetrized). The K integral is

$$K_{j,k} - \int \Psi_j(1)\Psi_k(2) \frac{e^2}{r_{12}} \Psi_k(1)\Psi_j(2)\, d\tau_1\, d\tau_2$$

The excitation energy for the $k \leftarrow j$ triplet state is

$$\Delta E = \varepsilon_k - \varepsilon_j - J_{j,k}$$

which is of lower energy than the corresponding excitation energy for the $k \leftarrow j$ singlet state. Thus it is the additional consideration of electron repulsion interactions which splits the singlet and triplet states that are degenerate by a simple Hückel LCAO-MO approach. It is also obvious that depending upon the value of the $-J + 2K$ term, the simple orbital energy difference, $\varepsilon_k - \varepsilon_j$, will give absolute energy differences quite different from those found experimentally that are improved by the addition of the J and K terms. Further, because the $-J + 2K$ term can vary in value, the order of energy levels determined by only $\varepsilon_k - \varepsilon_j$ may undergo changes when the $-J + 2K$ term is considered.

If ${}^1\Psi_1$, ${}^1\Psi_2$, ${}^1\Psi_3$, ${}^1\Psi_4$, ${}^3\Psi_1$, ${}^3\Psi_2$, etc., are solutions to the Schroedinger equation where each of these is the solution for a particular configuration, then the general solution is

$$\Psi = C_1\,{}^1\Psi_1 + C_2\,{}^1\Psi_2 + C_3\,{}^1\Psi_3 + C_4\,{}^1\Psi_4 + C_5\,{}^3\Psi_1 + C_6\,{}^3\Psi_2 + \cdots$$

with normalization requiring that

$$\sum_i |C_i|^2 = 1$$

and where the Ψ_i are orthogonal. If the molecular system were to be in a given stationary state—that is, completely described by a particular configuration and therefore a Ψ_i—then the corresponding coefficient would be equal to one while all others would be zero. However, our functions are not the true eigenfunctions and therefore all of the other coefficients may not vanish. Certain facts limit what interactions can occur: (a) states of different multiplicity cannot interact because of spin orthogonality, (b) states belonging to different irreducible representations of the group to which the molecule belongs cannot interact (see Chapter 3-A for a discussion of group theory). This configurational interaction results in a splitting apart of the energy of the individual configurations with the resulting production of new energy levels usually referred to as "states." Benzene provides a very good example of this (see Chapter 3-A and 5-B). Also, the excited singlet state corresponding to the ^1L_a transition in condensed alternant aromatic hydrocarbons is over 90 percent the $-1 \leftarrow 1$ configuration—that is, the configuration resulting from excitation from the highest filled to the lowest empty orbital. In view of this fact, and the fact that the $-J + 2K$ term is quite constant for these same hydrocarbons, Hückel theory is quite successful in estimating the relative location of this transition among these hydrocarbons. Except for the $-1 \leftarrow 1$ state, however, other states are composed of several excited configurations and Hückel theory is of doubtful value for estimation of such excited state energies.

Chapter 3 Absorption and Emission Processes

A. TRANSITION MOMENT INTEGRAL AND GROUP THEORETICAL CONSIDERATIONS

A transition between two stationary electronic states ψ_i and ψ_j can be induced by radiation provided its frequency $\nu = (E_j - E_i)/h$. In addition the probability of a transition occurring depends on whether the incident radiation can change the electronic distribution of the ground state to that which is necessary for the excited state. The expression for the probability of the transition is such that the probability depends on the square of a transition dipole moment integral

$$P \propto \left| \int \psi_i M \psi_j \, d\tau \right|^2 \tag{23}$$

where M is the dipole moment operator (a vector) and $M = M_x + M_y + M_z = \sum_i er_i$ where e is the charge on the N electrons and r_i is the vector distance of the ith electron from an origin of a coordinate system fixed for the molecule. The transition moment (or matrix elements of the electric dipole moment) between two states ψ_i and ψ_j is often denoted as R

$$R_{ij} = \int \psi_i M \psi_j \, d\tau \qquad \text{or} \qquad \langle \psi_i | M | \psi_j \rangle \tag{24}$$

where $d\tau$ is over the whole configuration space of the $3N$ coordinates. If $R = 0$, then the transition is forbidden as an electric dipole transition although the transition could occur very weakly if permitted by magnetic dipole or quadrapole radiation. Calculations show that magnetic dipole probabilities are 10^{-5} and quadrapole probabilities 10^{-8} of electric dipole transition probabilities; therefore, these are relatively highly forbidden transitions.

For a one-electron excitation in a homonuclear diatomic molecule, $R = \langle \psi_1 M_x \psi_2 \rangle$ where the electric dipole can be thought of as arising from the electronic charge times the length of the molecule which gives the dipole moment at the end of an oscillation of charge. Thus it can be seen that the dipole moment is not a permanent one but a transient one. The transition depends on the interaction of M with E, the electric vector of the incident radiation. The conceptualization of the integration and qualitative evaluation of R ($\neq 0$, $= 0$) can be seen in the following way. First consider an isotropic atomic case where we wish to see if the transition $2p \leftarrow 1s$ is allowed or forbidden.

$$(R_{p \leftarrow s})_x = \int_{-\infty}^{+\infty} \psi_{1s} M_x \psi_{2px} \, d\tau$$

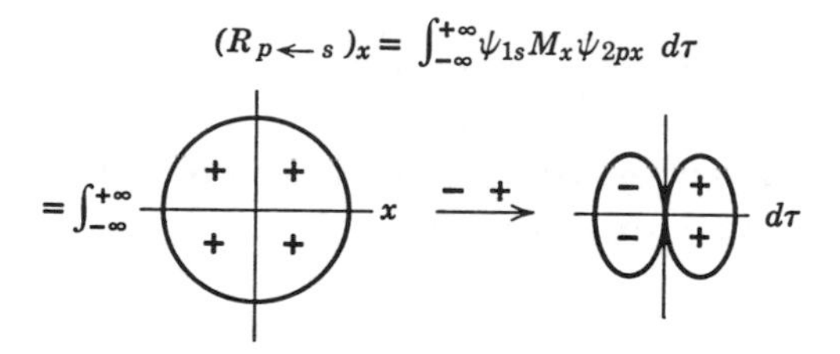

Although only two dimensions are shown it will not alter the conclusions. We now take the products in all quadrants as follows:

1. $+ - - = +$
2. $+ + + = +$
3. $+ + + = +$
4. $+ - - = +$

   Total $\overline{+}$

Thus R has some plus $(+)$ value and does not equal zero $(\neq 0)$ and therefore the transition is allowed. A similar procedure can be followed for $(R_{p \leftarrow s})_z$ with M_z. Note that a negative $(-)$ value would make no difference since $P \propto |R|^2$.

A more difficult situation arises for an anisotropic case such as ethylene (in a crystal where the carbon-carbon axis is parallel to the x direction).

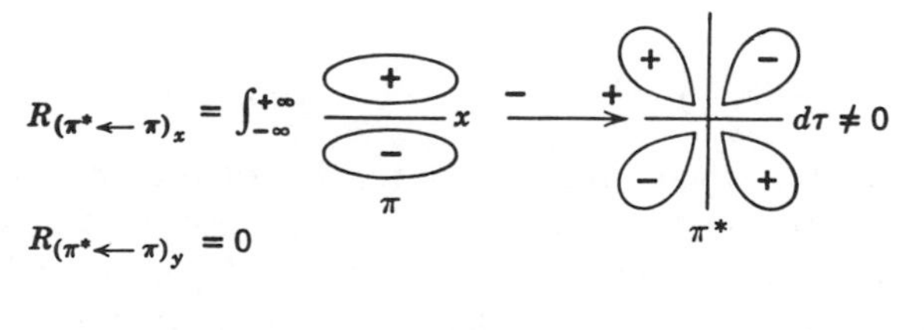

$R_{(\pi^* \leftarrow \pi)_y} = 0$

$R_{(\pi^* \leftarrow \pi)_z} = 0$

Thus only light polarized (see Section D) in the x direction will be effective in inducing the transition.

More sophisticated approaches can be used. These require the use of group theory. It must first be pointed out that this section presumes only to utilize group theory in a restricted way. It is beyond the scope of this book to develop the subject in detail or to the total breadth of its use. The principal areas of concern to us are state identification and evaluation of transition probability. The first can be accomplished because of the fact that the wave functions associated with the orbitals and states have symmetry properties determined by those of the molecule. The second can be solved because of the variance or invariance of the transition moment integrand to any appropriate symmetry operations and the meaning of this in terms of the zero or nonzero value of the integral; that is, we found if $R = 0$, then the transition was forbidden. We can now determine this property of R from group theoretical considerations. The benefit of this will become obvious since no extensive effort is required. This is in contrast to the situation that would develop if analytical evaluation of the integral were required.

The decision regarding the point group symmetry to which a molecule belongs depends on the nature and number of covering operations, such as rotation or reflection, that transform an object (e.g., a molecule) into a form indistinguishable from the original. The fundamental covering operations that develop the various point groups are the following:

1. Rotation about an axis through an origin.
2. Reflection in planes containing the origin.
3. Inversion that takes i to $-i$.
4. Improper rotation.
5. No operations that involve translational motion.

Operations of concern to us:

1. E: identity or the leave-it-alone operation.
2. C_n: an n-fold rotation and no other where the angle of rotation is $2\pi/n$ (about the principal or highest fold axis).
3. σ_h: reflection in a plane perpendicular to the principal axis.
4. σ_v: reflection in a plane containing the principal axis.
5. σ_d: reflection in a plane containing the principal axis and bisecting the angle between 2 twofold axes (c') perpendicular to the principal axis.
6. i: inversion through a center of symmetry.
7. S_n: rotation about an axis by $2\pi/n$ followed by reflection in a plane perpendicular to the axis of rotation.

The point groups of interest to us and the symmetry elements or operations involved (E is a symmetry element in all point groups) are the following:

1. C_n: n-fold axis of symmetry only. $C_1 = E$, $C_s = E + \sigma_h$, $C_i = i + E$.
2. C_{nv}: n-fold axis C_n plus σ_v plane.
3. C_{nh}: n-fold axis C_n plus σ_h plane.
4. D_n: n-fold axis C_n plus n-two-fold axes c' perpendicular to principal n-fold axis (C_n).
5. D_{nd}: n-fold axis C_n plus n-two-fold axes c' perpendicular to C_n and n-vertical planes σ_d bisecting angles between the c' axes.
6. D_{nh}: all operations of the D_n group plus a σ_h plane.

We have omitted the more special groups such as octahedral and tetrahedral.

It may be helpful to outline a procedure to determine the point-group assignment:

1. Find proper axes of rotation C_n.
2. Find the highest order axis of rotation, the principal axis.
3. Determine if you have n twofold axes (c') perpendicular to C_n.
4. If there are *no* n twofold axes perpendicular to C_n, then determine if there is a σ_h or σ_v plane. If there is a σ_h plane, then the group is C_{nh}. If there is a σ_v plane and no σ_h plane, then the group is C_{nv}.
5. If there are n twofold axes perpendicular to C_n, then determine if there are σ_h or σ_d planes. If there is an σ_h plane, then the group is D_{nh}. If there is a σ_d plane and no σ_h plane, then the group is D_{nd}.

Before proceeding, it will be worthwhile to utilize the procedure outlined above to actually assign several molecules to the proper point groups. All homonuclear diatomics belong to $D_{\infty h}$ and all heteronuclear diatomics belong to $C_{\infty v}$. Figure 5 shows some molecules with the appropriate point group to which they belong. Consider, for example, naphthalene:

1. There is a twofold axis, C_2, perpendicular to the plane (principal axis).
2. There are 2 twofold axes, c_2', in the plane perpendicular to the principal axis.
3. There are $2\sigma_v$ planes.
4. There is a σ_h plane that is the plane of the molecule.

These then generate a D_{2h} point group.

In group theory, letters such as A, B, C denote operators. Sets of operators are used in group theory and the members of the sets are called elements. The set of operations that sends a symmetrical figure or object into itself

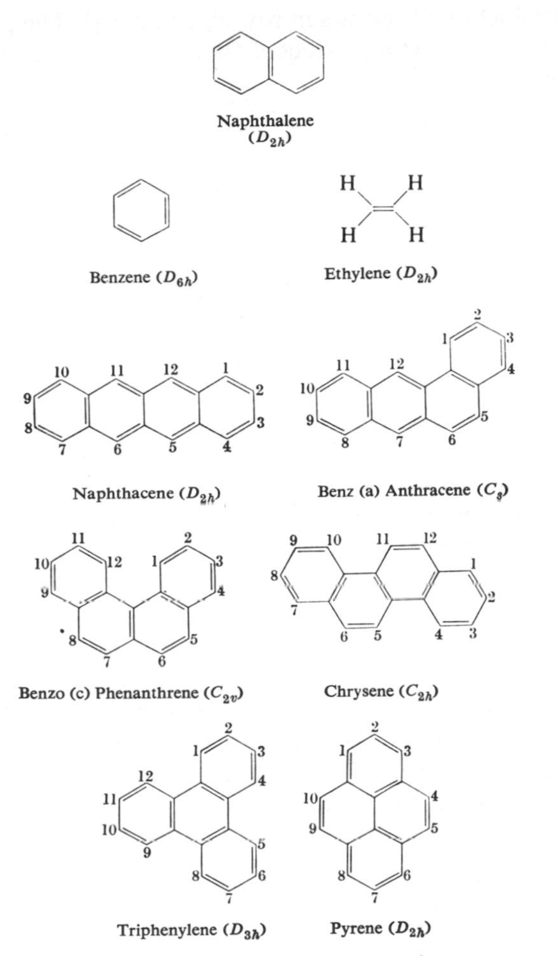

Figure 5 Some molecules and their point groups.

forms a group. Any set of elements forms a group if the following conditions are satisfied:

1. The product of any two elements in the set is another element in the set.
2. The set must contain an identity operator E such that $ER = RE = R$ where R is any element of the set.
3. The associative law of multiplication must hold; that is, $(RS)T = R(ST)$.
4. Every element must have a reciprocal or inverse such that if R is the inverse of S, then $SR^{-1} = R^{-1}S = E$, where the inverse is an element of the group.

Further, the product of PQ occurs in row P, column Q. The product of PQ may or may not be equal to the product of QP (if $PQ = QP$, then the elements P and Q commute).

Let us consider a group multiplication table of the C_{2v} group as follows:

	E	A	B	C
E	E	A	B	C
A	A	E	C	B
B	B	C	E	A
C	C	B	A	E

The set of elements form a group satisfying all the required conditions (see above). The elements E, A, B, C represent operations appropriate to the particular group such as rotation by 180° (A). Now a set of vectors can be assigned to the geometrical object and symmetry operations can be carried out. By doing this a set of transformation matrices are obtained that obey the same rules of combination as the elements themselves. Any set of elements, which may be numbers or matrices, form a *representation*, Γ_i, of a group provided the elements conform to the multiplication table of the group. From matrix algebra we can obtain the result that the traces of matrices combine in the same manner as the matrices themselves. The trace of the matrix is the sum of the elements on the principal diagonal. In group theory the trace is called the *character*. The tabulation of the symmetry elements and representations (given as sets of characters) for a group is called a character table. For example,

C_{2v}	E	$C_2(z)$	$\sigma_v(xz)$	$\sigma_v(yz)$	
A_1	+1	+1	+1	+1	T_z
A_2	+1	+1	−1	−1	
B_1	+1	−1	+1	−1	T_x
B_2	+1	−1	−1	+1	T_y

The different representations (A_1, A_2, B_1, and B_2) give the transformation properties of the object; that is, a set of characters for a representation typifies the covering operations. A *class* includes all operations of a given type, a total of four for C_{2v}—that is E, C_2, $\sigma_v(xz)$, $\sigma_v(yz)$. The order of the group h is equal to the total number of operations of the group (four for C_{2v}). If the object is a molecule, the representations could give transformation properties of molecular motions such as rotation or vibrations or of electronic wave functions, ψ_i. Since ψ_i is associated with a definite energy E_i, the energy states themselves may be classified by the symbols for a representation, Γ_i. The symmetry species for the representations are defined as follows:

1. A: symmetric with respect to rotation about the principal axis.
   B: antisymmetric with respect to rotation about the principal axis.
2. Single prime: symmetric with respect to reflection in the σ_h plane.
   Double prime: antisymmetric with respect to reflection in the σ_h plane.
3. Subscript 1: symmetric with respect to reflection in σ_v plane.
   Subscript 2: antisymmetric with respect to reflection in σ_v plane.
   Subscript g: symmetric with respect to inversion.
   Subscript u: antisymmetric with respect to inversion.
4. E: doubly degenerate.
5. T: triply degenerate.

We are now nearly ready to accomplish our principal goal—that is, to utilize group theory to decide whether $R = 0$ or not. The distinction between zero and nonzero values for R defines a selection rule. In order to proceed we must consider the determination of the direct product. This consideration arises when it is desired to evaluate integrals involving functions that are bases for representations of a group. An integral $\int \psi_i \psi_j \, d\tau$ differs from zero only if the integral is invariant under all operations of the group, therefore, totally symmetrical, or may be expressed as a sum of terms at least one of which is invariant. The integrand belongs to the representation $\Gamma_{\text{integrand}} = \Gamma_i \Gamma_j$, where $\Gamma_i \Gamma_j$ is the direct product of the representations of $\psi_i \psi_j$. The character of the direct product of two matrices is a product of the characters,

$$\chi_{\Gamma_i \Gamma_j} = \chi_{\Gamma_i} \chi_{\Gamma_j}. \tag{25}$$

Since a group table contains characters and the representations are basis functions, we can easily determine our $\Gamma_{\text{integrand}}$ by taking the direct product of the characters of Γ_i and Γ_j; for example, if $\Gamma_i = A_1$, and $\Gamma_j = A_2$, then $\Gamma_{\text{integrand}} = +1 \; +1 \; -1 \; -1 = A_2$. The integrand will be different from zero only if $\Gamma_i \Gamma_j$ is, or contains, the totally symmetrical representation. There is only one totally symmetrical representation and that one has characters of $+1$ for all operations (A_1 for C_{2v}). $\Gamma_i \Gamma_j$ may be reducible or irreducible but it is always irreducible if either or both of the two representations entering into the direct product are one-dimensional (nondegenerate). If both representations are of dimension greater than one, then $\Gamma_i \Gamma_j$ will be reducible—that is, expressible in terms of irreducible representations of the group: Γ_{red} or $\Gamma_R = \Gamma_i \Gamma_j = \sum_l a_l \Gamma_l$. Thus various direct products can results as follows:

$$\begin{aligned} \Gamma_{\text{irred}}^{\text{nondeg}} \Gamma_{\text{irred}}^{\text{nondeg}} &= \Gamma_{\text{irred}}^{\text{nondeg}} \\ \Gamma_{\text{irred}}^{\text{nondeg}} \Gamma_{\text{irred}}^{\text{deg}} &= \Gamma_{\text{irred}}^{\text{deg}} \\ \Gamma_{\text{irred}}^{\text{deg}} \Gamma_{\text{irred}}^{\text{deg}} &= \Gamma_{\text{red}} \end{aligned} \tag{26}$$

The number of times (a_l) the irreducible representation Γ_l occurs in the reducible representation Γ_R with compound character $\chi(R)$ is

$$a_l = \frac{1}{h} \sum_p g_p \chi_p(R) \chi_p(l) \tag{27}$$

where h is the order of the group, g_p is the number of operations in the class, $\chi(R)$ is the compound character of the reducible representation of the class, and $\chi(l)$ is the character of the irreducible representation of the class and the summation is over all classes, p.

A final point of interest concerns the representations to which M_x, M_y, and M_z belong. The individual axes as well as translations of a point in space transform in a manner such that they belong to a particular representation. In a group table this is denoted as T_x, T_y, T_z or X, Y, Z and is shown in the group table (see the table for C_{2v}, p. 24). Consider, for example, the x axis.

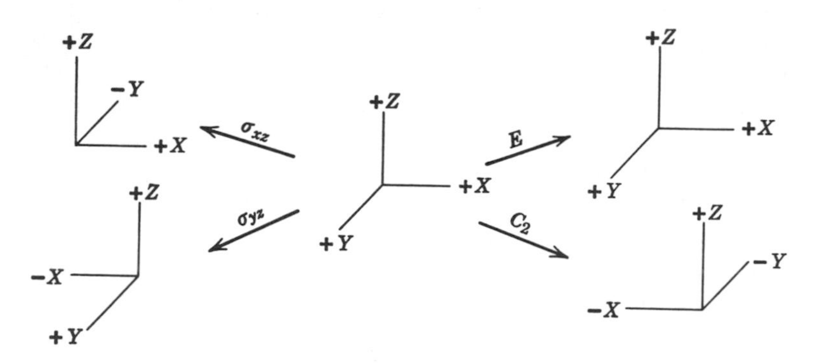

It does not change sign under $E(+1)$ but changes sign under $C_2(-1)$, it does not change sign under $\sigma_{xz}(+1)$ and does change sign under $\sigma_{yz}(-1)$ and, therefore, belongs to B_1 (see C_{2v} table, p. 24). Now M_x, M_y, and M_z are vectors along the coordinate axes and the transformation properties of the vectors are the same as the coordinate axes themselves.

Most often we are interested in triple direct products arising from integrals of the general type $\int f_a f_b f_c \, d\tau$. In our case the specific integral of interest is $R = \int \psi_i M_q \psi_j \, d\tau$ where ψ_i and ψ_j are molecular orbital or state functions and M_q is one of the dipole moment operators. The integral will not equal zero if the triple direct product is or contains the totally symmetric representation. This can occur only if the direct product of any two representations is, or contains, the representation to which the third belongs.

We are now ready to consider the possibilities for transitions in some particular cases. We will first consider a molecule belonging to point group C_{2v}. We shall consider the highest filled molecular orbital or ground state belonging to a_1 (capital letters are commonly used for states, A_1). We wish to decide if an electronic transition is allowed between a_1 and other molecular orbitals belonging to a_1, a_2, b_1, or b_2. Thus we need to evaluate integrals of the type $\int \psi_i M \psi_j \, d\tau$ where we now know ψ_i belongs to a_1 and M_x to b_1, M_y to b_2, M_z to a, and ψ_j is a_1, a_2, b_1, or b_2. The possible triple direct products, $\Gamma_{\psi_i}\Gamma_n\Gamma_{\psi_j}$, where $M = M_x$ are as follows:

$$a_1 b_1 a_1 = b_1$$
$$a_1 b_1 a_2 = b_2$$
$$a_1 b_1 b_1 = a_1$$
$$a_1 b_1 b_2 = a_2$$

Therefore, $b_1 \leftarrow a_1$ is allowed since the triple direct product, $a_1 b_1 b_1$, is the totally symmetric representation a_1. Since the representation for the highest filled molecular orbital is a_1, the direct product of the representation for M_q, Γ_M, and any molecular orbital to which a transition can occur, Γ_j, must be a_1. In general, the direct product $\Gamma_M \Gamma_j$ must equal Γ_i, the representation of the initial molecular orbital or state. That is:

$$\Gamma_i(\Gamma_M \Gamma_j) = \Gamma_i \Gamma_i = \text{total symmetric representation} \qquad (28)$$

This is only possible if $\Gamma_M = \Gamma_j$. Since Γ_M must equal Γ_j, it is only to those orbitals or states whose irreducible representations belong to M_x, M_y, and M_z that transitions are allowed from a totally symmetric orbital or state. Thus our problem is reduced to simply looking at the representations to which M_x, M_y, and M_z belong to determine to which orbital or state a transition can occur. In the case of a molecule of symmetry C_{2v} (see table on p. 24), we can have the following transitions:

Orbital Transitions	*State Transitions*	
$a_1 \leftarrow a_1$	$A_1 \leftarrow A_1$	z polarized
$b_1 \leftarrow a_1$	$B_1 \leftarrow A_1$	x polarized
$b_2 \leftarrow a_1$	$B_2 \leftarrow A_1$	y polarized

(More discussion concerning polarization will follow in Section D and Chapter 7.)

A much more complicated but highly instructive situation arises in the case of benzene (D_{6h}). The orbitals of concern are

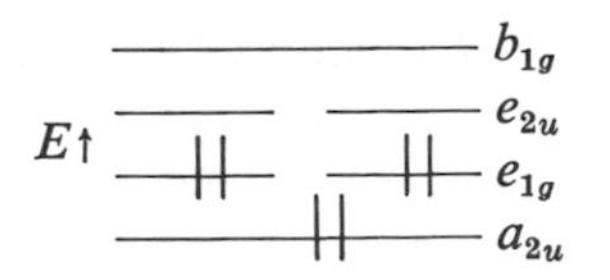

The lowest energy transition is $e_{2u} \leftarrow e_{1g}$. Immediately we know we will have

a reducible double direct product (see Section A, equation 26 last line). From a D_{6h} group table we find:

	E	C_6	C_3	C_2	C_2'	C_2''	i	S_3	S_6	σ_h	σ_d	σ_v
e_{1g}	2	1	−1	−2	0	0	2	1	−1	−2	0	0
e_{2u}	2	−1	−1	+2	0	0	−2	+1	+1	−2	0	0
$\chi_p(R)$	4	−1	+1	−4	0	0	−4	+1	−1	+4	0	0

Using (27) we can proceed to see how many times and which irreducible representations are contained in the reducible one. For example,

$$B_{1u}$$
$$\begin{aligned} &(E \quad 2C_6 \quad 2C_3 \quad C_2 \quad 3C_2' \quad 3C_2'' \quad i \quad 2S_3 \quad 2S_6 \quad \sigma_h \quad 3\sigma_d \quad 3\sigma_v) \\ &= \tfrac{1}{24}(4 + 2 + 2 + 4 + 0 + 0 + 4 + 2 + 2 + 4 + 0 + 0) = 1 \end{aligned}$$

Note that for several cases, as C_2', there are more than one operation of the same type in the class. A check of the other possibilities reveals that

$$e_{1g} \times e_{2u} = B_{1u} + B_{2u} + E_{1u}$$

In the LCAO-MO approximation with one-electron Hamiltonians, the three states are degenerate. However, with the inclusion of electron repulsion terms or the two electron operator e/r_{12} they do split. It is most important to point out that group theory can decide nothing concerning energetics. Nonetheless, since the ground state is A_{1g}, the only allowed transition is $E_{1u} \leftarrow A_{1g}$. On the basis of general orbital order we would suspect that the lowest three spin allowed transitions observed would be derived from the $e_{2u} \leftarrow e_{1g}$ transition and would be the symmetry allowed E_{1u} and the two symmetry forbidden B_{1u} and B_{2u} transitions. This is observed experimentally.

B. EINSTEIN COEFFICIENTS, ABSORPTION LAWS, AND *f* NUMBER

Other considerations in transition probability can be seen by considering the Einstein coefficients. The approach ideally applies to discrete states *in vacuo*. The coefficients are defined as follows:

E_2

B_{12} B_{21} A_{21}

E_1

B_{12}: induced absorption coefficient

B_{21}: induced emission coefficient

A_{21}: spontaneous emission coefficient

The rate of absorption from state 1 to state 2 is

$$n_1 B_{12} \rho_\nu \tag{29}$$

where n_1 is the number of molecules in state 1 and ρ_ν is the radiation density as given by Planck's black body radiation law. The rate of emission from state 2 to state 1 is

$$n_2(A_{21} + B_{21}\rho_\nu) \tag{29a}$$

where n_2 is the number of molecules in state 2. For a given radiation density the probabilities of induced emission and of absorption are assumed to be equal. At equilibrium (29) is equal to (29a). The number of molecules in the states is given by the Boltzman distribution law and ρ_ν from the black body law, from which the following equation results (*in vacuo*):

$$A_{21} = 8\pi hc\bar{\nu}^3 \frac{g_1}{g_2} B_{12} \tag{30}$$

where h is Planck's constant, c is the velocity of light, $\bar{\nu}$ is the energy of the transition in cm^{-1}, and g_1 and g_2 are the degeneracies of the lower and upper states. The coefficient B_{12} can be expanded in terms of experimental quantities to give

$$B_{12} = \frac{2303}{h\bar{\nu}N} \int \varepsilon \, d\bar{\nu} \tag{31}$$

where N is Avogadro's number, ε is the molar absorption coefficient, and $d\bar{\nu}$ represents the transition band width. More concerning the significance of A_{21} will be given in Chapter 7.

There obviously must be relationship among the intensity of a transition, I, B_{12}, and R. For the absorption coefficient B_{12} ($g_1 = g_2$)

$$B_{12} = \frac{8\pi^3}{3h^2c} |R_{12}|^2 = \frac{1}{8\pi ch\bar{\nu}^3} A_{21} \tag{31a}$$

and therefore I (absorption) $\propto \bar{\nu}_{12} |R_{12}|^2$. The intensity of a transition is commonly defined in terms of the f number or the oscillator strength. This measures the intensity of a given transition compared to that observed for an allowed transition at the same frequency for a three-dimensional harmonic oscillator,

$$\begin{aligned} f &= \frac{m_e hc^2\bar{\nu}}{\pi e^2} B_{12} \\ &= \frac{2303 m_e c^2}{N\pi e^2} \int_{\bar{\nu}_1}^{\bar{\nu}_2} \varepsilon \, d\bar{\nu} \\ &= 4.315 \times 10^{-9} \int_{\bar{\nu}_1}^{\bar{\nu}_2} \varepsilon \, d\bar{\nu} \end{aligned} \tag{32}$$

where the statistical weighting factors have been assumed to be equal. The value of f approaches one for highly allowed transitions. Considering that most absorption bands have half widths of several thousand cm^{-1}, the absorption coefficient for an allowed transition will be in the range of 10^5.

Finally it is possible to relate some extrinsic and intrinsic properties as

$$\log \frac{I_0}{I} = \varepsilon l c \tag{33}$$

where l = pathlength in cm, c = concentration in moles/l. and $\log I_0/I$ is known as optical density or absorbance.

In view of our present concern with transition probability it might be well to compare the types of forbidden transitions and the relative degree of forbidden character for each. The comparisons are relative to a fully allowed transition.

1. Spin forbidden: for second row atoms in molecules, the spin forbidden rule ($\Delta S = 0$) is the most rigorous—prohibition factor 10^{-4}–10^{-6}.
2. Space forbidden: cases in which the electron density in the region of space where orbitals overlap is very low—prohibition factor 10^{-1}–10^{-3}.
3. Orbitally forbidden: cases where there are large changes in the angular momentum about an axis—prohibition factor approximately 10^{-3}.
4. Symmetry forbidden: cases in which the allowed or forbidden character depends on symmetry considerations—prohibition factor 10^{-1}–10^{-2}.

C. FRANCK-CONDON PRINCIPLE

The electric dipole transition moment integral (equation 24) contains the total wave function for the system. The rotational portion may be neglected. This means that the wave function could be expressed as a product function,

$$\psi_{\text{total}} = \psi_e \psi_n \tag{34}$$

where ψ_e is the electronic part and ψ_n the nuclear coordinate or vibrational part. This is also true for the expression

$$H\psi_{\text{total}} = E\psi_{\text{total}} \tag{35}$$

The Born-Oppenheimer approximation permits us to write the complete wave function as a product of the nuclear and electronic part and therefore

$$H_e \psi_e = E_e \psi_e \tag{36}$$

$$(H_n + E_e)\psi_n = E_n \psi_n \tag{37}$$

Thus the Born-Oppenheimer approximation infers that ψ_e varies only slowly as a function of the nuclear coordinates and enables us to describe the

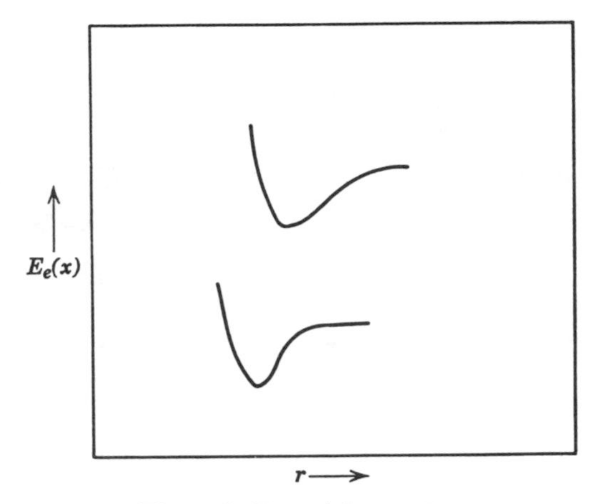

Figure 6 Potential energy curves.

electronic wave function independently of ψ_n. Therefore, the electronic energy is a function of the nuclear coordinates and is determined by solving the wave equation (36) for fixed positions of the nuclei. In the case of a diatomic molecule these solutions generate the familiar Morse potential energy curves, as in Figure 6. Although the nuclear portion does not depend upon the coordinates of the electrons, it will depend upon the kind of electronic state.

In the case of R expansion leads to

$$R = \int M_e \psi_{e_1} \psi_{e_2} \psi_{v_1} \psi_{v_2} \, d\tau + \int M_n \psi_{e_1} \psi_{e_2} \psi_{v_1} \psi_{v_2} \, d\tau \tag{38}$$

for a transition between vibrational levels of different electronic states. However, since M_n is not a function of the coordinates of the electron and since electronic eigenfunctions of different electronic states are orthogonal

$$R = \int \psi_{e_1} M_e \psi_{e_2} \, d\tau_e \int \psi_{v_1} \psi_{v_2} \, dr \tag{39}$$

where dr is the internuclear distance in a diatomic or of one of the normal coordinates in a complex molecule. With the Born-Oppenheimer approximation the first integral can be replaced by some average value $\bar{R}_e$ and

$$R_{v_2 \leftarrow v_1} = \bar{R}_e \int \psi_{v_1} \psi_{v_2} \, dr \tag{40}$$

Thus the relative intensities of vibronic (vibrational components of an electronic transition) transitions depend on the square of the overlap of the

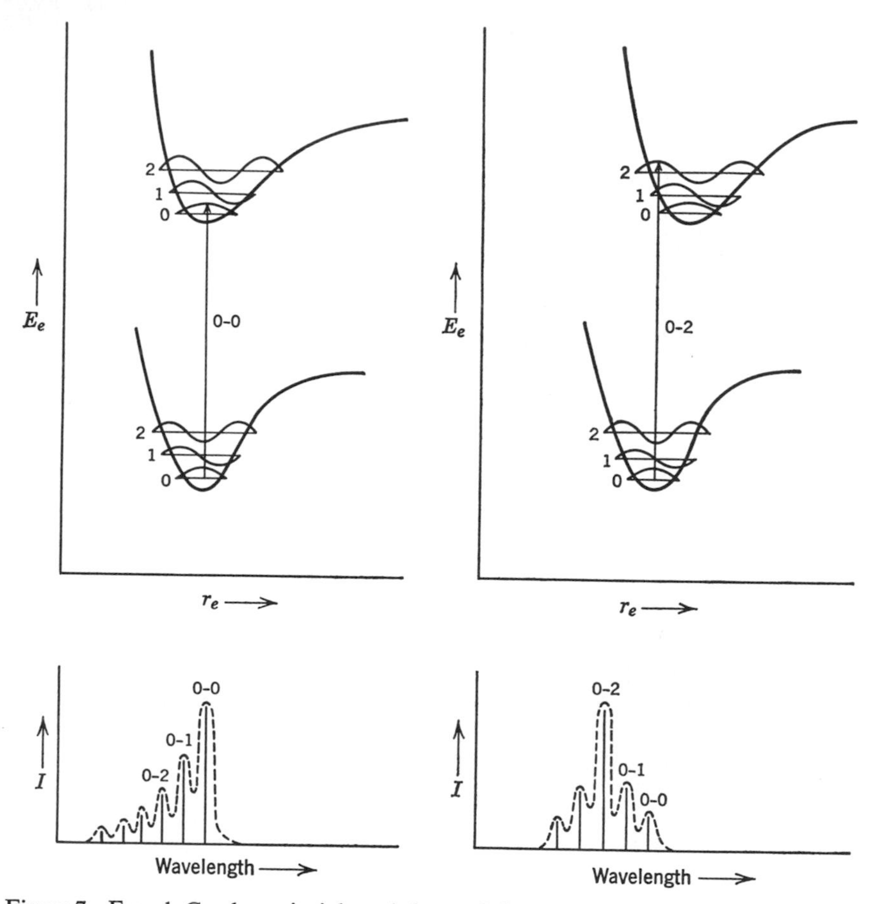

Figure 7 Franck-Condon principle and shape of absorption bands. The general shape of curves expected is indicated for vapor (———) and solution (- - - -).

vibrational functions of the two electronic states

$$I_{v_2 \leftarrow v_1} = \frac{8\pi^3}{3hc} I_0 l N_{v_1} \bar{R}_e^2 \left(\int \psi_{v_1} \psi_{v_2} \, d\tau \right)^2 \tag{41}$$

where I_0 is the intensity of the incident radiation, l the path length, and N_{v_1} is the number of molecules in the v_1 vibrational state of a particular electronic state (normally, the ground state).

There are no changes in the nuclear coordinates during an electronic transition and therefore the transitions are vertical. Thus it can be seen that the shape of an absorption band will depend largely on the relative location of the potential energy minima. The foregoing concepts in this section comprise what is known as the *Franck-Condon principle*. Examples of the relations described above are given in Figure 7. The schematic shows lines representing a vapor case and the dotted curve a solution result.

D. POLARIZATION

It will be recalled that the dipole transition moment integral contains M_x, M_y, and M_z components. Therefore, it is possible that with an oriented molecule a particular transition may occur only when the incident radiation is polarized along a certain direction. The molecular orientation can be accomplished by using crystals. In crystals it is possible to have a fixed relation between an axis of the molecule and the direction of polarization of the incident radiation. An example of the effect of this consideration is that previously shown for ethylene (Section A); that is, light polarized parallel to the long axis of ethylene could be absorbed with a resultant $\pi^* \leftarrow \pi$ electronic excitation whereas that parallel to the y or z direction could not be absorbed.

It is often easier to infer something about the relative polarization of absorption bands rather than the absolute polarization. This information is still very valuable in terms of characterizing the absorption band. This is accomplished by observing the direction of polarization of emission as a function of the wavelength of excitation. More concerning this will be given in Chapter 7, Section A4.

Chapter 4 Environmental Factors Affecting Absorption Spectra

A. GENERAL AND SPECIFIC SOLVATION EFFECTS

There has been considerable progress toward a better understanding of the effects of solvents on electronic transitions. The electronic transitions of particular interest are $\pi^* \leftarrow \pi$ and $\pi^* \leftarrow n$. The general nature of the former of these has been previously discussed (Chapter 2; see also Chapters 5 and 6). If a molecule contains atoms with nonbonding electrons, called n electrons, a transition is possible from the nonbonding orbital to the π^* or antibonding π orbital. More detailed consideration of the $\pi^* \leftarrow n$ transitions is given in Chapters 5 and 6. Although solvent effects on $\pi^* \leftarrow n$ transitions can be included in the discussion with those on $\pi^* \leftarrow \pi$ transitions, they represent cases where unusual interactions can occur and the quantitative order of prediction is less satisfactory. The effects of pressure can also be used to obtain data on solvent effects with particular reference to dispersion forces (Section B). Some of the major topics pertinent to the clarification of the effects of solvents on electronic spectra include (a) spectra of merocyanines [2], (b) general solvation theory with regard to $\pi^* \leftarrow \pi$ and $\pi^* \leftarrow n$ transitions [3–7], (c) general red shifts of nonpolar solutes [8], (d) dispersive interactions [9–12], nature of interactions of nitriles and chloroform with benzophenone [13], and (e) determination of excited-state dipole moments [6,14–16]. In addition expressions have been developed to deduce the effect of solvents on emission spectra [6,15,16]. Also, empirical wavelength corrections have been given for the $\pi^* \leftarrow n$ transitions in unsaturated ketones including steroids [17,18]. The latter data has been used to assign structures to unsaturated ketones [17,18]. For our general purposes it will not be necessary to be concerned with all of the foregoing topics.

In a discussion of solvent effects on transitions of organic molecules we should consider dipole forces as well as the Franck-Condon principle. Early work [3,4] had predicted a red shift (shift to longer wavelengths) for all transitions because of a transient polarization of the solvent molecules induced by the transition dipole moment of the solute (polarization effect). Although this prediction is general, other interactions may be dominant and either further increase the red shift or reverse the direction toward shorter wavelengths (blue shift). Therefore, the additional interactions of dipole-dipole, dipole-polarization, and hydrogen bonding should be considered [5]. Moreover, a consideration of the Franck-Condon principle leads to a classification of two components of strain—packing and orientation. The latter of these is normally considered to be the more important of the two. The orientation strain results from the nonequilibrium orientation of the solvent cage around the solute molecule in its excited state and includes dipole-polarization and dipole-dipole interactions (see Table 2). Orientation strain is particularly important when the solute and solvent are polar and when the permanent dipole moment changes upon excitation. Modification of the strength of any hydrogen bonding would also give importance to this type of strain. Packing strain results from an actual change in the geometric size of the molecule in the excited and ground states. Generally the percent change of the size of organic molecules is small and this strain can be neglected, except in specific instances.

The various interactions can be studied by considering various general combinations of solvent and solute molecules. Table 2 provides a summary of these and the expected results. Examples of some of these are as follows: I–$\pi^* \leftarrow \pi$ transition of benzene in hexane; II–$\pi^* \leftarrow \pi$ transition of naphthalene in chloroform; IIIA–$\pi^* \leftarrow n$ transition of acetone or crotonaldehyde in benzene or cyclohexane; IIIB–$\pi^* \leftarrow \pi$ transition of crotonaldehyde or nitrobenzene in cyclohexane; IVA–same solutes and transitions as in IIIA in ethanol; IVB–same solutes and transitions as in IIIB in ethanol.

The red shift in cases I and II occurs because of the polarization effect (Table 2). The form of the expression for the calculation of the shift depends on several factors including the solvent index of refraction, the frequency and intensity of the transition. The simplest equation is the following [3]:

$$\Delta\bar{\nu} = -1.07 \times 10^{-14}\left(\frac{1}{\bar{\nu}}\right)\left(\frac{f}{a^3}\right)\left(n^2 - \frac{n_2 - 1}{2n^2 + 1}\right) \tag{42}$$

where $\bar{\nu}$ and f are the frequency (cm^{-1}) and oscillator strength of the solute transition, respectively, n is the index of refraction of the solvent, and a is the radius of the spherical cavity of the solvent in which the solute resides. Other expressions have been derived for this shift that take into account other parameters. These will be discussed later. The energy diagrams appropriate

Table 2 Solute-Solvent Interactions and Nature of Spectural Shift

	I	II	III		IV	
			A	B	A	B
Solute[a]	NP	NP	P	P	P	P
Solvent[a]	NP	P	NP	NP	P	P
Change in solute dipole	None	None	Decrease	Increase	Decrease	Increase
Type interaction	Polarization[b]	Polarization	Polarization, dipole-polarization[c]	Polarization, dipole-polarization	Polarization, dipole-dipole, packing	Polarization, dipole-dipole, packing
Predicted shift[d]	Red	Red	Either direction	Red	Blue	Red
Example	$\pi^* \leftarrow \pi$, Benzene in hexane	$\pi^* \leftarrow \pi$, Naphthalene in chloroform	$\pi^* \leftarrow n$, Acetone in benzene	$\pi^* \leftarrow \pi$, Nitrobenzene in cyclohexane	$\pi^* \leftarrow n$, Acetone in ethanol	$\pi^* \leftarrow \pi$, Nitrobenzene in ethanol

[a] NP ≡ nonpolar, P ≡ polar.
[b] Polarization is equivalent to an induced dipole-induced dipole interaction.
[c] Dipole-polarization is equivalent to a dipole-induced dipole interaction.
[d] Absence of packing strain and compared to vapor.

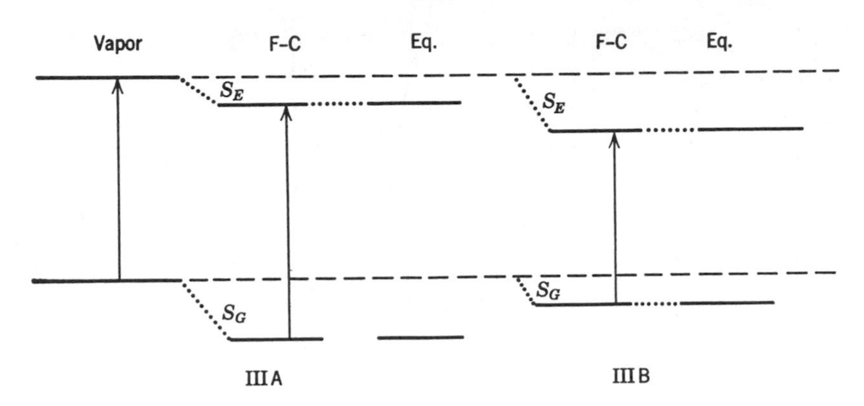

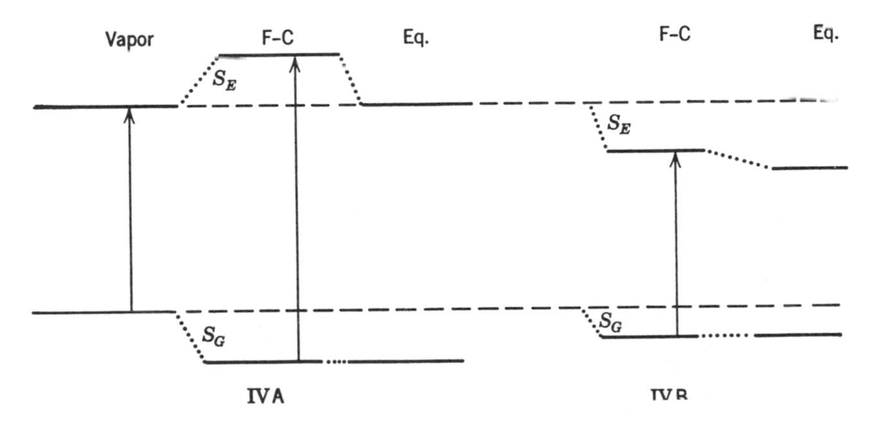

Figure 8 Energy diagrams for various solute-solvent interactions. F-C is the Franck-Condon state, Eq. is the equilibrium state, and S_E and S_G are the solvation energies in the excited and ground states, respectively.

to these and succeeding cases are shown in Figure 8. The Franck-Condon state is a nonequilibrium state formed immediately upon excitation. In the case of IIIA the $\pi^* \leftarrow n$ excitation results in a decrease of the dipole moment because of transfer of negative charge off the oxygen atom toward the carbon atom. Thus the dipole-polarization forces are less in the excited state than in the ground state and a blue shift results, (Figure 7 and Table 2). The degree of shift depends upon the relative change in the dipole moment. Additional superposition of the red shift (polarization effect) determines the ultimate shift. In the case of acetone the resultant shift is red while for crotonaldehyde it is blue. In case IIIB the increased dipole moment in the excited state causes an increase in the dipole-polarization forces in the excited state compared with ground state (Figure 7). This results in a greater lowering of the energy

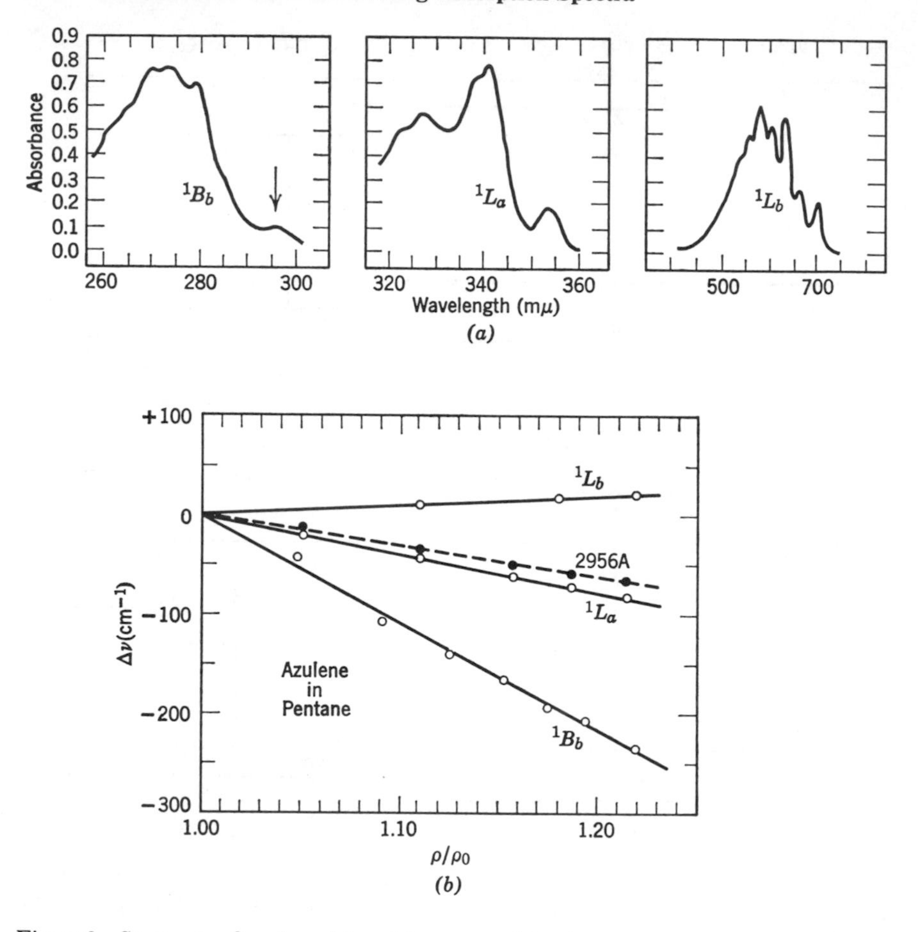

Figure 9 Spectrum of azulene (*a*) and frequency shift of certain bands as a function of solvent density (*b*).
Reprinted by permission from American Institute of Physics, W. W. Robertson and A. D. King, *J. Chem. Phys.*, **34,** 2190 (1961).

of the excited state than of the ground state and a red shift results (Table 2). The polarization red shift must be added (Table 2).

In case IVA there are the additional forces of dipole-dipole and possible hydrogen bonding compared with case IIIA (also packing strain). In the Franck-Condon state the less polar (decreased dipole moment) molecule is in an oriented solvent cage. The energy required to reorient the solvent molecules around the less polar molecule provides a negative energy term to the solvation energy (increase of the excited-state energy) (see Figure 8). Thus a blue shift results (Table 2). This will normally predominate over the polarization red shift (Table 2). The same type of argument applies in the case of problems involving hydrogen bonding if the strength decreases in the excited state. In case IVB the reverse of case IVA is found and the excited-state energy (in the Franck-Condon state) is lowered and further, more than the ground state by the interaction (Figure 8), and a red shift results (Table 2). The polarization red shift must be added (Table 2). The red shift will also occur in the hydrogen-bonding cases where the strength is increased in the excited state. It should be re-emphasized that in all these considerations the primary reference is the vapor.

Based on second-order perturbation theory, the following simplified equation can be developed to account for solvent interactions with solutes [6]:

$$\Delta\nu = (AL + B)\left(\frac{n^2 - 1}{2n^2 + 1}\right) + R\left(\frac{D - 1}{D + 2} - \frac{n^2 - 1}{n^2 + 2}\right) + S\left(\frac{D - 1}{D + 2} - \frac{n^2 - 1}{n^2 + 2}\right)^2 \quad (43)$$

where the terms A, B, R, and S are constants characteristic of a particular transition for an absorbing molecule, D is the dielectric constant, and n is the index of refraction of the solvent (at some particular visible or near ultraviolet frequency). The terms A and B are defined in more detail below. The terms R and S are defined as follows:

$$R = \left(\frac{2}{hc}\right)\left(\frac{m_g(m_g - m_e)}{a^3}\right) \quad \text{and} \quad S = \left(\frac{6}{hc}\right)\left(\frac{m_g(\alpha_g - \alpha_e)}{a^3}\right) \quad (44)$$

where m_g and m_e are the dipole moments of the solute in the ground and excited states, α_g and α_e are the molecular polarizabilities in the ground and excited states, and a is the cavity radius appropriate to the solvent.

The first term with A and L (weighed mean wavelength) [6] represents the induced dipole-induced dipole energy (or dispersive) portion. This corresponds to the polarization term and the general red shift expected that we discussed earlier in case I. The second term with the B represents the dipole-induced dipole (solute-solvent) energy portion. The third term represents the dipole-dipole interaction, and the fourth term represents the induced dipole-dipole (solute-solvent) energy portions. The L value generally increases in the same order as the solvent index of refraction. Most saturated hydrocarbons (as well as alkyl alcohols) should have about equal values (approximately 1000 Å) [6]. Benzene has a value that is approximately the same. The first term corresponding to the general red shift can be expanded for strong transitions in terms of the molecular constants to give

$$-2.14 \times 10^{-14} L\left(\frac{f}{a^3}\right)\left(\frac{n^2 - 1}{2n^2 + 1}\right) \tag{45}$$

where L is as before, f is the oscillator strength of the transition in question, n is the index of refraction of the solvent at the band frequency and a denotes the cavity radius. Often the solvent index of refraction at the sodium D line is used. It will be noted that this equation is similar to (42). If the transition is not strong, then other terms contribute to the shift.

The second term with B can be expanded to give

$$\left(\frac{m_g^2 - m_e^2}{hca^3}\right)\left(\frac{n_0^2 - 1}{2n_0^2 + 1}\right) \tag{46}$$

where n_0 is a solvent refractive index extrapolated to zero frequency (often n_D or n is used), m_g and m_e are the dipole moments of the solute in the ground and excited states respectively, a is as described earlier, and h and c are Plank's constant and the velocity of light. Several methods for evaluating excited-state dipole moments have been proposed [14,16].

In general, the first term contributes in case I, the first and fourth terms contribute in case II, the first and second terms in case III, and all four terms in case IV. Even though term four contributes in case II, this is usually much smaller than the first term. This is why the cause for the red shift given earlier in the discussion for both cases I and II was noted to be the polarization effect. In cases where the solvent is highly polar as in the case of water, however, the fourth term may contribute. In the cases of strong hydrogen bonding the equation does not apply although it may be capable of predicting the part of the shift exclusive of hydrogen bonding. The equation is semi-quantitative because of the model used in the development of the theory.

For low-energy transitions of nonpolar solutes in nonpolar solvents (a) the shifts are proportional to the polarizability of the solvent molecules, (b) the shifts increase as the mean separation of solute and solvent molecules decrease (thus the shift increases with increasing pressure), and (c) one energy term contributes to the shift for weak transitions but two terms contribute for strong transitions.

The large blue shifts that can occur in carbonyls for the $\pi^* \leftarrow n$ transition when the solvent is changed from a hydrocarbon to an alcohol have been shown to be mainly due to hydrogen bonding to the nonbonding (n) electrons of the carbonyl oxygen [19]. This interaction provides greater stabilization of the ground state compared with the excited state. In addition to ketones, N-heterocyclics show the same general behavior. The energy of the hydrogen bond formed would be the same as the magnitude of the blue shift providing a single hydroxylic molecule was bonding to the nonbonding electrons of the acceptor. In addition, the hydrogen bond energy can be evaluated from infrared data, and equilibrium constants can be calculated from ultraviolet absorption data.

The importance of the Franck-Condon principle and its role in the electronic transitions where hydrogen bonding is involved have been considered [20]. This is important in terms of the fact that early work [19] considered only the fact that the energy of stabilization by the hydrogen bond in the excited state was zero.

The alcohol-benzophenone system has been examined utilizing both ultraviolet and infrared [19,21]. From shifts and new band formation in the carbonyl region it can be deduced that monomer alcohol and dimer alcohol-benzophenone complexes predominate in solution [21]. There is also evidence for higher-order complexes. Of further significance is the existence of specific complexes between benzophenone and nitriles [21]. No obvious hydrogen bonding or other specific interaction appears to be possible. Several alkyl nitriles and benzonitrile, however, show well-defined isobestic points in the ultraviolet region (for the $\pi^* \leftarrow n$ transition) as the concentration of the nitrile is changed. An isobestic point occurs at a wavelength where two interconvertible species have an equal absorption intensity. In addition to the isobestic point there is a blue shift of the $\pi^* \leftarrow n$ transition. The interaction appears to be principally an electrostatic one involving the relatively electronegative oxygen of the carbonyl group and the relatively electropositive carbon of the nitrile group.

A final consideration of the effect of solvents concerns the retention or loss of the vibrational fine structure associated with the electronic transition. The Franck-Condon excited state is a very short-lived metastable state in fluid systems. This occurs because the solvent relaxation or orientation time is rapid and occurs in a time comparable to that required for molecular vibrations. This can result in the lack of vibrational quantization in the excited state with consequent blurring of the vibrational structure in the spectrum. This can occur in the cases in which (a) the solvent is polar, particularly if hydrogen bonding between solvent molecules can occur and (b) a polar solute and polar solvent are involved. In the first case the effect of any packing strain (for the solute) is important because with dipole or hydrogen bond forces between the solvent molecules, the reorientation time of the solvent cage will increase. In the latter case the presence of orientation strain is the predominant factor. It should be remembered that the orientation strain arises because of the difference in solvation energy between the Franck-Condon and equilibrium excited states.

All of these considerations point to the important roles general solvent as well as specific solvent interactions have on the spectra of solutes. Considerable care and attention to the points noted in this section must be given before valid and meaningful interpretations can be made.

B. PRESSURE AND TEMPERATURE EFFECTS

Of particular interest is the effect of high pressure (5–150 thousand atmospheres) on electronic transitions. Considerations include (a) spectral shifts [22], (b) London dispersion forces [12], (c) cavity sizes [23], (d) temperature-dependent spectral shifts [24], (e) crystal fields [25,26], (f) classification of optical spectra [23,25], (g) color centers in alkali halides [25], and (h) variables affecting charge transfer phenomena [27,28]. However, we will again be interested in only certain general features. Of particular interest relative to aromatic hydrocarbons are the facts that the wavelength of absorption in the vapor state and in solution are linearly related to the dielectric constant, and that a rough relationship exists between (Δ wavelength)/(Δ dielectric constant) and the oscillator strength or intensity of a transition [12,22]. The frequency shift of a nonpolar molecule in a nonpolar solvent is nearly a linear function of the solvent density. Moreover, the frequency shift increases with increasing intensity of a transition [24]. The effect of pressure is to increase the solvent density and consequently the dispersion forces that are responsible for the spectral shifts.

Another consideration is the relation between spectral shifts and the temperature. With prior knowledge of the effects of pressure on spectra and the use of a single solvent to avoid introducing solvent shifts, certain interesting factors evolve. A temperature lowering results in opposing shifts. The shift to longer wavelengths is caused by the increase in solvent density and the shift to shorter wavelengths is attributable to a change in band shape as a result of a decrease in the population of certain vibrational levels. The latter concept is supported by the fact that the residual temperature shift (that remaining after correction of total shift for density changes determined from pressure effects) is always to shorter wavelengths. Moreover, the temperature shift portion is independent of the intensity of the transitions and therefore of the dispersion forces, since the dispersion forces and resulting frequency shifts are directly related to the intensity or oscillator strength of a transition [12]. The final resulting shift depends on the relative orders of magnitude of the thermal or temperature effect (blue shift) and density effect (red shift). Thus in naphthalene the first relatively weak transition shows a resultant blue shift at $-80°C$, whereas two other stronger transitions show a resultant red shift at the same temperature and in the same solvent.

The usefulness of the pressure technique to distinguish electronic transitions is particularly valuable. This criterion has been applied in several instances [23,25]. For example [23] it is possible to show that a band thought to be associated with a particular electronic transition in azulene has a distinctively different characteristic shift as a function of pressure than do other bands of the transition. Figure 9 shows the spectrum and the frequency shifts versus solvent density for the three distinct transitions plus the questionable one at 2956 Å as indicated by an arrow. This definitively separates the 2956 Å band from the ^1B_b transition and makes it a separate electronic transition. In another case [23] a band in the spectrum of benz(a)anthracene undergoes a different shift than does the origin of a lower-energy transition with which it might be associated. This indicates the band to be the origin of a new electronic transition.

Pressure effects have been used to study ligand field and spin forbidden transitions [26]. In one particular case [26], K_2ReCl_6, as the pressure is increased not only do certain transitions undergo shifts but a new pressure-induced ligand field transition arises.

Finally, the effect of pressure on charge transfer complexes provides additional information on the transitions involved [27,28]. The nature of the acceptor levels in certain inorganic complexes has been proposed based on comparative shifts of the charge transfer transitions with pressure. Increase of pressure on aromatic hydrocarbon-tetracyanoethylene complexes causes large increases in the intensity of the charge transfer bands [28]. For the cases of benzene and hexamethylbenzene complexes there is a single band, whereas for biphenyl and naphthalene complexes the band is split, There are no general frequency trends in the bands of the complexes as a function of pressure. It is interesting that in the benzene complex even though the intensity increases with pressure, the equilibrium constant decreases. This is probably due to a new complex involving tetracyanoethylene and the solvent [28]. The increase in intensity with pressure presumably arises because of increased overlap between the π orbital of the acceptor and donor molecules with a resultant increased transition probability for the charge transfer band. [21].

Chapter 5 Effect of Substitution on Π Systems

A. ETHYLENE AND POLYENES AND SUBSTITUTED DERIVATIVES

In ethylene there is a strong absorption at 1610 Å corresponding to the lowest one-electron $\pi^* \leftarrow \pi$ transition. Progressive addition of double bonds to give polyenes causes the lowest transition to move to longer wavelengths (see Table 2-1). Neither the simple LCAO-MO nor FEMO, however, properly predict the fact that for large numbers of atoms (or double bonds) the wavelength of the lowest energy transition stays nearly constant. This can be corrected in the LCAO-MO approach by realizing all bond lengths are not equal; that is, all β terms are not equal in the calculation of the energy levels. In the FEMO theory modification of the potential function so that it is not constant (such as a sine curve potential) takes into account the bond alternation. With these changes both theories are in excellent agreement with experiment.

Certain polymethine dyes, such as symmetrical cyanines with nitrogen or carbon at the ends of a polyenic chain, can be treated by the simple LCAO-MO or FEMO theories. This occurs because all bonds have approximately the same double-bond character. Thus β is essentially constant or the potential is nearly constant. In certain cases, however, such as Ph-$(C{=}C)_n$-Ph, bond alternation again occurs.

The basis for the effect of substitutions for ethylene and polyenes, in general, is similar. Consequently the example of ethylene can be extended to other systems. The substituent influences the existing levels through both inductive and conjugative effects. The inductive effect acts through a modification of the potential field and is less consequential in terms of the magnitude of the modification of wavelengths of transitions of the parent or unsubstituted

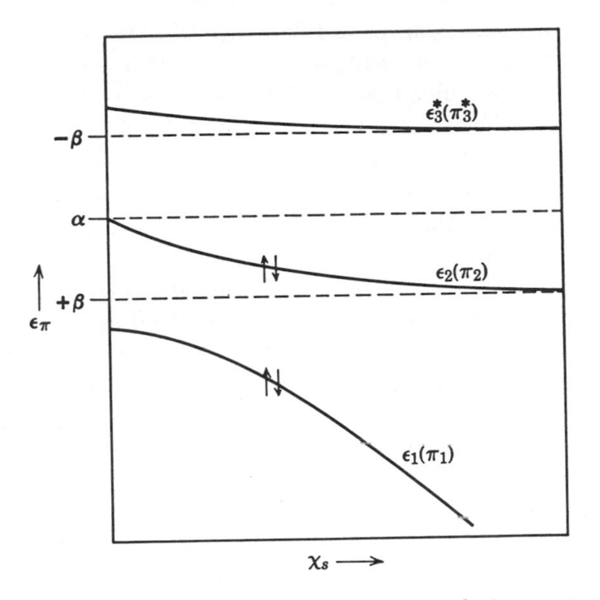

Figure 10 Variation in molecular orbital energies of a substituted ethylene as a function of the relative electronegativity of the substituent.

molecule. On the other hand, the conjugative effect increases the effective number of π electrons. The effect can be viewed as resulting in an increase in the effective path of the π electrons because of conjugation of the p electron of the substituent with the π electrons of the polyene. The result of this effect is the same as increasing the size of the box containing the electrons and causing a crowding of the π levels.

Using LCAO-MO theory the coulomb integral for a substituent, s, is $\alpha_s = \alpha + \chi_s\beta$ where α and β are the coulomb and resonance integrals for carbon and χ_s measures the relative electronegativity of s compared to carbon. If the electronegativity of the substituent is greater than that of carbon, χ_s is positive and the relative binding of the π electrons to the substituent is greater than for carbon. Thus one expects a relationship between χ_s and the ionization potential of the substituent. The new energy levels can be determined from the appropriate secular determinant. However, certain qualitative results can be given. If the electronegativity of the substituent is large, then the parent levels are essentially unaffected and therefore the location of the lowest energy transition will be essentially the same as without the substitution. Figure 10 gives a schematic picture of how the levels of a substituted ethylene vary as a function of χ_s. It can be seen that in the absence of consideration of inductive effects, the lowest energy transition, $\varepsilon_3^* - \varepsilon_2$ is lowered (red shifted) by conjugative substitution.

The inductive effect alone can have two consequences. If the substituent feeds in charge, a $+I$ or positive inductive effect, then a red shift results. On the other hand if the substituent withdraws charge, a $-I$ or negative inductive effect, a blue shift results. Generally the shift of a $\pi^* \leftarrow \pi$ resulting from inductive effects is much less significant than that from conjugative effects. This occurs because the α value of the carbon atoms is affected which modifies the energy of both the highest filled and lowest empty orbital in the same direction and with only a small differential.

The ionization potential and electron affinities are affected by both the conjugative and inductive effect. The conjugative effect decreases both the ionization potential and electron affinity. The $+I$ effect raises both ε_2 and ε_3^* thus decreasing both the ionization potential and electron affinity, while the $-I$ effect increases both the electron affinity and ionization potential.

The relative intensities of transitions are affected by both conjugative and inductive effects. Also particularly for the ionization potential and electron affinity, the change depends on the resultant of both the inductive and conjugative effects. For example, for chloroethylene the ionization potential is lower than for ethylene indicating the conjugative effect has more than compensated for the $-I$ effect. Further, from this we would expect the red shift from conjugation to predominate over the blue shift from $-I$, which is indeed the case.

Let us consider some specific cases of substitution. If the substituent is any halogen, hydroxyl, or amino group, there is a resultant red shift which indicates a dominance of the conjugative effect over the $-I$ inductive effect. The alkyl substituents present a somewhat more complex problem. The sp^3 carbons are less electronegative than are the sp^2 carbons because of the less effective shielding of three electrons compared to four (the π electron provides relatively little shielding). We can obtain similar results from energetics since sp^2 electrons have a higher ionization potential than do sp^3 electrons. This situation results in transfer of charge ($+I$, red shift) relative to the ethylene parent. The inductive effect is in the order *t*-butyl > *i*-propyl > ethyl > ethyl > methyl. The presence of a virtual p orbital on the alkyl groups permits interaction with the π electrons of ethylene giving a *hyperconjugative effect*. The order for this effect is methyl > ethyl > *i*-propyl. The observed red shift is in the order *i*-propyl > ethyl > methyl which indicates a predominance of the inductive effect for these cases. Ionization potential data indicate a similar result.

The addition of certain substituents causes not only the effects previously noted but also introduces additional molecular orbitals. Thus additional transitions are possible, and a marked alteration of the spectrum can occur. Examples of such groups are the aldehyde, keto, and carboxyl. For discussion purposes we shall consider the carbonyl case. The substitution of an aldehyde

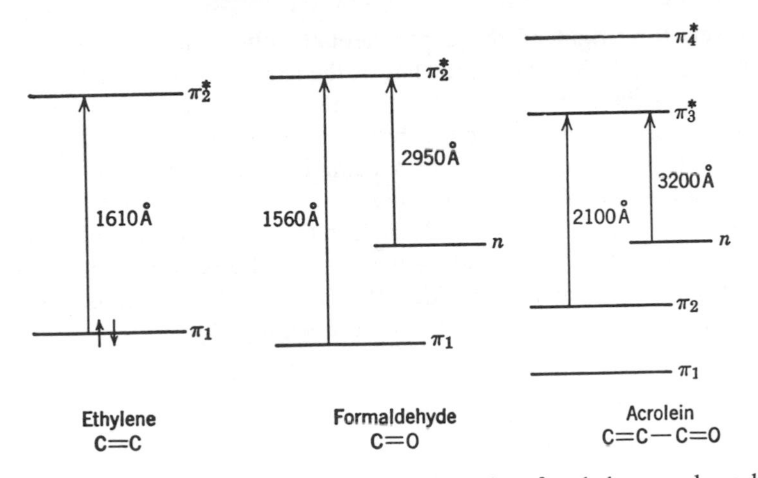

Figure 11 Schematic molecular orbital energy levels of ethylene and carbonyl compounds.

group on ethylene results in significant spectral changes. This occurs both because of a modification in the number of π levels and the addition of a nonbonding (n) level as shown in Figure 11. The latter level arises because of the presence of nonbonding p electrons on the oxygen atom. These orbitals lack the proper symmetry to interact with the $p - \pi$ electrons of carbon. As can be seen from Figure 11 a new low-energy transition arises, the $\pi^* \leftarrow n$ one, and because of the introduction of new π levels, the lowest energy $\pi^* \leftarrow \pi$ is dramatically red shifted from that of ethylene. In fact the lowest energy $\pi^* \leftarrow \pi$ transition is comparable in character and wavelength to that of butadiene (2170 Å) as might be expected (Figure 11).

The presence of the nonbonding level obviously can affect the ionization potential; that is, the first ionization potential would be from the n level and thus considerably lower than that of ethylene (or even lower than butadiene).

The foregoing results concerning the $\pi^* \leftarrow n$ transition require some further clarification. The basis given for the $\pi^* \leftarrow n$ transition being at lower energy in carbonyl substituted ethylene (and polyenes in general) was that the n level is at higher energy than the highest filled π level (see previous discussion and Figure 11). Similar considerations are also employed to explain parallel results for carbonyl substituted benzene derivatives as benzaldehyde (Section B-2). As a matter of fact the same type of arguments are used for the N-heterocyclics (Chapter 6-A). Although it is quite possible for the n level to be higher in energy than the highest filled π level, the difference in energy may be significantly less than the difference in energy between the lowest $\pi^* \leftarrow n$ and $\pi^* \leftarrow \pi$ transition. For example, in benzaldehyde the ionization potentials of the n orbital and highest filled π orbital are very nearly the same.

However, the origin of the experimentally observed $\pi^* \leftarrow n$ transition is approximately 1 eV lower than that of the lowest $\pi^* \leftarrow \pi$ transition.

In view of the above we must realize that it may be too naive to place the n level above the highest filled π orbital by the difference in the experimentally observed transition energies. Even schematically to place the n level above the highest filled π orbital may be incorrect. The salient point is that orbital energies alone may not properly reflect the experimental facts. There are several reasons for this including the neglect of electron repulsion and configurational interaction. Consider the lowest $\pi^* \leftarrow \pi$ transition between levels i and j and the lowest $\pi^* \leftarrow n$ between n and j.

$$\Delta E(\pi, \pi^*) = \varepsilon_j - \varepsilon_i - J_{i,j} + 2K_{i,j}$$
$$\Delta E(n, \pi^*) = \varepsilon_j - \varepsilon_n - J_{n,j} + 2K_{n,j}$$
$$\Delta E(\pi, \pi^*) - \Delta E(n, \pi^*) = \varepsilon_n - \varepsilon_i - (J_{i,j} - J_{n,j}) + 2(K_{i,j} - K_{n,j})$$

Commonly $\Delta E(\pi, \pi^*) - \Delta E(n, \pi^*) > 0$ as in the case of benzaldehyde. Now, since $\varepsilon_i \approx \varepsilon_n$, the difference must arise from the J and K terms and configurational interaction. Although it is difficult to precisely evaluate the J and K terms, $K_{i,j}$ is expected to $\geqslant 1$ eV, $K_{n,j}$ small (0.2–0.5 eV) and $J_{i,j} \approx J_{n,j}$. Thus the K terms would primarily contribute to the $\Delta E(\pi, \pi^*) - \Delta E(n, \pi^*)$. In addition, configurational interaction between the first and second π, π^* states is expected to be less than that between the first and second n, π^* states.

As noted in the preceding discussion it is difficult to evaluate the terms of concern. Nonetheless, the philosophical approach is sound regarding the frequent inadequacy of orbital energy considerations alone to explain experimental spectral data.

In those cases where $\pi^* \leftarrow n$ transitions are possible, we are particularly interested in the nature of the transition including symmetry characteristics, electronic-vibrational coupling, and the polarization expected for the transitions. The simplest case to consider is that of formaldehyde (C_{2v}), even though it presents certain complications that will be noted. The coordinate orientation is as follows

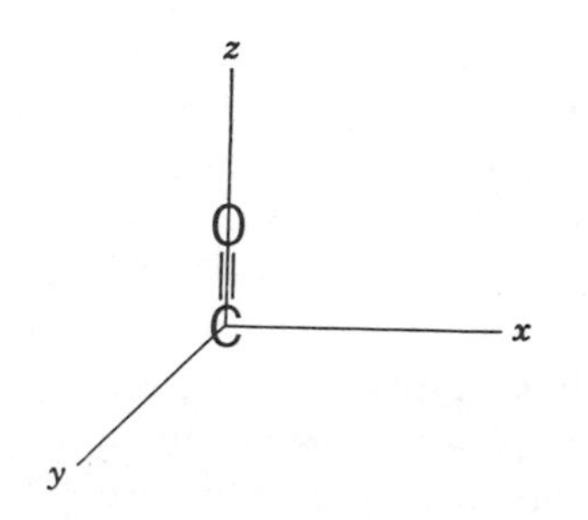

where z is the principal fold axis and long the C = O axis. The orbitals of concern are

$$b_2 = n[p_x(\mathrm{O})]$$
$$b_1^* = \pi^*[p_y(\mathrm{O}) - p_y(\mathrm{C})]$$

The transition moment integral will be

$$\int n[p_x(\mathrm{O})]M\pi^*[p_y(\mathrm{O}) - p_y(\mathrm{C})]\, d\tau$$

expansion of this will give two integrals, one of which, the product of p_x(O) and P_y(C), is small and can be neglected. This approach where the cross products are neglected is general for transitions from orbitals localized on the oxygen [29]. Thus the transition probability depends on integrals containing terms that are dependent on the local symmetry properties at the oxygen atom. In this case the integral is equal to zero and the transition is forbidden. It is generally true that $\pi^* \leftarrow n$ are forbidden when the n orbital is one of pure p character.

If any n orbital contains s character, then the integrals involving an s and p orbital do not vanish and the $\pi^* \leftarrow n$ transition is not formally forbidden (such as pyridine).

This general character of local symmetry dependence has led to the classification of n, π^* states as ^1U for those forbidden by local symmetry and ^1W for those allowed by local symmetry [30]. It should be noted that even though the transition may be allowed with respect to overall symmetry, as for acetaldehyde, it is still forbidden with respect to local symmetry and the transition is designated ^1U.

Despite the symmetry forbidden character of the $\pi^* \leftarrow n$, the transition does occur weakly in many molecules of the carbonyl type as formaldehyde, acetaldehyde, and benzophenone. Consider formaldehyde. The ground state is $^1A_1$ and the lowest n, π^* singlet state is $^1A_2$. The transition moment integral (selection rule for electronic transitions) is

$$\int {}^1A_1 M_q\, {}^1A_2\, d\tau = 0$$

for all M_q in C_{2v} symmetry. With vibronic perturbation such that nontotally symmetric vibrations distort the molecule, however, the integral may not vanish; for example, if an a_2, b_1, or b_2 vibration is excited, then a quadruple direct product is involved. It should be noted that in absorption the geometry of the ground state determines the vibronic selection rule, whereas in emission the geometry of the molecule in the excited state determines the vibronic selection rule. Specifically if a b_1 vibration is involved, then the integral will not vanish with M_q as M_y. Thus a b_1 vibration induces stealing of intensity from the allowed $^1B_2 \leftarrow {}^1A_1$ transition with resulting y polarization of the vibronic bands (see C_{2v} group table, p. 24). A b_2 vibration induces stealing from the allowed ${}^1B_1 \leftarrow {}^1A_1$ transition and an a_2 vibration from the allowed ${}^1A_1 \leftarrow {}^1A_1$ transition.

In the case of formaldehyde there is no a_2 vibration, it is a pseudorotation, and a prominent b_2 vibration is observed. A z component of polarization does occur and the O—O band is weak. These arise because the ${}^1A_2 \leftarrow {}^1A_1$ forbidden transition is allowed by a magnetic dipole transition [31].

Now if acetaldehyde or benzophenone are considered, either the reduction of the symmetry from C_{2v} or vibronic sources could be responsible for the increase in intensity of the $\pi^* \leftarrow n$ transition. Some factors complicate this picture, particularly for those cases that still appear to be of C_{2v} symmetry such as benzophenone and anthrone. The symmetry representations for C_{2v}, C_2, and C_s are as follows:

C_{2v}	C_2	C_s
$A_1(z)$	$A(z)$	$A'(x), (z)$
A_2	$A(z)$	$A''(y)$
$B_1(x)$	$B(x), (y)$	$A'(x), (z)$
$B_2(y)$	$B(x), (y)$	$A''(y)$

In anthrone the ${}^1U \leftarrow {}^1A$ shows a principal vibronic progression with polarization requiring b_1 or b_2 vibrations [32]. In benzophenone, however the principal progression is z polarized and therefore requires an a_2 vibration. In the latter case it is also possible that the molecule is effectively of C_2 symmetry and therefore a vibration of a symmetry in C_2 would account for the z polarization.

For the alkylphenones the $\pi^* \leftarrow n$ transition corresponds to a symmetry allowed $A'' \leftarrow A'$ with y polarization. An a'' vibration could cause stealing from the $A' \leftarrow A'$ transition that is polarized in the plane. The principal progression in alkylphenones (C_S symmetry) is z polarized making the principal source of intensity arise from an a'' vibration. In several aromatic aldehydes (C_S symmetry) a strong progression is polarized out-of-plane and therefore, represents the totally symmetric vibrational sequence. Other weak progressions exist in all cases that indicate other vibronic sources for the overall intensity increase of the $\pi^* \leftarrow n$ transitions [32].

Thus in benzophenone it is likely that the principal progression gains intensity from a totally symmetric a vibration in C_2 symmetry, or a_2 out-of-plane analog in C_{2v}. Therefore, intensity is stolen from the $A \leftarrow A$ allowed electronic transitions. The upper A state arises from splitting of the coupled ${}^1L_b$ states of the two benzene rings [32]. In conclusion, it appears that the source of intensity in the ${}^1U \leftarrow {}^1A$ bands is vibronic in origin which makes the molecule nonplanar. The vibrations would be a_2 in C_{2v}, a in C_2, and a'' in C_s.

The effect of additional substitution on the ketone is pertinent at this point. The substituents discussed previously as the halogens and alkyls still affect the π levels in essentially the same way. However, the energy of the n level, because it is nonbonding, is essentially unaffected. Thus, for example, a halogen would cause the π_2 and π_3^* levels of acrolein (Figure 10) to be pushed up in a manner as shown in Figure 9 for the π_2 and π_3^* levels of ethylene (π_2 affected more than π_3^*). Therefore, the $\pi^* \leftarrow n$ transition would be blue shifted compared to the parent ketone while the $\pi^* \leftarrow \pi$ transition would be red shifted. This difference will be seen again in the consideration of aromatic heterocyclics compared to carbocyclics. It is worth noting that there should be little alteration of the ionization potential. In both the spectral and ionization potential considerations, however, the inductive effect has been considered to be small compared to the conjugative effect. This is generally true from the spectral viewpoint since both levels are affected in the same manner; however, it is less true for the ionization potential consideration because most certainly the n level is specifically affected ($+I$ effect lowers the ionization potential).

It might be well to summarize some characteristics of and criteria for identification of $\pi^* \leftarrow n$ transitions. The $\pi^* \leftarrow n$ transitions are inherently weak because of the lack of spatial overlap between the orbitals. In addition, because of the different symmetry characteristics of the n and π orbitals relative to the molecular plane, the $\pi^* \leftarrow n$ singlet transition is out-of-plane polarized. Generally (but not always) the $\pi^* \leftarrow n$ transiton occurs at lower energy than any $\pi^* \leftarrow \pi$ transition. There is generally vibrational-electronic coupling in $\pi^* \leftarrow n$ transitions. This is characterized by the presence of either a particular stretching group frequency (as in the carbonyls) or a particular ring angular distortion (as in the N-heterocyclics). Finally, $\pi^* \leftarrow n$ transitions undergo blue shifts in hydroxyllic or hydrogen-ion containing solvents relative to their position in nonpolar solvents as hydrocarbons. Very commonly most of the vibrational fine structure becomes blurred in the hydroxyllic solvent system.

Another more or less special consideration is the formation of butadiene by substitution on ethylene. In this case the levels of the substituent are degenerate with those of the ethylene parent resulting in a splitting into four levels. Two of these are closely spaced and two are quite widely spaced (Figure 12). This should result, of course, in a large red shift and notable decrease in the ionization potential. This is verified experimentally where the red shift of the lowest energy transition is approximately 15700 cm^{-1}, and ionization potential is decreased by 1.4 eV.

In the polyene series the effect of substitutions on the spectra is essentially the same as that given above for ethylene, and for the same reasons. A new factor enters that is not present for ethylene—the existence of *cis* and *trans*

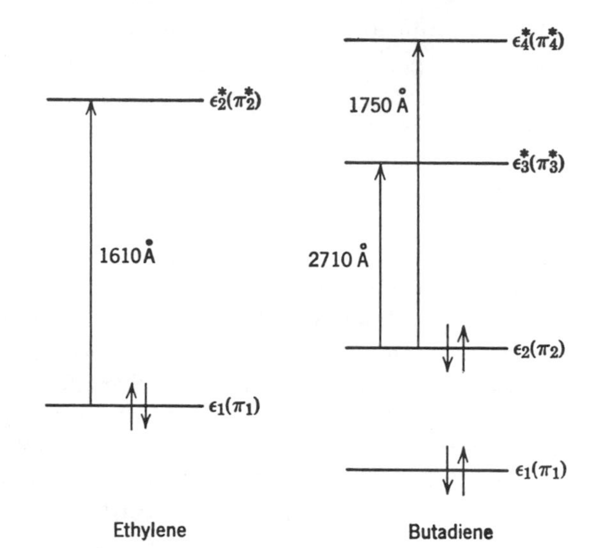

Figure 12 Schematic molecular orbital energy levels of ethylene and butadiene.

isomers. In the *trans* modification, $\varepsilon_4^* \leftarrow \varepsilon_2$ is forbidden (Figure 12). This corresponds to the $h \leftarrow f$ or ${}^1C \leftarrow {}^1A$ transition in the FEMO-Platt nomenclature system (see Figure 4). This transition is allowed in a *cis* form and has become known as the *cis* band. The $\varepsilon_3^* \leftarrow \varepsilon_2$ or $g \leftarrow f$ (${}^1B \leftarrow {}^1A$) transition is expected to be more intense in the *trans* form.

Finally, the problem of steric hinderance must be considered because of the possibilities of twisting about essential single and double bonds. Although it could be appropriate to consider it here, it will be more instructive to consider this factor later relative to the effect of substituents on transitions, (Chapter 5-D).

B. BENZENE AND DERIVATIVES

1. Benzene

In the discussion of group theory we deduced that the lowest molecular orbital transition, $e_{2u} \leftarrow e_{1g}$, was allowed since the double direct product contained a representation (E_{1u}) belonging to M_x, M_y. More specifically the transitions to the B_{1u} and B_{2u} states were forbidden and the transition to the E_{1u} state was allowed. Figure 13 schematically represents the situation in benzene. It can be seen that the $e_{2u} \leftarrow e_{1g}$ transition has four-fold degeneracy that is split up under electron repulsion to give the states noted. The energies for the original orbital scheme can be determined from the secular determinant (6 × 6) by the LCAO technique. The natures of the nodal

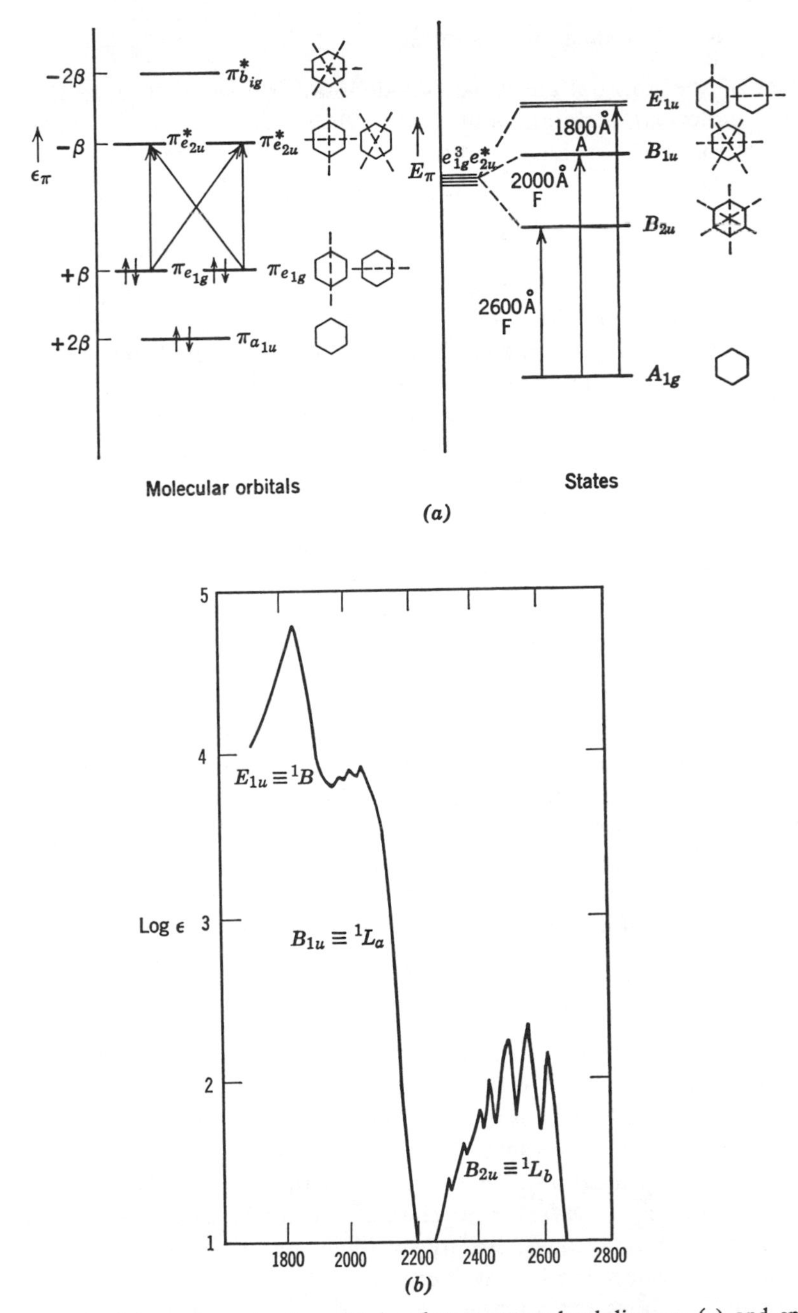

Figure 13 Schematic molecular orbital and state energy level diagram (*a*) and spectrum (*b*) of benzene, F = forbidden and A = allowed.
Reprinted by permission from American Institute of Physics, J. Petruska, *J. Chem. Phys.*, **34,** 1120 (1961).

properties of the appropriate wave functions are shown in Figure 12 for both the molecular orbitals and states. For example, the functions appropriate to the e_{1g} orbitals would be

$$\psi_{e_{1g}} = 2\phi_1 + \phi_2 - \phi_3 - 2\phi_4 - \phi_5 + \phi_6 \tag{48a}$$

$$\psi_{e_{1g}} = \phi_2 + \phi_3 - \phi_5 - \phi_6 \tag{48b}$$

The benzene absorption spectrum shows the presence of the B_{2u} and B_{1u} transitions despite the fact that they are forbidden by symmetry considerations (Figure 12). It must be remembered, however, that the forbiddeness refers to only the purely electronic transitions (zero vibrational level of A_{1g} to zero vibrational of B_{2u} or B_{1u}). If a vibrational quantum is added, the concern is no longer with a triple but a quadruple direct product; that is, it includes the representation to which the particular vibrational mode belongs. Thus we find that if the vibration belongs to e_{2g}, the direct product will transform properly and the vibronic (electronic-vibrational) transition will not be formally forbidden. In a vapor spectrum, however, the O—O band or purely electronic transition will not be seen in conformity with the prediction. The e_{2g} vibration is one that modifies the hexagonal geometry thereby modifying the previously rigorous D_{6h} symmetry. In other words, we could say quantatively that since the symmetry is no longer hexagonal (and, therefore, not D_{6h}), the transitions (vibronic) are no longer forbidden. The group theoretical approach provides the rigorous argument. In the vapor we then see a vibronic band corresponding to the e_{2g} vibration (520 cm^{-1}) in combination with a totally symmetrical ring-breathing vibration (a_{1g} 923 cm^{-1}). It should be noted that the B_{2u} and B_{1u} transitions (vibronic) are still relatively weak since the deviation from D_{6h} symmetry resulting from the e_{2g} vibrational coupling is not great. Thus the B_{2u} and B_{1u} transitions are nowhere as intense as the fully allowed E_{1u} transition (Figure 13).

2. Substituted Benzene Derivatives

Monosubstitution of benzene has several important consequences. First, the O—O band of the first transition can be seen. Second, there is red shifting of the spectrum and lowering of the ionization potential.

We shall first direct our attention to the lowest energy transition (the 2600 Å, B_{2u} or ^1L_b transition). The presence of the O—O band obviously indicates a relaxation of the forbidden character formerly associated with the transition. This occurs because, strictly speaking, the topological symmetry is reduced to C_{2v} where the lowest energy transition is not forbidden. It is vital to point out, however, that we are concerned about the symmetry of the electronic wave functions in the transition moment integral. Thus the important point is that although the topological symmetry is drastically altered, the electronic symmetry is not. Therefore, the O—O band appears, although comparatively weakly, and the ${}^1L_b$ transition (vibronic band system) is also still quite weak. It is obvious that the more the electronic symmetry is modified from D_{6h}, the stronger will be the previously forbidden transitions. Thus, the intensity will vary depending on the inductive and conjugative strengths of the substituents. Although no rules can be given, certain guiding principles are useful: (a) for forbidden transitions the intensity will increase upon substitution with the inductive effect more influential than the conjugative effect; (b) allowed transitions may increase or decrease in intensity, but the concentration of either plus or minus charge by induction generally increases the intensity.

The reason for and the amount of shifting of the ${}^1L_b$ transition again depends on the same factors and mechanisms previously discussed for ethylene; that is, the conjugative effect moves all the levels up but the highest filled more than the lowest empty with a resultant red shift. Table 3 gives some examples. The inductive effect has a more prominent influence on the spectra of the benzene derivatives than on the ethylene. For certain substituents such as amino and nitro, the effect of the substituent cannot be viewed as simply a perturbation on existing levels as was done for ethylene, etc., and implied above for benzene. This can be seen in the spectrum where extraordinarily large spectral changes occur. The transitions are not simply perturbed benzene transitions but may result, for example, from the interaction of benzene states and charge transfer states.

The effects of substituents on the ionization potential are parallel to that discussed for ethylene; that is, the ionization potential is lowered by the conjugative effects and $+I$ inductive effect and increased by the $-I$ inductive effect (Table 3). For alkyl groups it appears as if the influence of the inductive

Table 3 Change in Ionization Potential and Band Shifts for Some Derivatives of Benzene

Compound/Benzene	ΔIP[a] (9.24)(9.52)	$\Delta\bar{\nu}$ B_{2u} (38000 cm^{-1})	$\Delta\bar{\nu}$ B_{1u} (49500 cm^{-1})	$\Delta\bar{\nu}$ E_{2u} (54400 cm^{-1})
Fluorobenzene	−0.04 +0.15	−400	−1500	−1600
Chlorobenzene	−0.47 −0.10	−900	−2900	−1700
Bromobenzene	−0.11	−1600	−3000	−2000
Iodobenzene	−0.42	−1700	−5600	−3000
Methylbenzene	−0.42	−600		
Ethylbenzene	−0.47	−565		
i-Propylbenzene	−0.48	−350		
t-Butylbenzene	∼−0.68	−300		
Vinylbenzene	−0.64	−3800	−8900	

[a] The first column is photoionization data and the second is electron impact.

effect is not overcome by the hyperconjugative effect. This is supported by the ionization potential of the alkyl derivatives.

The substitution of certain groups as vinyl, to give styrene, cannot be treated as simply as above. This occurs because new (not just perturbed) π levels are introduced. Thus new transitions result.

Polysubstitution of benzene also has several interesting consequences. The most striking of these is the effect that the number and location of the substituents have on the intensity of the ^1L_b (B_{2u}) transition. The intensity is proportional to the square of the transition moment that can be represented by a vector whose direction is parallel to the plane of the ring and perpendicular to the substituent-benzene carbon bond. Another way to view this is that the B_{2u} state has dipolar symmetry and the vector sum of these is zero

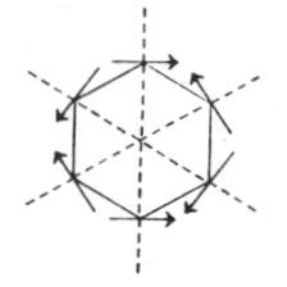

(thus the transition is forbidden) as illustrated. The intensity of the ^1L_b as implied above is vectorially additive,

$$\Delta I = M_i^2 + M_j^2 - 2M_iM_j \cos\theta_{ij} \tag{49}$$

which is appropriate for two substituents where θ is the angle between the substituents.

Substituents can be assigned spectroscopic moments [33] and these may be used in (49) to calculate the change in intensity relative to that of benzene itself. The sign will vary depending, in a general way, on whether the substituents are *ortho-para* or *meta* directing. Of great qualitative significance are the relative intensities expected for substitutions of the same kind; for example consider a 1, 2, a 1, 4, and a 1, 3, 5 substitution that give results as follows:

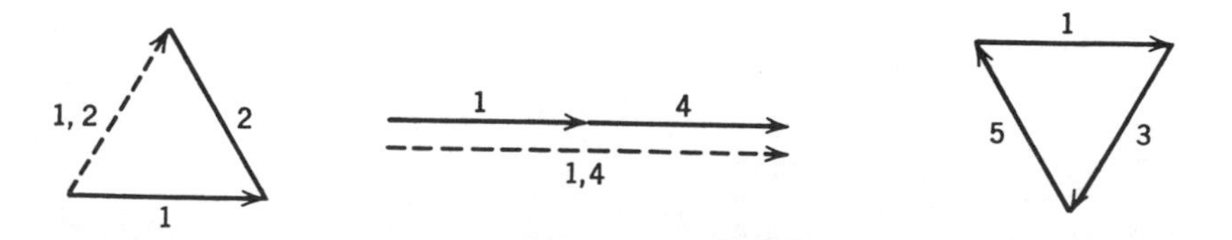

Consider a monosubstitution having a relative intensity of one or twice that of benzene, then 1, 2 is the same as that of the monosubstituted case, 1, 4 is two times that of the monosubstituted case, and 1, 3, 5, has a relative intensity of zero, or it is forbidden (like benzene itself).

From perturbation theory the wavelength shifts are predicted to be nearly additive for weakly interacting substituents; that is, the shift for a dichloro derivative would be twice that of a monochloro. As implied earlier, strongly interacting groups as amino and nitro cannot be treated in this manner. The experimental results are in good agreement with expectation.

Substitutions involving groups of the aldehyde and keto type have much the same consequences as for ethylene; that is, for example, a new transition of the $\pi^* \leftarrow n$ type arises. An example of this would be benzophenone. The intensities of such transitions are weak ($\varepsilon \approx 110$ for benzophenone). Additional substitution on the benzene ring affects the transitions in much the same way as noted for ethylene. Vapor phase studies for two aldehydes have indicated that the carbonyl bond is approximately 0.1 Å longer in the n, π^* excited state than in the ground state [34]. Benzoic acid has been shown to dimerize by hydrogen bonding in solution [35]. Also, benzoic acid hydrogen bonds with many other molecules, and the spectral shift (whether to the blue or to the red) depends upon whether benzoic acid acts as a donor or an acceptor in the complex.

The effect of substitution on the 2000 Å transition (^1L_a or B_{1u}) is similar in most respects to that for the 2600 Å one. The one difference is that the conjugative effect has a more marked influence on the shifting of this band than it did for the 2600 Å or ^1L_b transition.

C. CONDENSED AROMATIC HYDROCARBON

There are three subclasses within this group: the linearly condensed, nonlinearly condensed, and pericondensed. Examples of these are naphthalene, phenanthrene, and pyrene, respectively (Figure 14). In Chapter 2 certain theoretical considerations of the even and odd alternant hydrocarbons were discussed. Unfortunately a multitude of nomenclature systems exists for the bands of these molecules. Table 4 compares the most common systems. Here the ^1B_b transition has been correlated with Clar's strong band β since it would be expected that it would be polarized along the long axis. We will attempt to use the Platt system where possible and the group theoretical system as a supplement or replacement when necessary. In general there are three distinctive regions of absorption—the ^1L_b with $\varepsilon = 100$–1000, the ^1L_a with $\varepsilon = 1000$–$10{,}000$, and ^1B_b with $\varepsilon \sim 100{,}000$. Considerable care must be exercised, however, since these generalizations refer only to the lower-lying transitions (within the first four). In certain cases, such as anthracene, the first absorption region lies under the second and for the tetracyclics, at least, one other transition lies between the second and third regions of absorption referred to above. There are more transitions than those referred to

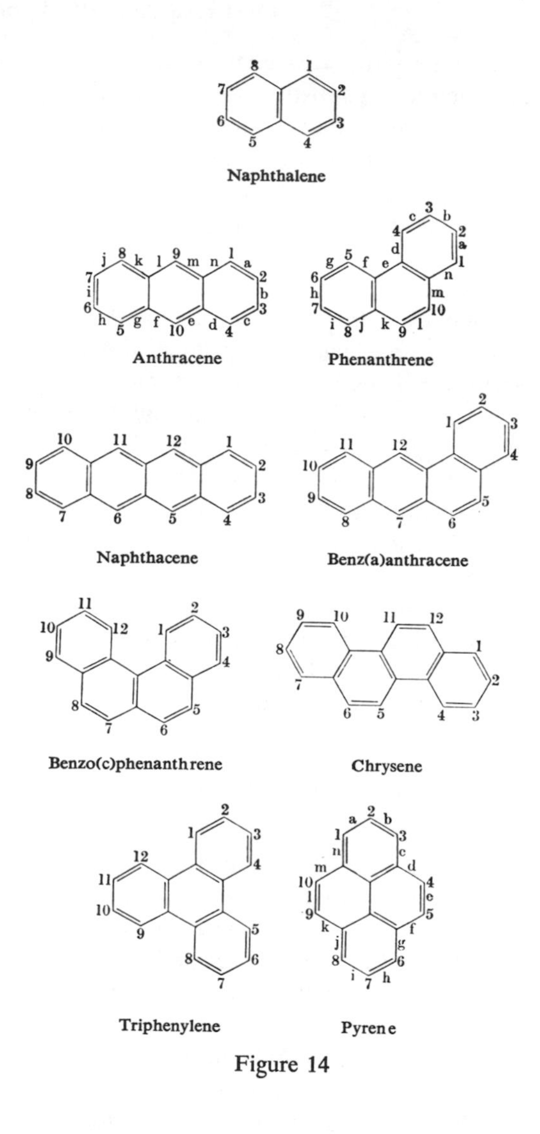

Figure 14

Table 4 Comparison of Nomenclature

Author				
Platt [36]	^1L_b	^1L_a	^1B_b	^1B_a
Clar [37]	α	ρ	β	
Moffitt [38]	V	U	X	Y

which occur at higher energies with $\varepsilon = 10^4 - 10^5$. More will be said on these points later.

The linearly condensed hydrocarbons, or polyacenes have planes of symmetry along both their short and long axes. This symmetry in the molecule dictates that the molecular orbitals must be symmetric or antisymmetric with respect to reflection through these planes. Therefore, the transition moment can induce the antisymmetry required for an allowed transition only if it is perpendicular to one of these planes. The transition moment for allowed transitions can then be either perpendicular ($\perp$) or parallel ($\|$) to the long axis of the molecule; or, alternatively, allowed transitions have either transverse (short axis) or longitudinal (long axis) polarization. Molecular orbital theory predicts that the long wavelength absorption has transverse polarization. Free-electron theory based on the perimeter carbon atoms [1], and another more detailed orbital theory based on the perimeter atoms [39] predict longitudinal polarization.

The orientation of the transition moment can be determined experimentally from the absorption of polarized light by suitably oriented crystals. The results from single crystals of pure naphthalene are not unambiguous [40] because of resonance interaction between the naphthalene molecules. In solid solution in durene this interaction is reduced and the crystal spectra show rather clearly longitudinal polarization of the long wavelength band [41]. The long wavelength band in the 3120 Å region results from a transition to the ^1L_b state. The next transition in the 2750 Å region has been assigned to the ^1L_a state and is short-axis polarized. The last and strongest transition of interest that occurs in the 2200 Å region is to the ^1B_b state and is long-axis polarized.

The orientation of the transition moment may also be determined by studying the effect of substitution. It has been suggested from this line of evidence that the long wavelength band in anthracene is short-axis polarized [29,42]. The increase in intensity in 1, 5- and 2, 6-dichloronaphthalenes has led to the suggestion that the long wavelength transition is an allowed one [43]. It has been shown that the longest wavelength band is the ^1L_a and is short-axis polarized [44,45]. Although this would seem to settle the issue the question as to the fate of the ^1L_b transition is not clear. This was the longest wavelength one in naphthalene. In general it has been assumed that it is under the ^1L_a transition in anthracene. However, both room temperature and particularly low temperature absorption spectra, where the resolution is excellent, apparently give no hint of the transition. Thus it has either become so weak that it has effectively disappeared or shifted by an extremely large amount compared to naphthalene. The latter does not seem reasonable. If it is in fact the former, no explanation has been offered for such a result.

The next member of the series is naphthacene with four linearly condensed

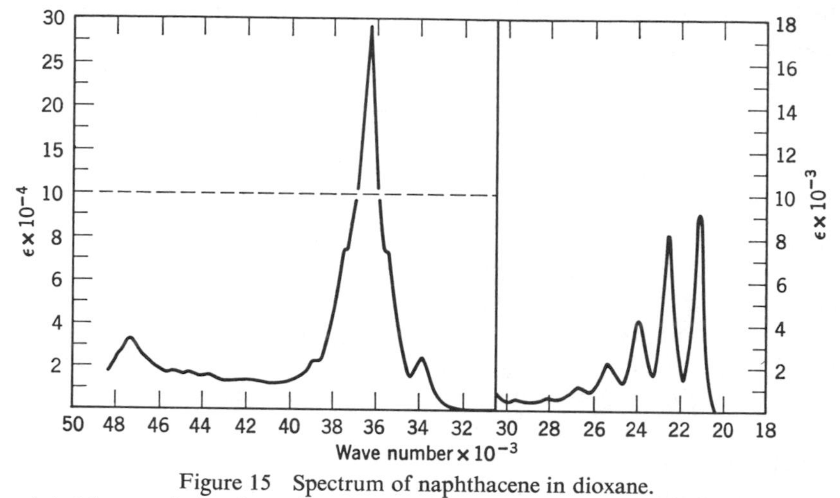

Figure 15 Spectrum of naphthacene in dioxane.
Reprinted by permission from American Institute of Physics, R. S. Becker, I. S. Singh, and E. A. Jackson, *J. Chem. Phys.*, **38**, 2144 (1963).

rings. This compound is deep orange whereas all of its predecessors have been colorless. This has occurred because the longest wavelength band is now moved into the visible to about 4700 Å (see Figure 15). There still is some question about the assignment of the longest wavelength band in naphthacene. Calculations are not in full agreement and neither is experiment. Assuming that this band is ^1L_a, then the fate of the ^1L_b band is again the problem. From polarization data the band has been assigned as ^1L_a [46]. On the basis of the anthracene data the ^1L_a assignment would seem to be most consistent. In any event it has been shown that another transition lies between the longest wavelength one and a very intense one in the region of 2750 Å (corresponding to the 2500 Å transition in anthracene) [47]. Apparently this is not the case in anthracene. It could be that this transition is the ^1L_b long-axis polarized transition. If this is true, however, then the ^1L_a transition has *red* shifted some 13,500 cm^{-1} while the ^1L_b has *blue* shifted some 1800 cm^{-1} proceeding from naphthalene to naphthacene and resulting in an unprecedented circumstance. It is apparent that some further investigation is needed to clarify the problem of assignment of the transitions in naphthacene.

Continued linear condensation progressively moves the longest wavelength band more into the visible. This is caused by the decrease in the energy difference between the energy of the lowest excited and ground states. The difference becomes progressively less as the number of condensed rings increases in the linear polyacene. Figure 16 shows the trend for the ^1L_b and ^1L_a transitions as a function of the number of rings. Pentacene (five rings) and hexcene (six rings) are almost black. Although some spectral work has been done on these [37], practically no assignments have been made.

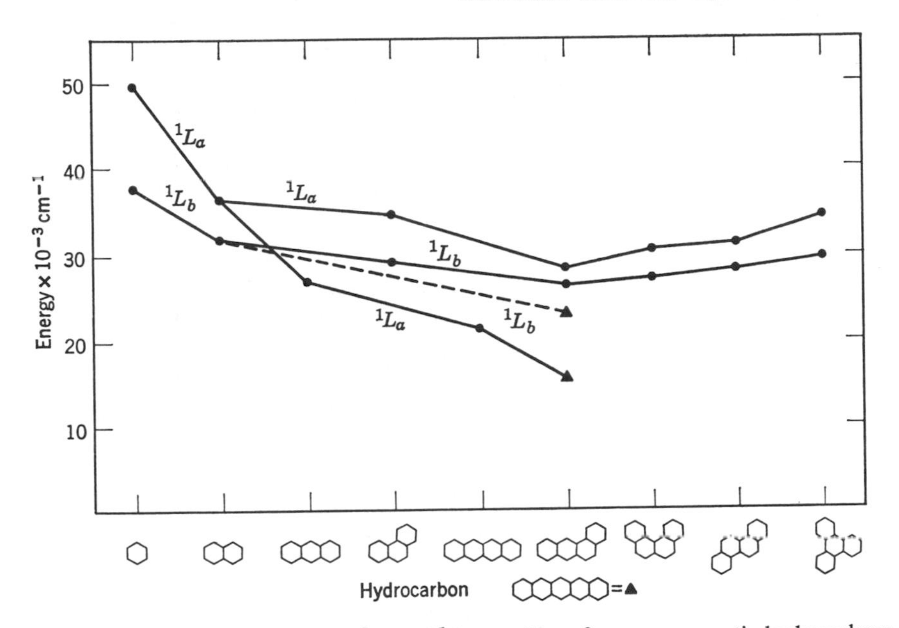

Figure 16 Trends in energy of the ^1L_a and ^1L_b transitions for some aromatic hydrocarbons.

The nonlinear hydrocarbons do not generally possess the well-defined axes of the linear hydrocarbons. The first member of this group is phenanthrene. The molecular orbital energies for this case were given in Figure 3. The molecular orbital theory correctly predicts that phenanthrene should absorb at longer wavelengths than naphthalene and at shorter wavelengths than anthracene. The compounds of general interest to us include a tricyclic phenanthrene (see above) and five tetracyclics. Although spectra are known for hydrocarbons with larger numbers of rings, little has been done relative to assignment of the transitions. In these cases the trend of energy difference between the first excited and ground states is not a simple one as in the case of the linear polyacenes. However, the energy difference in general increases as the number of "bends" in the molecule increases as shown in Figure 16.

There have been extensive theoretical and experimental studies on the nonlinear hydrocarbons [47]. It is found that in some cases a surprisingly large number of transitions exist between the onset of absorption and approximately 1950 Å; for example benz(*a*)anthracene has ten transitions (see Figure 17). Through the use of a broad variety of data including pressure, substitutional and solvent effects, band intensity, vapor spectra, and vibrational analyses, assignments have been made for all the tetracyclic hydrocarbons [47] except for the case of naphthacene in which problems still remain as previously noted. In this group the first two transitions are ^1L_b and ^1L_a, respectively. For pyrene, however, the first transition is short-axis

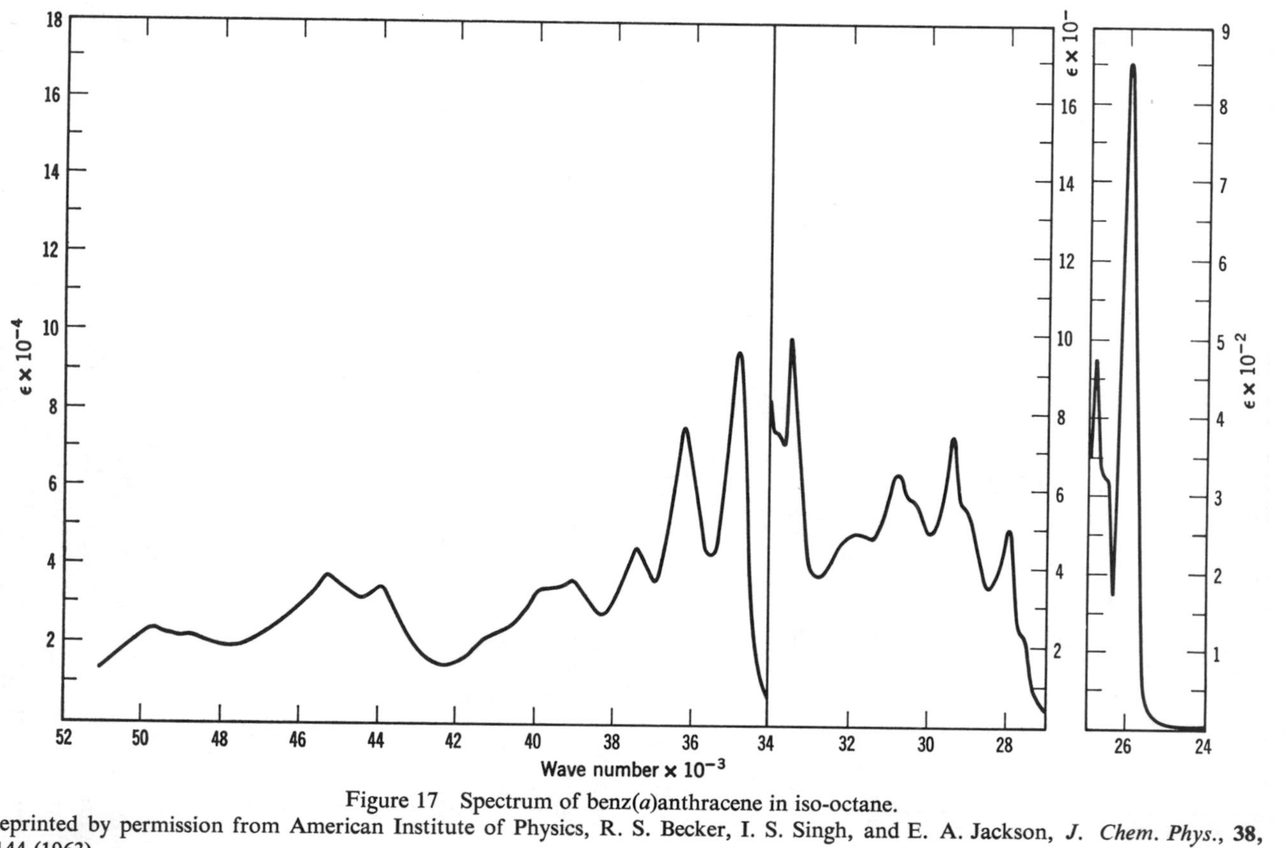

Figure 17 Spectrum of benz(*a*)anthracene in iso-octane.
Reprinted by permission from American Institute of Physics, R. S. Becker, I. S. Singh, and E. A. Jackson, *J. Chem. Phys.*, **38**, 2144 (1963).

and the second long-axis polarized. Except for chrysene and triphenylene there is a transition energetically lower than the ^1B_b. It is likely that this is a short-axis polarized transition of the ^1B_a type [47]. In triphenylene the ^1B_a and ^1B_b are degenerate and the first two transitions are forbidden but become vibronically allowed by an asymmetric vibration-electronic coupling.

Methyl substitution in any position generally red shifts the lowest four transitions. As implied by the use of the word "generally" there are a few exceptions. Multiple methyl substitution causes the second transition, ^1L_a, to undergo a substantially larger red shift than the first transition, ^1L_b. In fact, in one trimethyl substituted benz(*a*)anthracene, the ^1L_a shifts sufficiently for the ^1L_b to be no longer discernible. Fluoro substitution on benz(*a*)-anthracene causes both red and blue shifts of the first four transitions depending most strongly on the substituents position of substitution. For both methyl—and fluoro—the predominance of blue shifts occurs when substitution is at carbon number 1 (see Figure 14).

There are two other general areas of concern relative to aromatic hydrocarbons. These are the electron affinity and ionization potential values of the neutral molecules and the nature of the transitions in ions. Although ionization potential data have been known for some time, only recently has it been possible to determine electron affinities [48–51]. Table 5 shows both ionization potential and electron affinity data for a significant number of hydrocarbons along with some theoretically predicted values. These data are useful when considering charge transfer spectra.

The highest occupied orbital $\alpha + x\beta$, and lowest unoccupied orbital $\alpha - x\beta$, are arranged symmetrically about a midpoint α. The quantity $\alpha + x\beta$ is numerically equal to the ionization potential and $\alpha - x\beta$ is equal to the electron affinity. In addition because of the energy level arrangement, x is a linear function of the energy difference between the two orbitals. Further, the lowest molecular orbital transition energy is associated with the difference in energy between the $\alpha + x\beta$ and $\alpha - x\beta$ levels, ^1L_a. Therefore, a linear relationship should exist between the energy of this transition and the electron affinity and the ionization potential. The relationship between all the quantities discussed is shown in Figure 18. It should be emphasized that the lowest observed transition may not correspond to the ^1L_a transition [50,51].

Because of the pairing properties of the orbitals, the spectra of positive and negative ions should look quite similar. In addition, the spectra should be generally similar to the parent with the addition of a low energy transition. The orbital scheme would appear as follows for the neutral species:

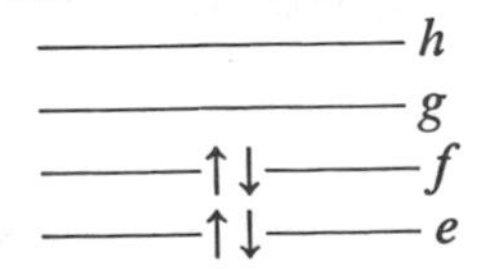

Table 5 Electron Affinities, Ionization Potentials, and Electronegativities for Some Aromatic Hydrocarbons[a]

	Electron Affinity (cV)	Ionization Potential[b] (eV)	$\chi = \frac{1}{2}$ (IP + EA)[c]
Naphthalene	0.15	8.26(W) 8.12(P.I)	4.20 4.13
Anthracene	0.56	7.55(W) 7.66(S)	4.11 4.05
Phenanthrene	0.31	8.03(W) 8.96(S)	4.18 4.17
Benz(*a*)anthracene	0.63	$(7.67 \pm 0.10)^d$ 7.54 ± 0.3^e	
Chrysene	0.40	$(7.90 \pm 0.10)^d$ 8.01 ± 0.3^e	
Benzo(*c*)phenanthrene	0.55	7.84(S)	4.19
Pyrene	0.59	7.72(S)	4.16
Triphenylene	0.29	$(8.01 \pm 0.10)^d$ 8.19 ± 0.3^e	
Benzo(*e*)pyrene	0.53	$(7.77 \pm 0.10)^d$	
Benzo(*a*)pyrene	0.68	$(7.62 \pm 0.10)^d$	
Dibenz(*a,h*)anthracene	0.60	$(7.70 \pm 0.10)^d$	
Dibenz(*a,j*)anthracene	0.59	$(7.71 \pm 0.10)^d$	
Picene	0.54	$(7.76 \pm 0.10)^d$	
Naphthacene	1.35^d	6.95 ± 0.3^e 7.15(S)	
	1.42^d	6.88^f	

[a] R. Becker and E. Chen, *J. Chem. Phys.*, **45**, 2403 (1966).

[b] (P.I.) photoionization with naphthalene, from H. Watanabe, *J. Chem. Phys.*, **22**, 1565 (1954); **26**, 542 (1957); (W) M. E. Wacks and V. H. Dibbler, *ibid.*, **31**, 1557 (1959); (S) D. P. Stevenson values, less than 0.56 eV (private communication to F. A. Matsen, 1957).

[c] The two values arise because of two sources of data for ionization potential, $\chi = 4.15 \pm 0.05$ (max deviation).

[d] Predicted values considering $\chi = 4.15 \pm 0.05$ and using the appropriate experimental electron affinity or ionization potential data given in the table. In the case of the electron affinity for naphthacene the three values result from the three ionization potential values.

[e] M. E. Wacks, *J. Chem. Phys.*, **41**, 1661 (1964).

[f] F. I. Vilesoc, *Dokl. Akad. Nauk. SSSR*, **132**, 643 (1960)

There would be one additional electron in level g for the anion and one less in level f for the cation. For the anion the additional transition $h \leftarrow g$ exists and for the cation an additional transition $f \leftarrow e$ exists. By simple molecular orbital theory these would have the same energy. Thus it would be expected that each ion would have a low-lying transition at nearly the same energy. This has been shown to be true for several hydrocarbons [52] including anthracene and pyrene, for example.

D. TWISTING AROUND ESSENTIALLY SINGLE AND DOUBLE BONDS

The principal molecules of concern here will be biphenyl, containing essentially a single bond, and ethylene. Other molecules such as stilbene and

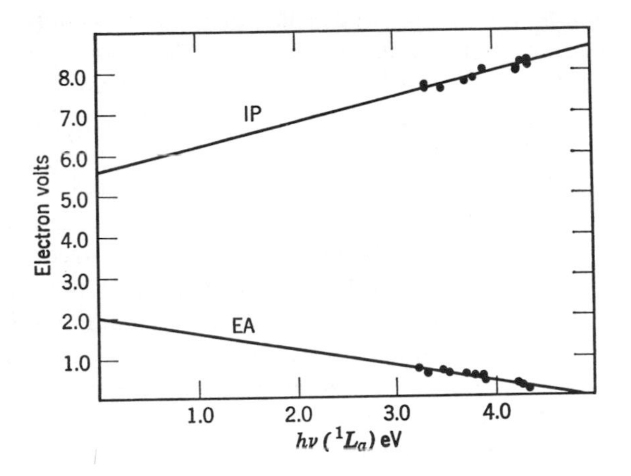

Figure 18 Correlation of electron affinities and ionization potentials with energies of ^1L_a transitions.
Reprinted by permission from American Institute of Physics, R. S. Becker and E. Chen, *J. Chem. Phys.*, **45**, 2403 (1966).

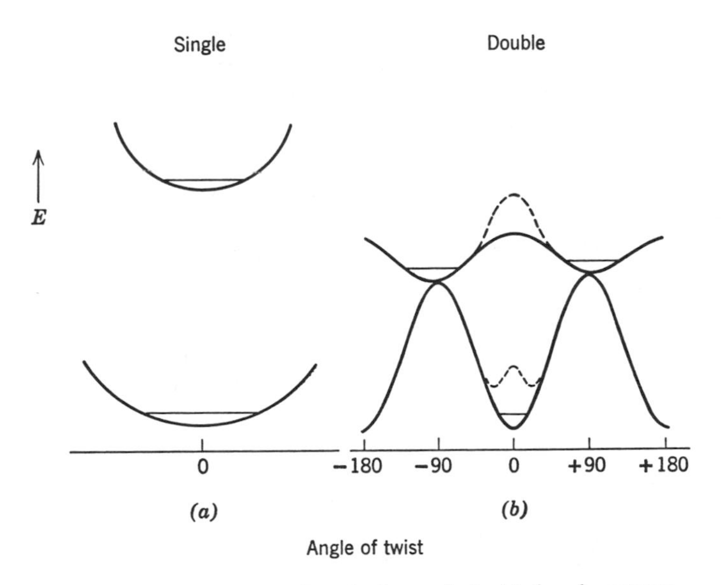

Figure 19 Potential energy curves for single- and double-bond systems.

azobenzene contain both types of bonds. The potential energy curves for these two main classes are shown in Figure 19.

In the case of biphenyl the highest occupied π orbital has a node between the two benzene carbon atoms that are connected by a "single" bond. The lowest empty orbital, however, has no such node. Thus in the excited state the connecting bond has more double-bond character than in the ground state resulting in the potential energy curves shown in Figure 19. Because of the differences in the shape of the potential energy curves, the O—O band will not be the strongest unless the angle of twist is very slight; that is, the band shape will be Franck-Condon forbidden in character. This is the case for biphenyl itself. When *ortho* substitution is made, a barrier is created at 0° where the height depends on the degree of hinderance and prevents planarity. Again, because of the relative shapes of the ground and excited-state potential energy curves, the most allowed transition is from the zero level to an even higher vibrational level in the excited state than before. Thus the maximum of the first transition is at shorter wavelengths than before (a blue shift).

In the case of ethylene, because the ground state is planar, the minimum in the potential energy curve occurs at 0°. The lowest empty π-molecular orbital has a node between the two atoms and this orbital is less stable than the $2p$ atomic orbitals from which it was made. For this and other reasons the geometric configuration of greatest stability is when the $2p$ orbitals are at right angles. In view of this the minimum of the potential energy curve in the excited state occurs when the angle is 90°. This relation results in a weak O—O vibrational band giving a Franck-Condon forbidden shape. Substitutions can lead to steric strain, depending on their size and their relative location. This results in a barrier at 0° (see dotted curve, Figure 19b.) Therefore, the energy difference between the ground and excited state will be decreased and a red shift will result. We should also note that because the minima are at angles different from 0° in the ground state there will likely be a change in shape of the band relative to the unsubstituted ethylene.

In the stilbenes substitution on the benzene portion in the *ortho* positions causes a blue shift. This is consistent with a twisting around the "single" bonds connecting the benzene rings to the ethylene carbon [29].

Chapter 6 Absorption Spectra of Heterocyclics

The aza aromatic cases have been most studied and are probably the best understood. Before proceeding to the aromatic cases, a few comments will be given on the five-membered ring cases. Table 6 compares the oxygen, sulfur, and nitrogen analogs of cyclopentadiene. There is considerable variation in data and interpretation of the spectra of these systems. Also certain notable variations in band intensities occur, for example, the furan band at 2500 Å has molecular absorption coefficient of only approximately 1 [53] while for the other compounds with bands in approximately the same region, the intensities vary from approximately $\varepsilon \sim 300$ to 520 [53,54]. Also there are some unexpected differences based on the solvent; for example, thiophene shows at maximum near 2350 Å with $\varepsilon \sim 4500$ in hexane [55] and near 2310 Å with $\varepsilon \sim 7100$ in isoctane [56].

A qualitative interpretation nevertheless shows that the spectrum of thiophene most closely resembles benzene. This might be expected since the ionization potential of the sulfur lone pair electrons would be the least and

Table 6 Bands of Some Five-Membered Ring Compounds

Compound	λ (Å max)	λ (Å max)
Cyclopentadiene	2390	1910
Furan	2500	2070
Pyrrole	2400	2070
Thiophene	2310	2280

thus show the most participation in delocalization. Other factors, such as variation in nature of hybridization of the heteroatom, must play a role.

A. PYRIDINE AND OTHER AZA HETEROCYCLICS

Substitution of N for C results in two principal new features: (1) reduction in the symmetry and (2) introduction of lone pair electrons. The first of these is responsible for the increase in intensity of the formerly forbidden bands in benzene (B_{2u} and B_{1u}, see Figure 11). These do not become fully allowed despite the obvious reduction of skeletal symmetry from D_{6h} to C_{2v}. This indicates and emphasizes the fact that skeletal symmetry is not the factor of prime importance but it is the change in the symmetry of the electronic functions that is of concern. Thus the π-electron functions or charge cloud has been distorted from D_{6h} symmetry but only to a relatively minor degree. Another way to consider this is (similar to the case of benzene) to utilize the vector diagrams for the B_{2u} (^1L_b) transition (Chapter 5-B). In the case of pyridine the vector length at N is different than those at the carbons and a residual moment remains (contrary to benzene where the vector sum equals zero). The principal band assignments are undoubtedly similar to benzene.

The introduction of n orbitals permits an additional transition of the $\pi^* \leftarrow n$ type to occur. This is expected to be at longer wavelengths than the lowest $\pi^* \leftarrow \pi$ ($^1B_{2u}$) transition because the n orbital is at still higher energy than the highest π orbital. In pyridine it is difficult to resolve the $\pi^* \leftarrow n$ transition in the tail of the lowest $\pi^* \leftarrow \pi$ but it does occur as a shoulder with a maximum near 2700 Å. The transition is symmetry allowed although only the s part of the nonbonding orbital contributes to the transition probability or intensity of the transition. As noted previously $\pi^* \leftarrow n$ transitions are generally weak. This occurs because of poor or negligible overlap of the n and π^* orbitals [58]. The $\pi^* \leftarrow n$ transition of substituted pyridines also cannot be clearly identified and is some sixteen times weaker than the adjacent $\pi^* \leftarrow \pi$ [59]. Calculations on the location of the lowest n, π^* states generally agree with experiment [60].

The $\pi^* \leftarrow n$ transitions are predicted to blue shift with a purely conjugative substitution. This is *opposite* to that expected for substituents of the same nature for the alternant aromatic hydrocarbons. The reasons for this result are parallel to those given for the effect of substitutions on the $\pi^* \leftarrow n$ transition of carbonyls (Chapter 5-A); that is, the π and π^* levels are affected in the usual manner while the n level is less affected.

Replacement of another of the CH groups by a nitrogen atom in the *ortho*, *meta*, or *para* positions to form pyridazine, pyrimidine, or pyrazine, respectively, moves the $\pi^* \leftarrow n$ transition to longer wavelengths (3400 Å, 2980 Å and 3400 Å, respectively). The unusually large red shift in pyridazine arises because of significant overlap of the adjacent n orbitals giving combinations

$$n_1 + n_2$$
$$n_1 - n_2$$

and a splitting such that the $\pi^* \leftarrow (n_1 - n_2)$ transition occurs at significantly lower energy than the usual $\pi^* \leftarrow n$. The lowest $\pi^* \leftarrow \pi$ transition intensity of pyrazine would be expected to be considerably more intense than that of pyridine based on the vector diagram model. Indeed this is the case (approximately three times greater compared with the theoretical expected value of four times if the two nitrogen atoms were independent). Also, the $B_{2u}(^1L_b)$ transitions are blue shifted in two cases and red shifted in the other (pyrazine).

Further substitution gives molecules such as *s*-triazine and *s*-tetrazine. The $\pi^* \leftarrow n$ transition of *s*-triazine is near the diazines (~2700 Å) and is quite intense ($\varepsilon \sim 1050$, twice that of pyridine). The ^1L_b transition at 2220 Å would be expected to be of comparable intensity to that of benzene since in the vector diagram model, the vector sum is zero. This is the case ($\varepsilon \sim 150$ versus $\varepsilon \sim 110$). In *s*-tetrazine there are two $\pi^* \leftarrow n$ transitions located at approximately 5500 Å and 3200 Å with the latter being very weak. These most likely are transitions from the $n_1 \pm n_2$ nonbonding orbital combinations. The ^1L_b transition is very similar to that of pyridine in energy and intensity.

The N lone pairs are adequately described by assuming sp^2 hybridization in the ground state. On excitation, however, rehybridization occurs and the remaining electron resides in an orbital with approximately only 10 percent s character [61].

For the azanaphthalenes [62] the lowest $\pi^* \leftarrow n$ transitions are at considerably longer wavelengths than those of pyridine and diazines but at shorter wavelengths than *s*-tetrazine. For example, benzpyrazine (quinoxaline) shows an $\pi^* \leftarrow n$ transition as a shoulder at 3500–3800 Å. This transition blue shifts in water while the $\pi^* \leftarrow \pi$ transitions are little shifted. The spectra of the mono-aza anthracenes and phenanthrenes bear resemblance to those of the alternate hydrocarbons, although the fine structure is decreased and relative band intensities are changed. In particular, acridine is significantly different than anthracene in the 2600–3800 Å region. Acridine has two clearly defined transitions while anthracene does not. Diazoanthracenes, such as phenazine, show only a long wavelength tail that contains the $\pi^* \leftarrow n$ transition.

Hydrogen bonding affects the absorption spectra of N-heterocyclics [63,64]. There is evidence that the hydrogen bond does not exist or is very weak in the excited singlet and triplet states [63,64]. This leads to the proposition that the blue shift of the $\pi^* \leftarrow n$ transition should be greater than the energy of the hydrogen bond in the ground state by an amount equal to the Franck-Condon destabilization energy. In addition, in the case of emission spectra only the Franck-Condon destabilization energy is important in determining the shift. Moreover these shifts should be small and to the *red* of that for nonhydrogen bonded (or protonated) molecules of the N-heterocyclic type. The long wavelength bands of acridine change as a function of the age of ether in a solvent (at 77°K) [65]. This can not be seen at room temperature. The bands presumably arise from complex formation with a peroxide impurity. The complex can further coordinate with alcohol or hydrogen ion giving trimolecular complexes. Proton addition to acridine causes a very large shift of the lowest energy transition as is typical of many of the N-heterocyclics. With the addition of supporting data from emission spectra [65], it seems clear that acridine does not have an n, π^* singlet state lowest but, rather, a π, π^* state.

Some data also exists on molecules of particular interest in biological systems, such as the purines. For example, 9-methylpurine shows an $\pi^* \leftarrow n$ transition as a marked shoulder in the 2910–3200 Å region. This shoulder undergoes a strong blue shift in a water-acid solution (pH = 5); however, it appears still to be present in the 2900 Å region. There have also been studies on DNA. With the help of optical rotatory dispersion data it is evident that an $\pi^* \leftarrow n$ transition exists in the long wavelength tail of the DNA spectrum at around 2900 Å.

Finally, we can make some generalizations concerning the effects of substitution on the $\pi^* \leftarrow \pi$ of aza aromatics. The first $\pi^* \leftarrow \pi$ transition of pyridine undergoes red shifts with almost all substitutions. There may be a change in the direction of the shifts, however, depending on the position of substitution. In general, substituents cause red shifts of the first $\pi^* \leftarrow \pi$ transition of the diazines. Most substituents produce red shifts in the aza naphthalenes with the exception of certain positions where the fluorine atom is involved. The same is generally true for the aza tricyclics as well.

B. PORPHYRINS, CHLOROPHYLLS, AND DYES

A significant amount of investigation of these compounds has taken place in recent years regarding both electronic absorption and emission (Chapter 8-A). No attempts will be made to discuss these compounds in terms of their significance in biological processes. Figure 20 gives structures for some compounds to be considered.

1. Porphyrins

All free base porphyrins show similar four-band visible spectra of moderate intensity (ε = 5000–30,000) in the region of 5000–6000 Å and a notably

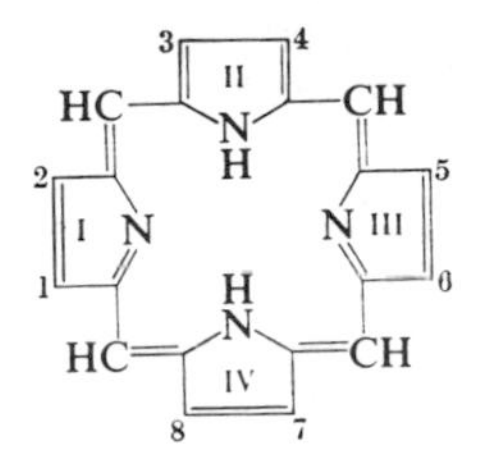

Porphyrin nucleus (porphin)

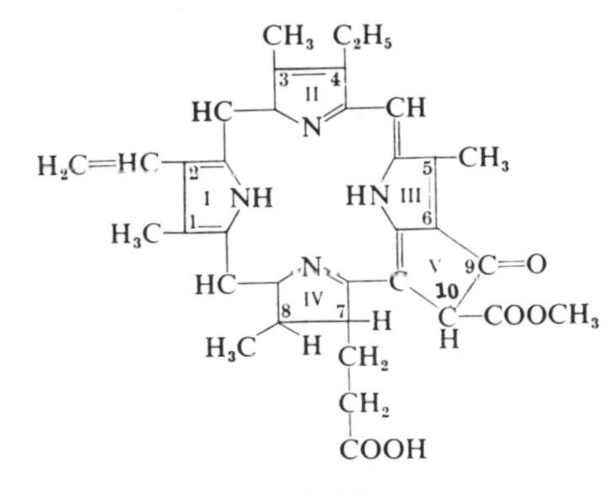

Pheophorbide *a*

1. Mesoporphyrin IX, dimethyl ester-1,3,5,8-tetramethyl-2,4 diethyl-6,7-di-(methyl propionate) porphyrin.
2. Metallo-porphyrin: same as porphyrin with metal complexed in center.
3. Phthalocyanine: same as porphyrin except methine carbons are replaced by nitrogen atoms and four phenyl groups are fused at positions 1, 2; 3, 4; 5, 6; 7, 8.
4. Chlorophyll *a*: same as pheophorbide *a* except magnesium complexed in center of ring and propionic acid residue in position 7 is esterified with phytol alcohol ($C_{20}H_{39}OH$).
5. Chlorophyll *b*: same as chlorophyll *a* except that a formyl group is substituted for the methyl group at position 3.

Figure 20 Structures of porphyrins and chlorophylls and their derivatives.

intense band (ε = 300,000–600,000), called the *Soret band*, in the 4000 Å region. The substitution of a metal into the porphyrins affects the spectrum most strikingly by altering the four-band visible region into a predominately two-band spectrum. Different metals primarily affect the intensity of the long wavelength bands and produce a shifting of the entire spectrum.

There have been several theoretical attempts made to explain these spectra [66–71]. These have varied in their approach beginning with the relation of porphin to an 18-membered cyclic polyene [66], LCAO-MO calculations [67,69], a vector model [68], SCF-MO calculations with configuration interaction and, finally, a type of combination approach derived from the theory

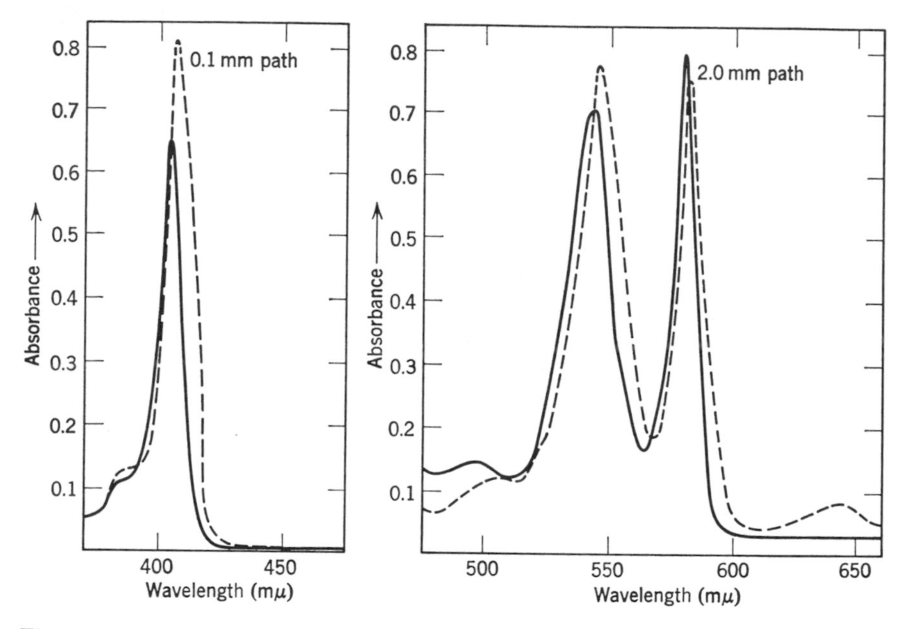

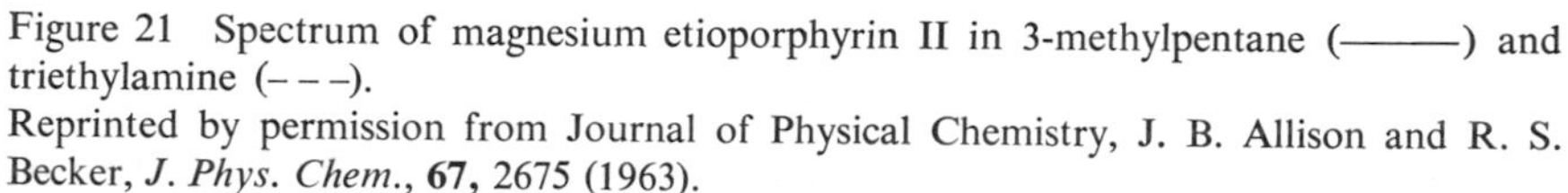

Figure 21 Spectrum of magnesium etioporphyrin II in 3-methylpentane (———) and triethylamine (– – –).
Reprinted by permission from Journal of Physical Chemistry, J. B. Allison and R. S. Becker, *J. Phys. Chem.*, **67**, 2675 (1963).

of cyclic polyenes and employing configurational interaction and empirical parameters [20]. Extensive studies have been carried out relative to the affect of metal substitution and solvents on the porphyrin spectra in both absorption [72] and emission (Chapter 8-A). In the metallo-porphyrin spectra the moderately intense visible transition (in the green-yellow region) and the very intense Soret band (in the blue region) are different electronic transitions to states of E_u symmetry (Figure 21) [70]. Each contains some vibrational components (Figure 21). The metal porphyrins are considered to have D_{4h} symmetry (any peripheral substitutions are considered as perturbations). In the case of the unsubstituted parent, porphin, the spectrum is complicated in the 5000–6600 Å region by the presence of extra bands compared with the metal derivatives giving a four-band structure, as in Figure 22, called Q bands. In this case the hydrogen atoms can be considered as perturbations occurring on opposite nitrogens. This presumably stabilizes the structure to resemble an 18-member polyene. Porphin can be considered as a superposition of a cyclic polyene perturbation into the original states resulting from D_{4h} symmetry (metal porphyrins) where the Q and B states were degenerate. In a true 18-member polyene the lower states (Q) are not degenerate but the next upper state (B) is still degenerate [70]. Therefore the nominally two-band structure (one electronic transition with vibrational structure in the 5000–6600 Å region) of the metal porphyrins (D_{4h}) increases to four in the parent porphin containing the original degenerate pair of Q states (plus vibrational structure). The B state does not split, keeping the nominally single band in the 4000 Å region.

Barium, lead, and mercury porphyrins show unusual spectra [72]. In particular, the mercury and lead spectra appear abnormal compared with usual metallo-porphyrin spectra, as the magnesium derivative, whereas the barium spectrum appears normal [72,73]. This was considered unusual, since the ionic or covalent size of barium is greater than either lead or mercury. This can be explained in terms of the electronegativity difference in these cases [72]. The barium-nitrogen bonding can be considered highly ionic and therefore the barium porphyrin would have a geometry essentially the same as that of the porphyrin anion (D_{4h}) and thus would have a normal spectrum compared with other metal porphyrins (e.g., magnesium). On the other hand, the mercury and lead derivatives can be considered to be largely covalently bonded and because of the large size of the metal, geometrical distortion from the normal coplanar structure can occur. This, of course, alters the molecular geometry (symmetry) with consequent abnormal spectra compared to normal porphyrins (e.g., magnesium).

Finally, the degree of configurational interaction is essentially constant for the D_{4h} metal derivatives [72] as had been predicted [70]. It also has been shown however, that there is no simple correlation between the energy of the first excited state and the electronegativity or electronic configuration of the central ion [70] as had been previously supposed [72].

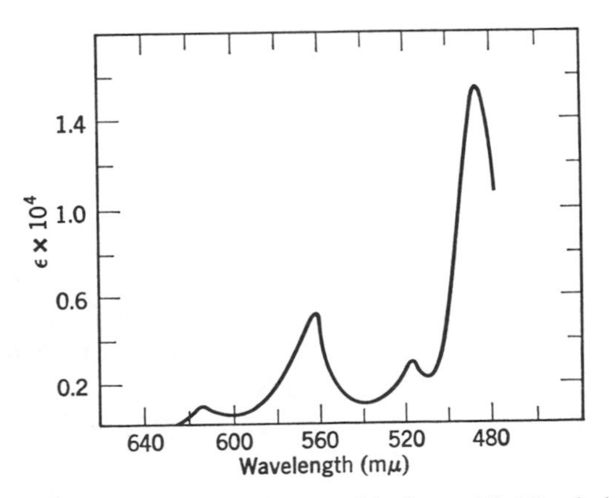

Figure 22 Spectrum of porphin in dioxane [A. Stern, H. Wenderlein, and H. Molvig, *Z. Physik. Chem.*, **177A**, 40 (1936)].

2. Chlorophylls and Derivatives

Chlorophylls are considered to be of particular importance because of their significance in photosynthesis. Crystalline chlorophyll has been prepared and the spectra studied [74].

The spectra of the chlorophylls and derivatives show two principal regions of absorption. One of these is in the 6600 Å region with high intensity ($\varepsilon =$ 60,000–80,000) and the other is in the 4300 Å region and is even more intense ($\varepsilon =$ 110,000–160,000). These are considered to be two separate transitions with the one at the longer wavelength containing vibrational structure on the short wavelength side. The spectra of derivatives such as pheophytins pheophorbides, and chlorophyllides are quite similar to the chlorophylls [74,75]. Several other chlorophylls besides the *a* and *b* derivatives (Figure 20) exist and these are called chlorophylls *c* and *d*, for example. The spectra of these look similar to the usual chlorophylls except that in the case of chlorophyll *c* the long wavelength band is appreciably weaker than the shorter wavelength band [75].

There is evidence for chlorophyll-ether and alcohol complexes in various solvents and at various temperatures to 77°K [76]. Significant differences exist in the long wavelength absorption band of chlorophyll in dry hydrocarbons and in hydrocarbons with certain amines or alcohols [77,78].

3. Dyes

Although certain of the compounds to be discussed are not normally considered as dyes, they are nevertheless colored and are therefore included in this category.

Many naturally occurring pigments are of interest not only because of their relation to biological phenomena, but also because they are a type of molecule for which special theories apply quite well (FEMO, Chapter 2). The area of particular interest at present is that of the linear conjugated polyenes.

The naturally occurring pigments such as carotenes (yellow), lycopene (red), and carotenols (yellow) are long-chained polyenes with alkyl substitution in various locations. The terminal carbons may be attached to unsaturated six-membered rings or to substituted alkyl chains. The spectrum of the carotenoids (carotenes and carotenols) is characterized by two or three intense bands ($\varepsilon \sim 100,000$) in the region from approximately 5000–4000 Å (Figure 23). Because of the possibility of geometrical isomers, there is some modification of the spectrum depending upon the particular isomer chosen. FEMO theory predicts a series of transitions from the ground state (^1A) to various excited states (^1B, ^1C, etc.). The *all trans* form has a center of symmetry such that the ^1A to ^1C transition would be forbidden while the

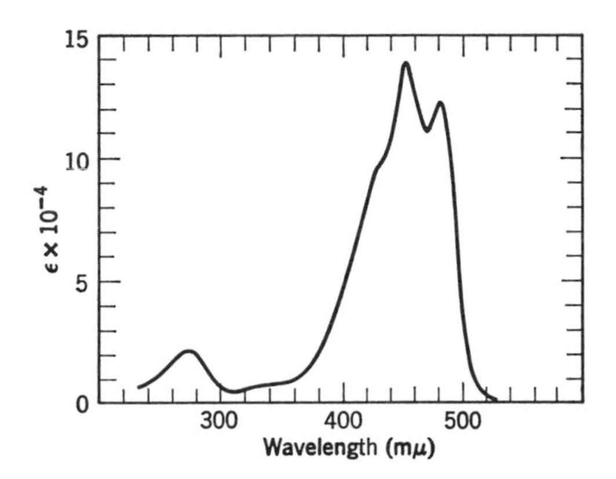

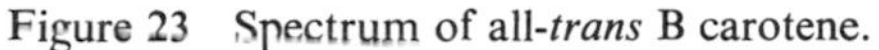
Figure 23 Spectrum of all-*trans* B carotene.
Reprinted by permission from Journal of the American Chemical Society, L. Zechmeister and A. Polgar, *J. Amer. Chem. Soc.*, **80**, 4826 (1958).

^1A to ^1B and ^1D should be allowed. In a *cis* isomer, the center of symmetry no longer exists so the ^1A to ^1C transition should increase in intensity while the ^1A to ^1B and ^1D should decrease. Thus in essence a new band will grow in with an intensity proportional to the bend of the molecule (a function of the location of the *cis* band). This has been confirmed for several molecules [79].

An important group of molecules that is in the dye category are the phthalocyanines. These are the tetraazatetrabenzporphins (Figure 20). Metal derivatives of these exist and the variety possible should be comparable to the porphyrins. However, not so many have been studied either in absorption or emission. In general the absorption spectra show a moderately intense band ($\varepsilon \sim 50{,}000$–$80{,}000$) in the ultraviolet ($\lambda \sim 3500$) and one or two very intense bands ($\varepsilon \sim 150{,}000$–$350{,}000$) in the deep yellow and red regions ($\lambda \sim 5900$–7100) [80]. The metallo derivatives are generally very stable thermally. The compounds are green, blue, or blue-green in color.

Chapter 7 Characteristics of Excited States

There are four characteristics that can be associated with a molecular emission. These are (a) energy, (b) lifetime, (c) quantum yield, (d) polarization.

Before proceeding to a detailed discussion of each of these, it will be helpful to qualitatively consider some aspects of the emission processes. On electronic excitation to any state above the first excited singlet, commonly de-excitation occurs to the lowest excited singlet state. From this state there can occur fluorescence emission, continued de-excitation to the ground state with no emission, crossing to the triplet state, or any combination of these (Figures 1 and 24). If the triplet state is occupied, phosphorescence, or de-excitation to the ground state, or both, can occur. Because the lowest triplet is below the lowest excited singlet state, phosphorescence occurs at longer wavelength than does fluorescence. The shape of the fluorescence spectrum often resembles that of the absorption band of longest wavelength, with approximately a mirror-image relationship between the emission and lowest energy absorption band. Frequently the fluorescence spectrum is located immediately adjacent to the longest wavelength absorption band on the long wavelength side; sometimes a larger wavelength interval is found between the absorption and emission spectrum. The wavelength displacement of the fluorescence with respect to the absorption is referred to as the "Stokes shift." Further details regarding shape and intensity of emissions and the nature and magnitude of the shift are discussed in subsequent sections.

A. ENERGY

The energy associated with an emission process is the easiest of the characteristic qualities to determine. In essence the fluorescence spectrum is

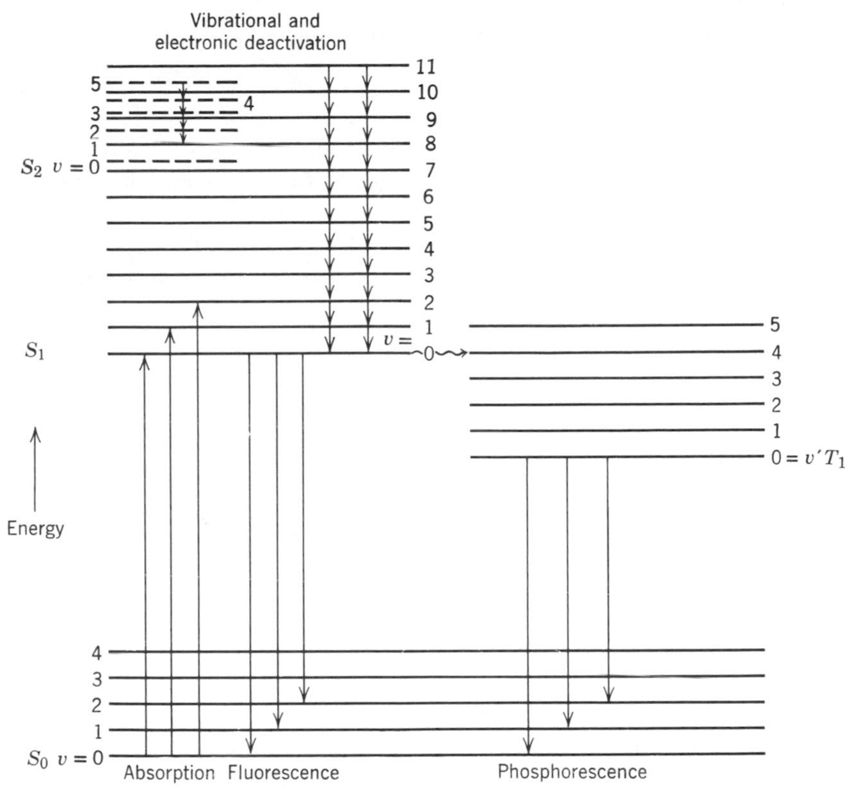

Figure 24 Schematic energy-level diagram showing electronic states and associated vibrational levels.

produced by exciting the molecule (with radiation from some external energy source) followed by emission from the excited molecule. The energy E associated with any wavelength of the fluorescence spectrum is given by the relation $E = h\nu$, where ν is the frequency of the emitted radiation. Of course that vibronic band with highest energy determines the energy of the excited state (relative to the ground state). This vibronic band corresponds to the emission origin and represents the transition from the zero vibrational level of the excited state to the zero vibrational level of the ground state. Any other energy would include vibrational energy as shown in Figure 24.

B. LIFETIME

The second characteristic of an excited state, its mean lifetime, is somewhat more difficult to measure, particularly when it is short (10^{-7}–10^{-10} sec). Observations show that the fluorescence intensity decays after the withdrawal of the exciting source according to a first-order rate equation, that is, exponentially with the time

$$I = I_0 e^{-t/\tau} \tag{50}$$

where I_0 is some initial intensity at an arbitrary time zero, I is the intensity at some later time t, and τ is a constant. When the time t over which the decay is measured is equal to τ, the intensity has fallen to $1/e$ of its initial value. τ is defined as the mean decay time for the emission process or the mean lifetime of the excited state. Of course there are relations between the rate constants and the lifetime τ. One relation is

$$k_F = \frac{1}{\tau} \quad \text{or} \quad \tau = \frac{1}{k_F} \tag{51}$$

where k_F refers to the rate constant for fluorescence and τ is the corresponding lifetime. The τ will equal τ_0, the intrinsic or natural lifetime, *only* in the absence of any deactivational processes, both internal and external. The observed lifetime τ is determined by all the deactivational processes and can be expressed as $\tau = 1/(\sum_i k_i)$. Thus the observed mean lifetime may differ from the natural lifetime. The relationship between τ and τ_0 for fluorescence is $\tau_0 \Phi_F^0 = \tau$, where τ is the observed lifetime of the particular emission in the absence of external quenching, and Φ_F^0 is the quantum yield of the fluorescent emission in the absence of any external quenching processes (Section C). External quenching results from collisions between the solute molecule and any other molecule. It is known that internal deactivational processes must shorten the lifetime from what it would be in the absence of such deactivational processes. One way this can be observed is from the effect of substitution of deuterium for hydrogen on certain aromatic hydrocarbons. In the deuterium compounds the phosphorescence lifetimes are notably longer than those containing hydrogen (for more details see Chapters 9 and 11).

In the case of phosphorescence the decay also follows the first-order rate equation. The intrinsic or natural lifetime for phosphorescence emission is also defined in the same way as for fluorescence,

$$\tau_0 = \frac{1}{k_P} \tag{52}$$

where k_P is the rate constant for phosphorescence. However, the relationship between the quantum yields and the intrinsic and observed lifetimes is not quite so simple as in the case of fluorescence. If we define k_{IC} as the rate constant for internal conversion (vibrational deactivation of electronic energy) between S_1 and S_0, and k_{IS} as the rate constant for intersystem crossing between S_1 and T_1 (see Figure 21), then the intrinsic lifetime of phosphorescence is related to the observed lifetime as follows:

$$\tau_0 = \tau\left(\frac{k_{IS}}{k_{IC} + k_{IS}}\right)\frac{1 - \Phi_F^0}{\Phi_P^0} \tag{53}$$

where Φ_F^0 and Φ_P^0 are the intrinsic quantum yields of fluorescence and phosphorescence, respectively (see Section C). It is often assumed that internal conversion occurs only from the triplet state and then the relationship is simplified to

$$\tau_0 = \tau\left(\frac{1 - \Phi_F^0}{\Phi_P^0}\right) \tag{54}$$

The experimental procedures for measuring decay times vary according to the order of magnitude of the decay. Shutter-type arrangements are the simplest. The emission can be detected by the use of an arrangement allowing periodic entrance of the exciting radiation to the sample through one shutter and passage of the emitted radiation to a detector through a second shutter. The latter shutter normally opens and closes with the same frequency as the first but is delayed in time. If the luminescence persists for a period up to 10^{-4} sec after the removal of the exciting radiation, mechanical shutters consisting of slots in rotating disks or cylinders can be used, as in the Becquerel phosphoroscope [81]. The fluorescence of dyes decays in some 10^{-8} sec after removal of the exciting radiation, and for such periods mechanical shutters are replaced by electro-optical shutters without moving parts. One such device is a Kerr cell placed between crossed Nicol prisms. The cell contains a liquid, such as nitrobenzene, of high dipole moment and high polarizability, which becomes doubly refracting in an electric field. The arrangement transmits light when the field is applied and is opaque when the field is zero, and therefore acts as a shutter of very low inertia. The light exciting the fluorescent system is passed through one such shutter and the fluorescence passes through a similar shutter that can be actuated a fraction of a microsecond later. It is sometimes not necessary to have a second Kerr cell.

Another procedure to measure lifetimes employs a method known as *phase shift*. In this method the exciting light is sent through a water tank in which standing waves are being generated by a vibrating quartz crystal. The wave pattern diffracts the light and the diffraction image of one split is focused on another. The emergent modulated light is split so that a small part goes to a reference detector and the remainder to the sample. The lifetime is determined by first scattering the light normally going to the sample to a second detector and the phase relation between the signals is measured. The sample is then put into place. The emitted fluorescence is modulated in intensity according to the excitation modulation but is delayed in phase by an angle that is related to the observed fluorescence lifetime by [82]

$$\tan \Delta\phi = 2\pi f\tau \tag{55}$$

where f is the frequency of modulation and $\Delta\phi$ is the angular delay in the phase. This technique permits measurement in the nanosecond (10^{-9} sec) range.

C. QUANTUM YIELD

The quantum yield is the most difficult characteristic to measure. This is particularly true when absolute quantum yields are desired. It is somewhat more simple to obtain information on relative quantum yields. In general the quantum yield of an emission is defined as the number of quanta emitted per exciting quantum absorbed. Therefore

$$\Phi_F = \frac{\text{number of fluorescence quanta emitted}}{\text{number of quanta absorbed to a singlet excited state}} \tag{56}$$

The quantum yield of phosphorescence is defined as

$$\Phi_P = \frac{\text{number quanta phosphorescence quanta emitted}}{\text{number of quanta absorbed to a singlet excited state}} \tag{57}$$

In the case of phosphorescence because of the forbidden character of the direct absorption $T_1 \leftarrow S_0$ (see Figure 21), excitation usually occurs via a singlet state. The quantum yields become intrinsic, Φ_F^0, Φ_P^0, when they are measured in the absence of external quenching by collisional processes.

In the absence of photochemical reaction (such as self-decomposition) the following expression is valid:

$$\sum_i \Phi_i^0 = \Phi_F^0 + \Phi_P^0 + \Phi_{IC}^0 = 1 \tag{58}$$

where Φ_{IC}^0 is the quantum yield for internal conversion by vibrational deactivation (radiationless deactivation between electronic states, see Figures 1 and 24). In many cases Φ_{IC}^0 may be small and for the sake of discussion, let it be considered to be zero. Then

$$\Phi_F^0 + \Phi_P^0 = 1 \tag{59}$$

This equation points up the important fact that these are *complementary* in nature. Even though phosphorescence may not be seen (as in a fluid solution) the triplet state *is still occupied* to a degree dependent on the intrinsic quantum yield. This has important significance if the triplet state is involved in a subsequent reaction.

The quantum yield of fluorescence can be related to several rate constants in the following way:

$$\Phi_F^0 = \frac{k_F}{k_F + k_{IC}^S + k_{IS}} = k_F \tau_F \tag{60}$$

where k_{IC}^S is the rate constant for internal conversion from the first excited singlet. In the case of phosphorescence, if $k_{IS} \gg k_F + k_{IC}^S$, then

$$\Phi_P^0 = \frac{k_P}{k_P + k_{IC}^T} = k_P \tau_P \tag{61}$$

where k_{IC}^T is the rate constant for internal conversion from the lowest triplet state. If the above condition between k_{IS} $k_F + k_{IC}^S$ is not met, then

$$\Phi_P^0 = \left(\frac{k_{IS}}{k_F + k_{IC}^S + k_{IS}}\right)\left(\frac{k_P}{k_P + k_{IC}^T}\right) \tag{62}$$

The quantum yield of intersystem crossing is

$$\Phi_{IS}^0 = \frac{k_{IS}}{k_{IS} + k_F + k_{IC}^S} \tag{63}$$

In the case of collisional bimolecular quenching (external quenching) the general relation between lifetime and quantum yields is

$$\frac{\tau_Q}{\tau} = \frac{\Phi}{\Phi^0} \tag{64}$$

where τ_Q is the observed lifetime in presence of external quenching, τ is the observed lifetime in the absence of external quenching, Φ^0 is as previously defined, and Φ is the observed quantum yield in the presence of external quenching. Thus the observed τ_Q is

$$\tau_Q = \frac{\Phi}{\Phi^0}\tau \tag{65}$$

For example, if $\Phi^0 = 0.3$ and $\Phi = 0.03$, then the observed lifetime will be 0.1 of the lifetime observed in the absence of collisions. Collisions effectively shorten the lifetime from that obtained in the absence of collisions.

Relative quantum yields are commonly measured with a rhodamine *B* quantum counter [83]. One method employs a rhodamine dye that is excited by the fluorescence of the sample and the emission from the dye is fed to a photomultiplier whose signal is compared with a standard signal from a photocell. The intensities are converted to relative quanta per second. Further details are available in the literature [83–85]. The problem of reabsorption and re-emission of the primary fluorescence may have to be considered [83]. Absolute quantum efficiencies [83,86–88] can be determined by measuring the sample emission with a calibrated photomultiplier and comparing this with the response to a known portion of the light scattered from a block of magnesium oxide substituted for the sample and illuminated by the same source.

In the remainder of this chapter all superscripts 0 for Φ and subscripts Q for τ will be dropped since it will be assumed that competitive external quenching is absent, except when stated otherwise.

D. POLARIZATION

Emission polarization can be utilized to establish the relative orientation of the emission transition dipole to that of particular absorption bands. From this it is possible to determine the relative orientation of the transition dipoles for each of the absorption bands. Further, with a knowledge of the polarization characteristics of absorption bands it is possible to assign the nature of the state from which the emission occurs. More will be said concerning these shortly.

The molecules in solution are randomly oriented and are held stationary during the excitation-emission process by making the solution rigid (as an

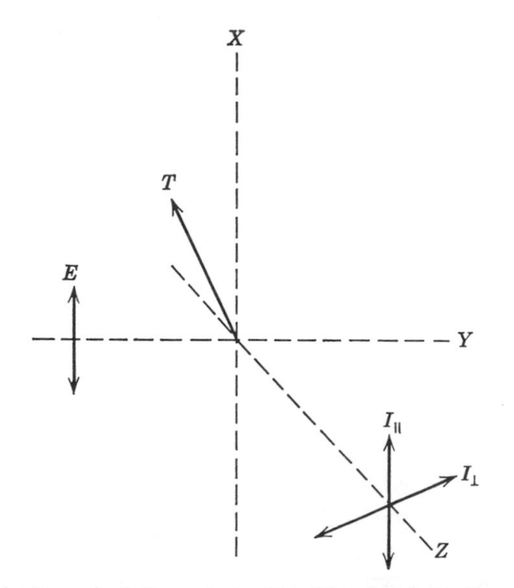

Figure 25 Determination of relative polarization. The electric vector of the exciting light E is plane polarized in the XY plane, the absorption transition moment T is at some angle to the electric vector of the exciting light, and observation of emitted polarized components $I_{\perp}$ and $I_{\parallel}$ is made along Z.

organic solvent system at low temperature). Figure 25 shows a schematic of the problem. The molecules whose absorption transition moments are parallel with the electric vector of the light will be excited the most, but less preferred orientations will contribute less. The degree of polarization is defined as follows:

$$P = \frac{I_{\parallel} - I_{\perp}}{I_{\parallel} + I_{\perp}} = \frac{(3\cos^2\alpha) - 1}{(\cos^2\alpha) + 3} \tag{66}$$

where $I_{\parallel}$ and $I_{\perp}$ are the intensities of the observed parallel and perpendicular components, respectively, and where α is the angle between the emission and absorption transition moments. Commonly the α is either 0 or 90° so that P takes on a maximum value of $+\frac{1}{2}$ and a minimum value of $-\frac{1}{3}$, respectively. These values are most often not found in practice because of some depolarization resulting from external factors such as strains in the rigid glass matrix. Nonetheless the method is useful by noting changes in P as a function of the exciting wavelength or evaluating the sign of the polarization. Care must be exercised relative to the last procedure since various vibronic bands may be of different polarizations (see Chapter 9-D).

Figure 26 shows an example of the polarization of the emission changing sign as a function of wavelength of excitation. The values would indicate that the transition moments of the corresponding absorption bands are mutually perpendicular.

In aromatic hydrocarbons the allowed transitions are in-plane polarized. Thus it is possible to determine something about the direction of the emission. Particularly, it has been found that for several hydrocarbons, phosphorescence is polarized perpendicular to the plane of rings. In addition to this fact, such information can aid in determining state assignments and the nature of perturbing states (see Chapters 11 and 12).

Molecules with $\pi^* \leftarrow n$ transitions provide a good example of the method of assignment of the nature of the triplet state from which emission occurs utilizing polarization data. The $\pi^* \leftarrow n$ singlet transition is out-of-plane polarized while the $\pi^* \leftarrow \pi$ singlet transitions are in-plane polarized. Further, phosphorescence from n, π^* triplet states is in-plane polarized while that from π, π^* triplets is out-of-plane polarized. Thus if a phosphorescence is negatively polarized when irradiating into the $\pi^* \leftarrow n$ singlet transition and positively polarized when irradiating into the $\pi^* \leftarrow \pi$ singlet, the emission is polarized perpendicular to the $\pi^* \leftarrow n$ absorption and parallel to the $\pi^* \leftarrow \pi$ absorption; that is, it is in-plane polarized and originates from an n, π^* triplet. This is the case for pyrimidine and pyrazine. On the other hand, for example, quinoxaline (1,4 diazanaphthalene) exhibits out-of-plane polarization of the phosphorescence and therefore originates from a π, π^* triplet. Nitro containing compounds such as 4-nitrobiphenyl also show a

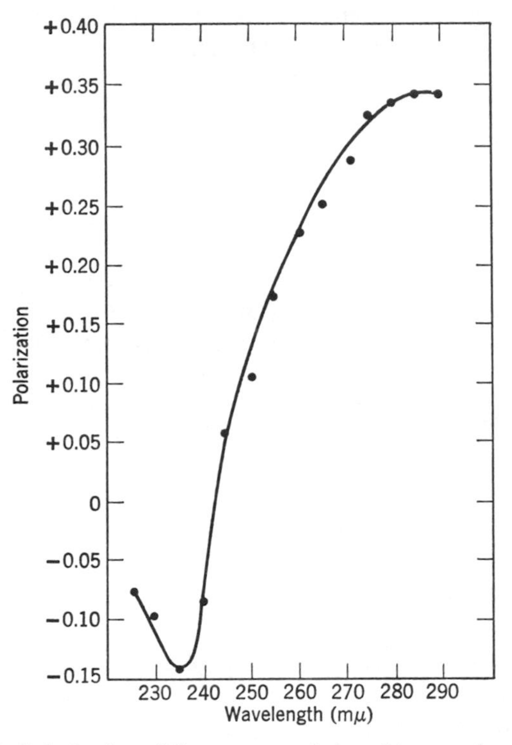

Figure 26 Polarization of fluorescence of phenol in propylene glycol at −70°C. Reprinted by permission from Biochemical Journal, G. Weber, *Biochem. J.*, **75**, 335 (1960).

phosphorescence that is in-plane polarized and therefore is consistent with assigning the triplet as n, π^* [88a]. Other examples are discussed in succeeding chapters.

E. RELATIONSHIP BETWEEN ABSORPTION INTENSITY AND LIFETIME

We deal here with the fundamental relation between the transition probabilities for induced plus spontaneous emission. Although the results have been expressed in a variety of manners, the following is a general expression:

$$\frac{1}{\tau_0} = A_{21} = \left(\frac{8\pi 2303 c\bar{\nu}^2 n^2}{N}\right)\left(\frac{g_1}{g_2}\right)\int \varepsilon \, d\bar{\nu} \tag{67}$$

where the expansion for A_{21} results from substitution of (28) into (27) except that n^2 arises when there is a diffracting medium. The integration is taken over an entire absorption band. Again the actually observed lifetime τ will agree with τ_0 providing there are no deactivational modes present other than emission. In the expression for A_{21} the term $\bar{\nu}^2$ appears. For atomic transitions of small half width this is accurately determined by the absorption maximum. In molecular absorptions, however, the half width is large and no $\bar{\nu}$ really exists. In this case, either the 0–0 frequency is used for $\bar{\nu}$ or the band maximum is used for $\bar{\nu}$; however, see the following discussion and (68).

In the case of molecular systems the electronic states are not discrete because of vibrations and, moreover, the vibrational fine structure has breadth. Therefore, transitions can occur over a range of frequencies or energies. Additional modifications are thus required. The final equation for an allowed transition can generally be written [82] as

$$\frac{1}{\tau_0} = A_{21} = \left(\frac{8\pi(2303)cn^2}{N\langle\bar{\nu}_f^3\rangle_{\text{av}}^{-1}}\right)\left(\frac{g_1}{g_2}\right)\int \varepsilon \, d \ln \bar{\nu} \qquad (68)$$

where the integral is over the entire electronic absorption band and $\langle\bar{\nu}_f^{+3}\rangle_{\text{av}}^{-1}$ is the reciprocal of the mean value of $\bar{\nu}^{+3}$ in the fluorescence spectrum. If the absorption were sharp and the fluorescence and absorption occurred at the same wavelength, the $\bar{\nu}$ could be considered to be constant. Therefore a factor of $1/\bar{\nu}$ could be extracted from under the integral, $\langle\bar{\nu}_f^{+3}\rangle_{\text{av}}^{-1}$ would

Table 7 Comparison of Observed and Calculated Lifetimes (nanoseconds)

Compound	Solvent	Observed	Calculated	Error (percent)
Perylene[a]	Benzene	4.79	4.29	−10
Perylene[b]	Benzene	5.65	5.06	−10
Acridone[a]	EtOH	11.80	12.06	+2
Acridone[b]	Absorption EtOH	15.1	15.9	+5
9-Aminoacridine[a]	EtOH	13.87	15.43	+11
9-Aminoacridine[a]	EtOH-HCl	14.07	14.19	+1
9-Aminoacridine[a]	H_2O	16.04	16.40	+2
9-Aminoacridine[a]	H_2O-HCl	15.45	16.67	+8
9-Aminoacridine[b]	Absorption EtOH	15.3	15.60	+2
Fluorescein[a]	H_2O-NaOH	4.02	4.37	+9
Rhodamine B[a]	EtOH	6.16	6.01	−2
Rubrene[a]	Benzene	16.42	22.42	+37
Anthracene[b]	Benzene	16.7	13.52	−19
9,10-Diphenylanthracene[b]	Benzene	8.9	8.77	+1
9,10-Dichloroanthracene[b]	Benzene	10.4	11.0	−29
Quinine sulfate[b]	1 N H_2SO_4	3.59	27	−25
Fluorescein[b]	0.1 N NaOH	4.96	4.70	−5

[a] Reference 82
[b] Reference 89

become $\bar{\nu}^{-3}$, and then the equation would become identical to (67). The $\langle \bar{\nu}_f^{+3} \rangle_{\text{av}}^{-1}$ can be most accurately evaluated by plotting intensity (relative quanta at each frequency) vs frequency in (cm^{-1}) and $\bar{\nu}^{-3}$ times the intensity vs frequency (cm^{-1}). The ratio of the areas under these curves gives $\langle \bar{\nu}_f^{-3} \rangle_{\text{av}}^{+1}$ with good accuracy [82]. Also a mean frequency can be estimated from the emission curve, the value cubed and taken as $\langle \bar{\nu}_f^{-3} \rangle_{\text{av}}^{+1}$.

The formula generally holds for moderately strong transitions ($\varepsilon \approx 8000$) and where there is not a large change in configuration in the excited state. The latter implies ideally that the maximum intensity lies at the 0–0 band. If it is near this (up to 0–3 band), the formula is still valid. In the case of weak (but not symmetry forbidden) transitions the correct order of magnitude may be expected from the equation. It would appear that the error may range from 1 to 30 percent [89] and τ_0 is generally low, particularly in the absence of a mirror-image relationship between the absorption and emission [89]. In the case of the first absorption band of benzene (symmetry forbidden) the calculated lifetime is estimated to be approximately 16 percent too long [82].

Table 7 gives some fluorescence lifetimes and compares the observed values with those calculated from (68). The calculated values include the consideration of the fluorescence quantum yields [82,89]. It is worthwhile to emphasize that both the absorption and emission must be measured in the same solvent in order to compare observed and calculated lifetimes.

Chapter 8 Nature of Emission Processes

A. FLUORESCENCE

The simplest examples of fluorescence are afforded by monatomic vapors at low pressure. Sodium atoms, for example, in their ground state, $^2S_0$, absorb the well-known D lines and are thus excited to the states $^2P_{1/2}$ and $^2P_{3/2}$. These revert, after a mean lifetime of 1.6×10^{-8} sec, to the ground state. This type of fluorescent emission in which the wavelengths of the emission and absorption lines coincide is called *resonance radiation*. The term arises from the classical (prequantum) concept that absorption resulted from a resonance between the frequency of the exciting light and an electronic frequency within the atom.

Resonance spectra can be observed in diatomic molecules at very low pressure in the gaseous state. On excitation by monochromatic radiation, emission is not confined to the absorbed frequency. Transitions occur from the excited state (a particular rotational state of a particular vibrational state of the excited electronic state) to rotational states of a number of vibrational states of the ground electronic state. Thus the "resonance spectrum" of a diatomic molecule consists of a series of lines, one coinciding with the absorption line, the others at longer wavelengths representing the return to higher vibrational levels of the ground state. The intervals between the principal members of the series therefore represent energy differences between vibrational levels in the ground state.

An important question arises regarding the number of possible excited singlet states from which the fluorescence may originate. Absorption can occur to any of the excited states S_1, S_2, S_3, etc. Fluorescence emission occurs, however, in almost every case from the lowest excited state S_1 (Figures 1 and 24). Moreover, the vibrational level origin within S_1 is the zero level. Thus the lowest energy band in absorption and the highest energy band in emission should coincide. This is in fact quite true for vapor but may not be true for solutions (see Section 10–A). An exceptional case is that of azulene, for which the fluorescence emission is from the second excited singlet level rather than from the first.

In general the lifetimes of excited states of singlet character can be expected to range from 10^{-7} to 10^{-10} sec. However, associated with each state there are vibrational transitions whose lifetimes can be expected to range from 10^{-11} to 10^{-14} sec. Moreover, in a complex molecule there are large numbers of excited states that it may be recalled are actually potential energy surfaces. These may be represented in a cross-sectional sense as shown in Figures 4 and 5 of Chapter 3-C. Once excitation has occurred to a vibrational level of some relatively high singlet state, say S_3, the molecule immediately begins vibrating and an extremely rapid vibrational cascading process starts through the vibrational levels associated with the excited state (Figures 1 and 24).

It may well be that the potential energy surface for S_3 is crossed or cut by that for S_2. Some vibrational level of S_2 will then be isoenergetic or nearly so with that of S_3 (e.g., level 8 of S_1 and level 1 of S_2 in Figure 24). In this case there is a high probability that during the next and succeeding vibrations, the molecule will be vibrating in a manner appropriate to that for S_2. As the cascading continues S_1 may cross S_2 and the above process is repeated until finally the vibrational cascading or *internal conversion* process has carried the molecule to the lowest vibrational level of the lowest excited electronic state S_1. From here the molecule emits to various vibrational levels of the ground state and the typical fluorescence band results as in Figure 27. It is important to note that the potential energy curves need *not* actually cross but that they be sufficiently close to interact strongly and "mix," thereby ultimately providing a pathway for the energy loss just described. From experimental results, it is quite evident that the probability of the deactivational process, as described above, occurring is extremely great since emissions other than from S_1 are almost nonexistent. Thus electronic energy is transformed and lost as vibrational energy by internal conversion. Also it is highly likely that part of the energy loss associated with internal conversion occurs during collisions of the excited molecules with other molecules. The proportion of the energy lost in this way is not known. The transformation from high to low excited electronic states, however, occurs not only in fluid media but also in rigid media and the gas phase in which the number of collisions must be severely diminished or even reduced to zero. More concerning this will be given in Chapter 9-B.

If it happens that the ground state and lowest excited state potential energy curves cross or mix, the fluorescence intensity may be weak or absent. The molecules can also lose energy from the first excited state by collisions. For

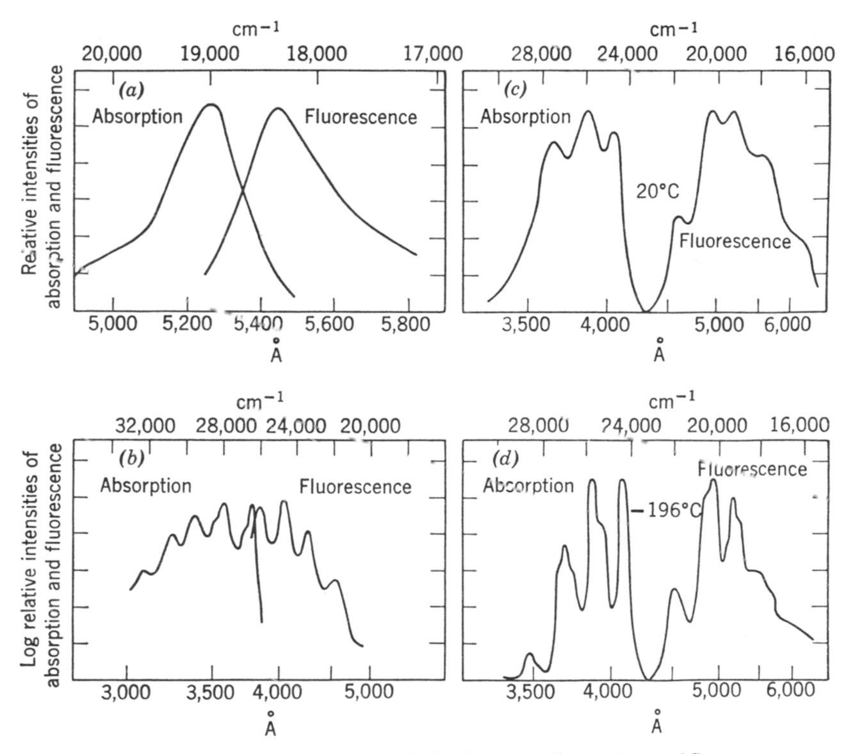

Figure 27 Approximate mirror-image relation between absorption and fluorescence spectra: (*a*) fluorescein (E. L. Nichols and E. Merritt, *Phys. Rev.*, **31,** 376 (1910); (*b*) anthracene in dioxane (G. Kortum and B. Finckh, *Z. Physik. Chem.*, **B52,** 263 (1942); (*c*) diphenyl-octatetraene in xylene at 20°C (K-W. Hausser, R. Kuhn, and E. Kuhn, *Z. Physik. Chem.*, **B29,** 417 (1935)); and (*d*) same as (*c*) at −196°C.

example the fluorescence intensity or quantum yield often increases with an increase in the solvent viscosity. The implication is that fewer collisions will decrease the number of molecules whose excitation energy is degraded before emission can occur.

The rules relating the shape of an absorption spectrum and the Franck-Condon principle also apply to the general shape of fluorescence spectra. There have been numerous accounts in the literature of the approximate mirror-image relationship between absorption and emission for example (see Figure 27). This can occur if the vibrations in the excited state are comparable to those in the ground state. Differences in the dipole orientation, in hydrogen bonding and in geometry between the two states can modify this relationship. The mirror-image relationship is therefore not to be considered as general and its existence will be one of the characteristics of the emission useful in the interpretation of the process.

Finally it is worthwhile to consider the case of azulene. Excitation of azulene results in fluorescence emission from the second excited singlet state and moreover, no phosphorescence occurs [90,91]. This can be interpreted in several ways. Excitation into the second excited state is not followed by internal conversion to the first state because of the difference in energy and symmetry characteristics between them; that is, they do not cross or mix. Excitation into the first excited state results in internal conversion (thus no fluorescence) to the ground state because of crossing of the potential curves [90]. Also it is possible that as an alternative explanation, crossing occurs to the triplet level [90] and thus the population in the potentially fluorescent state is converted into the equivalent triplet state occupation. It is possible that phosphorescence has not been observed because of a difficulty of observing the expected long wavelength emission [91]. Also it is possible that the lowest triplet state lies above the lowest excited singlet; hence occupation of the former is vanishingly small [90]. The observations of emission are inadequate to clarify completely the peculiarities of the azulene fluorescence. Recent information based on oxygen-perturbed singlet-triplet absorption studies (Chapter 14) indicates that the lowest triplet state is either higher than the first excited singlet state or only very slightly lower [92].

B. PHOSPHORESCENCE

1. Spin-Orbit Coupling and Transition Probability

In order to have a better understanding of phosphorescence and triplet states we shall first discuss spin-orbit coupling. As inferred earlier, in light atoms or molecules containing light atoms the spin angular momentum of the electrons can be considered as a quantized independent quantity. Its value must be conserved during a transition (pure $L - S$ coupling). The total wave function for a system can be written

$$\psi_i = \phi\Gamma \tag{69}$$

where ϕ and Γ are the space and spin parts of the total function. The ordinary Hamiltonian does not contain any terms involving the spin. The transition probability between two states based on the transition moment integral is

$$P = \left| k \int_{-\infty}^{+\infty} \psi_n M \psi_m \, d\tau \right|^2 \tag{70}$$

where the integration is over all space and M is the electric dipole operator (er). If the complete wave function [69] is substituted into [70], the integral vanishes (=0) if states m and n have different multiplicities. This happens because integration over the spin coordinates when the spin functions Γ are different, gives zero. In other words the spin functions are orthogonal.

$$\int \Gamma_1 \Gamma_2 \, d\Gamma = 0 \tag{71}$$

Spin-orbit coupling effectively mixes some singlet character into triplet states and some triplet character into singlet states thus destroying their pure character and orthogonal nature. The spin-orbit interaction causes an additional small term to appear in the Hamiltonian operator as follows:

$$H_{\text{total}} = H_0 + H_{\text{rep}} + H_{\text{SO}} \tag{72}$$

where the terms in the total Hamiltonian are respectively the ordinary Hamiltonian, one including electron repulsions and the additional one for spin-orbit interaction. The spin-orbit interaction part is introduced as a relativistic correction. The greatest contribution to spin-orbit interaction occurs when an electron is closest to the nucleus and is moving with relativistic velocity.

The properties of the spin-orbit operator can be deduced for any nuclear potential field but the properties are most easily seen for a spherical field, which is used as an approximation for molecules. The spin-orbit Hamiltonian has the form

$$H_{\text{SO}} = \sum_{i}^{n} \sum_{j}^{m} C r_{ij}^{-1} \frac{dV(r_{ij})}{dr_{ij}} (l_{xi} s_{xi} + l_{yi} s_{yi} + l_{zi} s_{zi}) \tag{73}$$

where r_{ij} is the distance of the ith electron from the jth nucleus, l and s are the orbital and spin operators of the ith electron, and n and m are the number of electrons and nuclei, respectively, and C is a constant. The sums are over all electrons and nuclei. The magnitude of spin-orbit interaction depends on an average value of $(1/r)(dV/dr)$ for the electron. Moreover, for a bare nucleus $V = (Ze/r)$ and $(1/r)(dV/dr) = (Ze/r^3)$. For a nucleus surrounded with electrons, the electrostatic potential changes more rapidly with r because of the fast change in shielding by the electrons as the nucleus is approached. Thus the spin-orbit interaction tends to be large in atoms of high atomic number and for orbits that penetrate the core more deeply.

One component, x, of the spin-orbit Hamiltonian for molecules (assuming a central field potential) may be written [92a]

$$H_{\text{SO}}^{x} = \frac{1}{2n} \sum_{i}^{n} \sum_{k}^{n} (A_i l_{xi} + A_k l_{xk})(s_{xi} + s_{xk}) + \frac{1}{2n} \sum_{i}^{n} \sum_{k}^{n} (A_i l_{xi} - A_k l_{xk})(s_{xi} - s_{xk}) \tag{74}$$

where

$$A_i = \sum_j^m \left(\frac{1}{r_{ij}}\right)\left(\frac{dV(r_{ij})}{dr_{ij}}\right), \qquad A_k = \sum_j^m \left(\frac{1}{r_{kj}}\right)\left(\frac{dV(r_{kj})}{dr_{kj}}\right) \tag{75}$$

and where j is a particular nucleus, r_{ij} is the distance from the electron i to the nucleus j, and the sum is over all electrons and nuclei. This equation (74) for one component is more complete than (73), since it includes spin-orbit coupling of electron i with the repulsive field of electron k. Both terms of (74) are symmetric to electron exchange. Therefore, antisymmetry is retained when H_{SO}^{x} operates on any wave function. The spin parts, however, contain $(s_{xi} + s_{xk})$ and $(s_{xi} - s_{xk})$ which are symmetric and antisymmetric, respectively, with respect to spin exchange. Therefore $(s_{xi} + s_{xk})$ operating on a wave function that is antisymmetric with respect to spin exchange will generate a wave function that is still antisymmetric in the spin part. However, operating with $(s_{xi} - s_{xk})$ will give a wave function that is symmetric in the spin part. The factor that differentiates wave functions of different multiplicity is the symmetrical (triplet) or antisymmetrical (singlet) behavior with spin exchange. Thus the second term in (74) is responsible for mixing states of different multiplicity and thereby increases the $S \leftrightarrow T$ transition probability. Since the A_i are totally symmetric [92a], it is the l_i that determine the orbital part of the states that may mix.

The perturbed triplet state wave function is

$${}^3\psi = {}^3\phi^0 + \sum_i \left[\frac{\langle {}^1\phi_i^0 |\, H_{\mathrm{SO}}\, | {}^3\phi^0 \rangle}{|{}^3E^0 - {}^1E_i^0|}\right] {}^1\phi_i^0 \tag{76}$$

where ${}^3\phi^0$ is the unperturbed triplet state function, ${}^1\phi_i^0$ are the perturbing singlet functions and ${}^3E^0$ and ${}^1E_i^0$ are the energies of the unperturbed triplet and perturbing singlet states, respectively. The terms in brackets are the mixing coefficients.

The perturbed ground-state singlet wave function, ${}^1\psi_0$ is

$${}^1\psi_0 = {}^1\phi_0^0 + \sum_i \left[\frac{\langle {}^1\phi_0^0 |\, H_{\mathrm{SO}}\, | {}^3\phi_i^0 \rangle}{|{}^1E_0^0 - {}^3E_i^0|}\right] {}^3\phi_i^0 \tag{77}$$

where ${}^1\phi_0^0$ is the unperturbed ground singlet state function, ${}^3\phi_i^0$ are the perturbing pure triplet functions, and ${}^1E_0^0$ and ${}^3E_i^0$ are the energies of the unperturbed ground singlet state and perturbing triplet states, respectively. The terms in brackets are the mixing coefficients. The transition moment integral, M, between ${}^1\psi_0$ and the lowest triplet ${}^3\psi_1$ is

$$\langle {}^1\psi_0 |\, er\, | {}^3\psi_1 \rangle = \sum_i \left[\frac{\langle {}^1\phi_0^0 |\, H_{\mathrm{SO}}\, | {}^3\phi_i^0 \rangle}{|{}^1E_0^0 - {}^3E_i^0|}\right] \langle {}^3\phi_i^0 |\, er\, | {}^3\phi_1^0 \rangle + \sum_i \left[\frac{\langle {}^1\phi_i^0 |\, H_{\mathrm{SO}}\, | {}^3\phi_1^0 \rangle}{|{}^3E_1^0 - {}^1E_i^0|}\right] \langle {}^1\phi_i^0 |\, er\, | {}^1\phi_0^0 \rangle \tag{78}$$

Thus the phosphorescence will have a polarization which is dependent on that of the $S_i \leftrightarrow S_0$ and $T_i \leftrightarrow T$ transitions.

Further, if it is assumed that it is only one particular singlet, ${}^1\phi_p^0$, that mixes strongly with the lowest pure triplet ${}^3\phi_i^0$ and the ground state is unperturbed, then

$$\langle {}^1\psi_0 | \, er \, | {}^3\psi_1 \rangle = \frac{\langle {}^1\phi_p^0 | \, H_{\rm SO} \, | {}^3\phi_1^0 \rangle}{|{}^3E_1^0 - {}^1E_p^0|} \langle {}^1\phi_p^0 | \, er \, | {}^1\phi_0^0 \rangle \tag{79}$$

where ${}^1\phi_p^0$ is the perturbing singlet and ${}^1E_p$ is the energy of the perturbing singlet. Thus it can be seen that the intensity of the $T \leftrightarrow S$ transition is "borrowed" from the $S_p \leftrightarrow S_0$ transition. The part $\langle {}^1\phi_p^0 | \, H_{\rm SO} \, | {}^3\phi_1^0 \rangle$ is proportional to $\zeta_{n,1}$, the spin orbit coupling factor which for hydrogenic-like atoms is equal to

$$\left[\frac{e^2h^2}{2m^2c^2a_0^3}\right]\left[\frac{Z^4}{n^3(l+1)(l+\frac{1}{2})l}\right] \tag{80}$$

where Z is the atomic number of the atom and n and l are the principal and orbital angular momentum quantum numbers respectively of the electron of concern. Since the transition probability is proportional to $\langle {}^1\psi | \, er \, | {}^3\psi \rangle^2$, the $S \leftrightarrow T$ probability is dependent on Z^8 and $1/r^6$. This particularly strong Z and r dependence applies to molecules as an approximation only, since a central field potential was assumed. Nevertheless there is a strong dependence on Z and r. The development of $H_{\rm SO}$ for fields other than spherical has been made [93].

2. Paramagnetism of the Phosphorescent State

The essential problem is to prove that the phosphorescent state is indeed the spectroscopic triplet state. Earlier work gives reasonably clear indication of this. Lewis and Calvin [95] and Lewis, Calvin, and Kasha [96] showed that fluorescein in a boric acid glass at 25°C became paramagnetic on exposure to light, returning to its normal diamagnetic state in the dark. In this medium fluorescein is present as the cation, I, with an absorption maximum at 4390 Å and a strong yellow phosphorescence with a mean life of about 2 sec. In contrast, the well-known anion has an absorption maximum

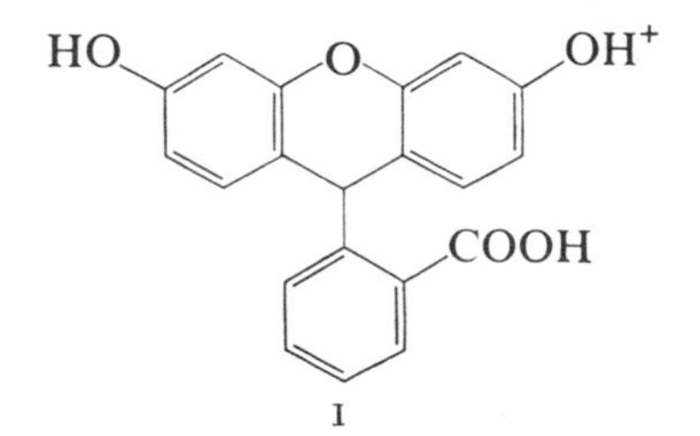

I

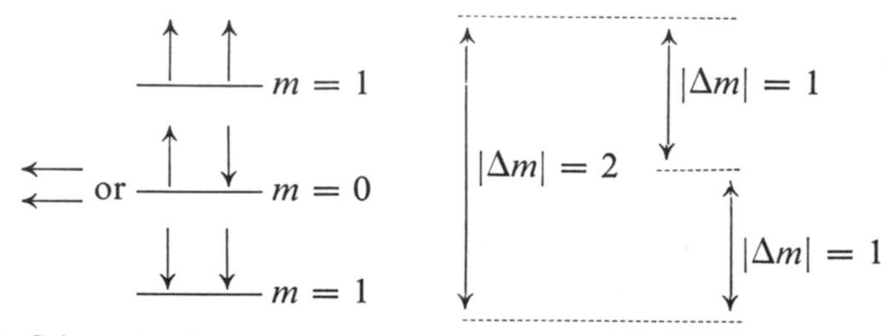

Figure 28 Schematic diagram showing components of a triplet level and relative spin orientations.

at 4920 Å, with a high quantum yield of green fluorescence. As a result of the long life of the phosphorescent state before emitting, and the high quantum yield, indicating little radiationless deactivation, a relatively large fraction of the fluorescein molecules can be maintained in the excited phosphorescent state. As many as 88.5 percent of the molecules can be in the excited state. The magnetic measurements were made by means of a modified Theorell magnetic microbalance and it was shown that the illuminated phosphor became paramagnetic, reverting to the diamagnetic state in the dark. From estimates of the concentration of dye molecules in the excited state the molar paramagnetic susceptibility of the excited state was found to agree, within an estimated maximum experimental error of 13 percent, with the observed susceptibility of the ground state of the oxygen molecule (known to be $^3\Sigma$).

Further evidence of the paramagnetic nature of phosphorescent state, however, has been sought by the search for paramagnetic resonance of this state which would of course provide additional unequivocal evidence for its triplet nature. An electron paramagnetic resonance (epr) absorption of the phosphorescent state of naphthalene has been observed [97]. This was obtained from single crystals of a solid solution of naphthalene in durene. The spectra are due to transitions between neighboring sublevels of the triplet. The triplet electronic degeneracy is removed by magnetic dipole-dipole interaction. In the presence of an external magnetic field there are different energy levels and transitions are possible between them (Figure 28). The levels result from the different possible spin orientations relative to the external field. Further, epr absorption by the triplet occurs not only in single crystals but in rigid glass solutions of some aromatic hydrocarbons. Triphenylene, 1,3,5-triphenylbenzene, coronene, and naphthalene all exhibit epr spectra in their phosphorescent state in rigid glasses of primary alcohols at 77°K [98]. The decay times are 13.3, 5.1, 7.9, and 3.0 sec, respectively. These should be the same as the lifetimes of the phosphorescences. They are in fact similar, though not identical (see Table 8, Chapter 11-A), which may be the result of the difference in solvents.

In addition to hydrocarbons, other molecules show epr spectra in their phosphorescent state. Several ketones such as aminobenzophenone, phenylbenzophenone, and others in rigid glasses of alcohols at 77°K show such absorption [99]. The decay time for aminobenzophenone is close to that observed for phosphorescence (0.41 sec). Also N-ethylacridone exhibits epr absorption in the phosphorescent state at 77°K in a rigid glass [100]. A single crystal solid solution of quinoxaline (1,4-diazonaphthalene) in durene at 77°K shows an epr absorption in its phosphorescent state [101]. The lifetime of the epr signal is estimated to be 0.2 ± 0.1 sec and that for phosphorescence is 0.25 ± 0.02 sec. The epr spectra of the triplet states plus phosphorescence data for 2-phenyl-S-triazene, 2,4,6-triphenyl-S-triazene, 2,4-diphenyl-6-methyl-S-triazene, biphenyl, *m*-terphenyl, and 1,3,5-trimethyl benzene show that triplet state for the phenyl-S-triazenes is of π, π^* type [102].

It is now obvious from the preponderance of evidence that the phosphorescent state and triplet state are in fact one and the same.

3. Intercombinations

In Figure 24 the essential intercombination processes involving states of different multiplicity are shown. The term "intersystem crossing" has been coined to describe the internal conversion (or radiationless transition) process that depends on spin-orbit coupling [94]. A more detailed discussion of radiationless processes will be given in the following chapter. This crossing is commonly between the first excited singlet and the lowest triplet (the ($T_1 \leftsquigarrow S_1$) transition, see Figure 1). The radiative intercombination process is phosphorescence and involves the transition $S_0 \leftarrow T_1$ (Figure 1). The intersystem crossing ratio $\chi = (\Phi_p/\Phi_F)$ has been used as a measure of the population of a triplet state relative to the lowest excited singlet state (S_1) [94]. The fraction of molecules leaving S_1 and going to T_1 will depend on the ratio of the appropriate rate constants and can be expressed as

$$\Phi_{IS} = \frac{k_{IS}}{k_F + k_{IS} + k_{IC}^S} \tag{81}$$

Often k_{IC}^S is considered small compared to other rate constants. Also the ratio (81) is not equal to χ.

It will be recalled that a direct transition from $T_1 \leftarrow S_0$ is forbidden in the absence of spin-orbit coupling because of spin orthogonality. Of course in the presence of spin-orbit coupling this process becomes less forbidden, but it is still an extremely unlikely process compared to any transition involving no change in multiplicity. It is even considerably weaker than any symmetry-forbidden process involving no change in multiplicity. Thus it is not normally feasible to attempt to excite phosphorescence emission by utilizing the $T_1 \leftarrow S_0$ transition.

As has been implied earlier the procedure to excite phosphorescence is to utilize $S_1 \leftarrow S_0$ or higher transitions, followed by intersystem crossing to the triplet, followed by phosphorescence. The question may arise of why this method is more successful than $T_1 \leftarrow S_0$ since without spin-orbit coupling they are both forbidden. At first glance one might expect that even with spin-orbit coupling $T_1 \leftarrow S_0$ and $T_1 \leftsquigarrow S_1$ would be about equally probable. One important factor which this neglects is that the mixing between states (now our S_0, S_1, and T_1) also depends on the energy difference between the triplet and singlet states concerned. Therefore, since $T_1 - S_0$ energy differences are commonly of the order of 20,000 cm^{-1} and $S_1 - T_1$ energy differences are commonly of the order of 3000–9000 cm^{-1}, the intersystem crossing probability ($T_1 \leftsquigarrow S_1$) is considerably greater than the direct transition from the ground state to the triplet state.

It is generally true that radiative processes involving intercombinations (phosphorescence) are of the order of 10^{-6} of those involving no multiplicity differences. This means that the lifetimes for phosphorescence are as a general rule some million times slower than for fluorescences. The rate constant for the intersystem crossing process is of the order of 10^7 sec^{-1} (or has a lifetime of 10^{-7} sec). Considering that the general order of fluorescence lifetimes is 10^{-8}–10^{-9} sec, one can see the considerable difference in the intersystem crossing $T_1 \leftsquigarrow S_1$, and phosphorescence $S_0 \leftarrow T_1$ transition probabilities. Although the $T_1 \leftsquigarrow S_1$ process is slower than the fluorescence $S_0 \leftarrow S_1$ process, it is fast enough to compete with the fluorescence process and give triplet-state occupation. It is important to point out that in many cases the relative rates for the $S_0 \leftarrow S_1$ and $T_1 \leftsquigarrow S_1$ processes can be altered to be strongly in favor of the $T_1 \leftsquigarrow S_1$ process with consequent high or even total occupation of the triplet state and a resulting complete absence of fluorescence. This can be accomplished by the introduction of a high atomic numbered center, a paramagnetic center, or by having a molecule in which the $S_1 \leftarrow S_0$ transition is weak, where the fluorescence will have a long lifetime. In any of these cases the intersystem crossing rate constant may be equal to or considerably larger than that for fluorescence. More details and examples on these points will be given in later sections.

Finally, the effect of paramagnetic ions or molecules should be considered. In the case of paramagnetic ions, the systems of particular interest are the metallo-organic complexes containing such ions. For ions, the spin momentum alone need be considered. Because of this spin momentum, there exists an extremely strong inhomogeneous magnetic field. This would be as strong, if not stronger, than any similar field created by a high atomic numbered center. Thus it would be expected that complete or nearly complete spin-orbit coupling would occur in complexes containing paramagnetic ions. This would in turn increase the probability of intersystem crossing to such a degree that it would dominate the probability for $S_0 \leftarrow S_1$ emission and fluorescence would be quenched. Thus phosphorescence might be expected to be the sole emission. Generally this is borne out by experiments (see Chapter 13).

The interest in the effect of molecules that are paramagnetic primarily centers around their effect as external perturbing agents on the $T_1 \leftarrow S_0$ absorption of other molecules. Oxygen (O_2) represents the best known paramagnetic molecule. Although the known perturbing influence of O_2 is probably not solely the result of a magnetic effect, this factor appears to be of consideration (see Chapter 14).

Chapter 9 Nature of Radiationless Processes

A. VIBRONIC PERTURBATIONS

In addition to direct spin-orbit coupling of the triplet to the singlet manifold, (see the previous chapter) there are other coupling mechanisms that are possible [103]. Mechanisms include direct spin-vibronic coupling which is very weak but a first-order term, spin-orbit coupling and vibronic coupling in the singlet manifold, and finally vibronic coupling in the triplet manifold and spin-orbit coupling [103]. In the previous section the H_{so} was purely electronic. If nontotally symmetric vibrations exist, then vibronic terms can also introduce mixing of the triplet and singlets.

The total electronic Hamiltonian can be written for the zero order Born-Oppenheimer approximation as follows:

$$H = H_0 + H_{so} \tag{82}$$

where H_0 contains the kinetic and potential energies for the electrons as well as the spin-energy operators and H_{so} is the spin-orbit operator. Both terms contain the nuclear coordinates as parameters. The eigenfunctions of H_0 and H depend on the nuclear coordinates, which is of prime significance for vibronic perturbation. The Hamiltonian can be expanded in a Taylor series in the normal coordinates of the triplet state for the emission process (ground state for absorption). To first order, the nuclear coordinate dependence can be written as follows:

$$H = H_0^0 + H_{so}^0 + \sum_a \left[\left(\frac{\partial H_0}{\partial Q_a}\right)_0^0 + \left(\frac{\partial H_{so}}{\partial Q_a}\right)_0^0 \right] Q_a + \cdots \tag{83}$$

where the 0 superscript refers to the equilibrium position of the nuclei and Q_a are all the $3N$-6 normal coordinates belong to the appropriate electronic state.

If spin-orbit coupling is weak, then multiplicity classifications of the eigenfunctions of H_0 are appropriate to a good approximation for H. These could be called nearly pure singlets and triplets versus pure singlets and triplets of H_0. The first-order expression provides vibronic mixing of the nearly pure singlets and triplets. It is further possible to then consider H_{so}^0 as a perturbation on the nearly pure states and express them in terms of the H_0^0 and pure states at equilibrium nuclear positions [103,104].

Direct second-order perturbation theory on eigenfunctions of H_0^0 gives three perturbing terms,

$$\begin{aligned} H(1) &\equiv H_{so}^0 && \text{spin-orbit coupling} \\ H(2) &\equiv \sum_a \left(\frac{\partial H_0}{\partial Q_a}\right)_0^0 Q_a && \text{vibronic coupling} \\ H(3) &\equiv \sum_a \left(\frac{\partial H_{so}}{\partial Q_a}\right)_0^0 Q_a && \text{spin-vibronic coupling} \end{aligned} \tag{84}$$

with H_0^0 as the zero-order Hamiltonian. The perturbed pth triplet function ${}^3\psi_p$ to second order is [103]

$$\begin{aligned} {}^3\psi_p = \text{pure triplet terms} &+ \sum_{k\neq p} ({}^3E_p^0 - {}^1E_k^0)^{-1} \Bigg\{ H_{kp}(1) + H_{kp}(3) \\ &+ \sum_{n\neq p} \left[\frac{H_{kn}(1)H_{np}(2) + H_{kn}(2)H_{np}(1)}{|{}^3E_p^0 - {}^{3,1}E_n^0|} \right] \Bigg\} {}^1\phi_k^0 \end{aligned} \tag{85}$$

where E_p^0 is the energy associated with the zero-order triplet wave function p, ${}^{3,1}E_n^0$ is the zero-order energy of electronic state ϕ_n^0 (singlet or triplet) which is coupled to the pth triplet or to ${}^1\phi_k^0$ vibronically or spin orbitally, and ${}^1\phi_k^0$ is the zero-order singlet wave function for any of the excited singlet states. The matrix elements H_{rs} (also see equations 84 and 85) have the following further significance:

$H_{kp}(1)$	spin-orbit coupling	
$H_{kn}(2)\ H_{np}(1)$	vibronic coupling among singlets and spin-orbit coupling	
$H_{kn}(1)\ H_{np}(2)$	spin-orbit coupling and vibronic coupling among triplets	(86)
$H_{kn}(3)$	spin-vibronic coupling	

If the perturbed triplet p is the lowest triplet, then

$$^3\psi_1 = {}^3\phi_1^0 + \sum_k ({}^3E_1^0 - {}^1E_k^0)^{-1}\Bigg\{H_{k1}(1) + H_{k1}(3) + \sum_n \left[\frac{H_{kn}(1)H_{n1}(2) + H_{kn}(2)H_{n1}(1)}{|{}^3E_1^0 - {}^{3,1}E_n^0|}\right]\Bigg\}\,{}^1\phi_k^0 \tag{87}$$

where ${}^3\phi_1^0$ and ${}^3E_1^0$ are the zero-order wave function and energy for the lowest triplet, the H_{rs} are matrix elements as defined above, ${}^{3,1}E_n^0$ is the zero-order energy of the electronic state ϕ_n^0 (singlet or triplet) which is coupled to ${}^3\phi_1^0$ or ${}^1\phi_k^0$ either spin orbitally or vibronically, and ${}^1\phi_k^0$ is any of the zero-order singlet states.

Equation 87 does not contain $H(2)$ as a first-order term since from (84) it can be seen that only $H(1)$ and $H(3)$ directly connect electron spin and configuration space; that is, they contain H_{so}. In second order, all H's are contained but only two with $H(1)$ and $H(2)$ remain, either because the others vanish or because they are relatively small [103]. For molecules with hydrogen, carbon, or nitrogen atoms, $H(2) \gg H(1) > H(3)$. In spin-orbit perturbations, however, $H(2)$ enters only in the second-order term and therefore is less important than $H(1)$ which enters in the first-order term. This assumes $H(1)$ does not vanish. If there are heavy atoms in the molecular system, $H(1)$ and $H(3)$ would be expected to increase in importance relative to $H(2)$. Both $H(2)$ and $H(3)$ are energy expressions and therefore are totally symmetric. Thus the partial derivatives contained in each term should have symmetry properties in the electronic coordinates identical to those of Q_a in the nuclear coordinates such that total symmetry results. The direct product of the representations of the two states that are mixed by $H(2)$ or $H(3)$ must be or contain the representation to which the perturbing vibration Q_a belongs.

It is also worthwhile to evaluate the above considerations in terms of the transition moment between the ground state and lowest triplet, assuming no mixing of triplet states into the ground state. For convenience in nomenclature we make the following modification in symbolism:

$H(1)$ to H_{so}	spin-orbit coupling	
$H(2)$ to H_v	vibronic coupling	(88)
$H(3)$ to H_{sv}	spin-vibronic coupling	

Thus for molecules containing carbon, hydrogen, and nitrogen atoms $H_v \gg H_{so} > H_{sv}$. The first- and second-order contributions to the ground-state singlet to triplet transition moment M_{ST} are

$$M_{ST}(\text{1st order}) = \sum_k ({}^3E_1^0 - {}^1E_k^0)^{-1}\langle {}^1\phi_k^0|\,H_{so}\,|\phi_1^{30}\rangle M_{0k} \tag{89}$$

(compare with 79) and

$$
\begin{aligned}
M_{ST}&(\text{2nd order}) \\
&= \sum_{k,n} ({}^3E_1^0 - {}^1E_k^0)^{-1}({}^3E_1^0 - {}^1E_n^0)^{-1} \langle {}^1\phi_k^0 | H_v | {}^1\phi_n^0 \rangle \langle {}^1\phi_n^0 | H_{so} | {}^3\phi_1^0 \rangle M_{0k} \\
&+ \sum_{k,n} ({}^3E_1^0 - {}^1E_k^0)^{-1}({}^3E_1^0 - {}^3E_n^0)^{-1} \langle \phi_k^0 | H_{so} | {}^3\bar{R}_n^0 \rangle \langle {}^3\phi_n^0 | H_v | {}^3\phi_1^0 \rangle M_{0k}
\end{aligned} \tag{90}
$$

for which spin-vibronic contributions are neglected. All terms are as previously defined and M_{0k} is the zero-order transition moment of the perturbing transition ${}^1\phi_k^0 \leftarrow {}^1\phi_0^0$, where ${}^1\phi_0^0$ is the zero-order wave function for the ground state (subscript 0). The first term in (90) represents the vibronic coupling of the perturbing singlet ${}^1\phi_k^0$ with an intermediate singlet state ${}^1\phi_n^0$ which is itself spin-orbitally coupled to the lowest triplet ${}^3\phi_n^0$. The second term represents spin-orbit coupling between the perturbing singlet ${}^1\phi_k^0$ and an intermediate triplet ${}^3\phi_n^0$ which is itself vibronically coupled to the lowest triplet ${}^3\phi_1^0$.

The above considerations have been applied to molecular systems including benzene [103] where it is found that the matrix element $H(1)H(2)$ is dominant, which leads to an assignment of the triplet as ${}^3B_{1u}$. More concerning these considerations is given in later discussions on benzene and N-heterocyclics [105].

B. INTERNAL CONVERSIONS

There are three principal theories concerning radiationless transitions.

1. The excited molecular system (electronically and vibrationally) interacts with the time-dependent phonon field originating from the solvent [106].
2. Nonstationary electronic states of the system (molecule plus solvent) are connected by time-independent perturbations of electrostatic or spin-orbit type [107,108,109].
3. The process is truly intramolecular and no interactions (of any kind) with a solvent are necessary [110,111].

In addition to these there is an important consideration regarding the effect of solvent viscosity on both the radiative and radiationless transition probabilities from the triplet. A discussion of this will be postponed until the theories concerning radiationless transitions have been examined.

The considerations in the phonon field theory for radiationless transitions are parallel with those of radiative transitions [106] and the theory applies to a solute molecule in a crystalline solvent. That is, in radiative transitions the molecule interacts with the photon field, whereas in nonradiative transitions the molecule interacts with the phonon field. Phonons arise from fundamental excitation of lattice modes. Just as in radiative processes, it is possible to deduce Einstein probability coefficients that can relate the probability of phonon absorption and emission to characteristics of the solute molecule. There are two Einstein coefficients that are important for the nonradiative process: one is spontaneous and the other is induced. The rate constant for the nonradiative process would be proportional to the sum of the two as

$$k \propto \frac{4\omega^3}{3\hbar c_s^3} |\mu^{rl}_{1\leftarrow 2}|^2 + \frac{4\omega^3 |\mu^{rl}_{1\leftarrow 2}|^2}{3hc_s^2(e^{\hbar\omega/kT-1})} \tag{91}$$

where $\omega/2\pi c$ is equal to the energy difference between states 1 and 2 in cm^{-1}, ω is the resonance frequency between states 1 and 2, c_s is the velocity of sound, and $\mu^{rl}_{1\leftarrow 2}$ is the matrix element connecting states 1 and 2 radiationlessly [106]. The matrix element is composed of two parts, nuclear and electronic. The nuclear part connects vibrational states of a given electronic state while the electronic part connects different electronic states. The latter, in general, will depend on the vibrational overlap between the different vibrational levels of the two electronic states. This is generally the greatest between states of identical vibrational quantum number.

The second term in (91) refers to the induced radiationless emission process. It can be seen that this portion is strongly temperature dependent.

Phonon emission occurs if the final molecular state is below the initial one and phonon absorption occurs if the reverse is true. Thus, if a radiationless process should require phonon absorption, the observed rate would be temperature dependent (i.e., there would be no spontaneous emission). On the other hand, if phonon emission occurs, two temperature sensitivities can result: first, for low-energy phonons there would be a significant temperature dependence since the rate of induced emission dominates the spontaneous rate near room temperature but at lower temperatures they are approximately equal; second, for high-energy phonons the emission rate may change by a factor of approximately 10 and therefore is relatively temperature insensitive.

A comparison between the Einstein coefficients for spontaneous radiative emission and radiationless emission shows that radiationless transitions would be some 10^{15} times faster than radiative ones. This occurs because in the ratio

$$\frac{A_{\text{radiationless}}}{A_{\text{radiative}}} = \left(\frac{c}{c_s}\right)^3 \frac{|\mu^{r1}_{1\leftarrow 2}|^2}{|\mu^{r}_{1\leftarrow 2}|^2} \tag{92}$$

the ratio $c/c_s \approx 10^5$ where c is the velocity of light, c_s is as previously defined, see equation (91), and μ^{r1} and μ^{r} are the matrix elements connecting the states of radiationless and radiative transition, respectively. A more detailed consideration of the ratio of $|\mu^{r1}_{1\leftarrow 2}|^2/|\mu^{r}_{1\leftarrow 2}|^2$ introduces some further modifications. For nonradiative transitions within an electronic state and between states within a given manifold this factor would appear to reduce the ratio of the A's by a factor of 10^2–10^3 for a 100 cm^{-1} phonon. Further, in the case of internal conversions between states the ratio of A's must be taken times the vibrational overlap (i.e., the Franck-Condon factor or integral).

Presumably, for fast intersystem crossing (a nonradiative process) the transition needs to be between the lowest singlet and triplet states higher than the lowest one [106].

The second theory pertinent to radiationless transitions involves time-independent perturbations whereby the states are coupled by electrostatic or spin-orbit interactions [107]. Essentially there is coupling between the electronic states of the solute and solvent.

Several factors are important before proceeding. The initial and final states of the molecule actually refer to nonstationary states of the overall system; that is, the initial state is rigorously the zero-vibrational level of the initial electronic state of the system molecule plus lattice. The final state is comprised of the final molecular electronic state, molecular vibrations, and lattice vibrations. Thus radiationless transitions occur between an initial state and a final state that are degenerate when molecular and lattice vibrations are added to the final state through time-independent perturbations of the type previously described, see theory 2. Therefore, it can be seen that the principal role of the solvent is to increase the degeneracy of the system through coupling of its electronic and vibrational states which superpose with those of the solute. Finally, it should be stated that since a perturbation approach is used, $\tau_{\text{electronic}} \gg \tau_{\text{vibrational}}$, that is, vibrational energy returns to thermal equilibrium far more rapidly than does electronic energy.

A critical point is that there are vibrational states of the final state that are nearly degenerate with the initial state in the free molecule. These vibrational states are directly coupled to the initial state through mechanical coupling with the lattice. The solute-solvent interaction is considered to be weak such that there is little distortion of the free molecule states. Figure 29 shows a schematic of the couplings involved. In Figure 29 only one vibrational and lattice mode is considered. In general in a lattice there will be many modes, and a large number of final states consisting of a singular vibrational mode with the lattice modes will exist. Further, α represents a matrix element coupling together zero-order final states that differ by only a small number of vibrational quanta; and β represents a matrix element connecting the initial state and one of the directly coupled final states as ψ_0' (Figure 29) where ψ consists of electronic (ϕ) and vibrational (γ) part. In this case, $\alpha \gg \beta$ which infers weak coupling between the initial state, ψ_0, and one of the final states, ψ_0', but that strong coupling exists between ψ_0' and the various indirectly coupled final states, ψ_0^{I}, ψ_0^{II}, etc. There are a large number of states contributed by the lattice vibrations. These will be degenerate with a final

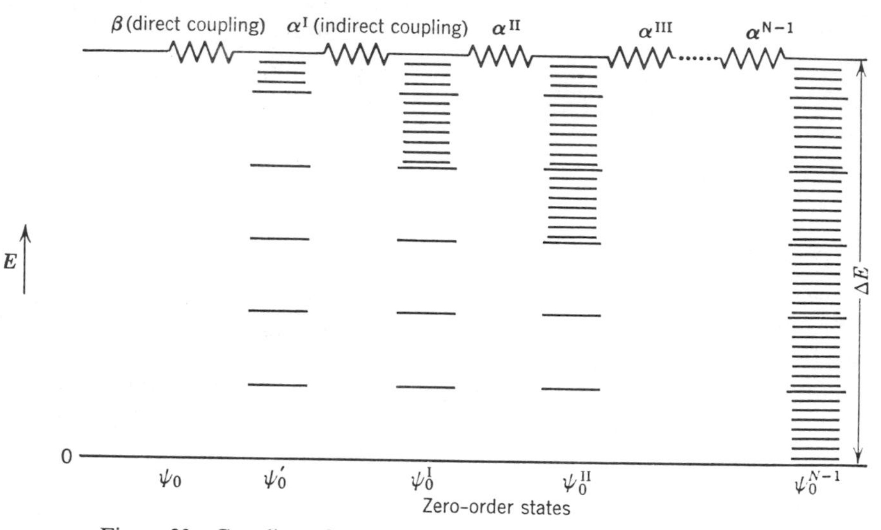

Figure 29 Coupling of states of the system molecule-medium [397].

state and form a band coupling the zero-order states differing by one or a small number of vibrational quanta (see Figure 29).

For those cases where the purely electronic state energy difference is very large, vibrational overlap (Franck-Condon integrals) becomes small and so does the transition probability. The probability depends on vibrational overlap in the following manner:

$$k = \frac{2\tau_v}{\hbar^2}\,[\langle\gamma_n\gamma_n'\rangle]^2\,[\langle\phi H'\phi\rangle]^2 \tag{93}$$

where τ_v is the vibrational relaxation time (10^{-11}–10^{-13} sec), γ_n and γ_n' are respectively, the 0 level vibrational wave functions in the initial electronic state, ϕ, and the vibrational wave function at about ΔE, Figure 29, in the final electronic state (ϕ'), and H' is a perturbing Hamiltonian. If the process is $S_0 \leftsquigarrow T_1$ or $T_1 \leftsquigarrow S_1$, then H' is the spin-orbit coupling operator but if it is $S_0 \leftsquigarrow S_1$, then H' is the electron repulsion or vibronic operator. In addition, the $\langle\gamma_n\gamma_n'\rangle$ integrals are summed over all directly coupled states. It is worthwhile to point out that actual crossing of the potential energy surfaces is not necessary since the vibrational wave functions extend beyond the potential surface. In the case of crossing, the vibrational overlap would be large; that is, approximately one. Obviously in other cases (no crossing) it could be expected to be considerably less than one.

In many ways the general approaches of Hunt et al. [108], Bryne et al. [109] are similar to that of Robinson and Frosch [107], although differences are also present. These principally involve the determination of the Franck-Condon factors as a function of the energy separation of the electronic states.

More discussion concerning Franck-Condon factors is given later in this section. Earlier consideration [108] treated internal conversion as a tunneling process between two potential energy surfaces that did not cross or intersect. Potential energy surfaces are constructed utilizing relations between bond length and bond order and the rate of internal conversion is derived empirically based on the distance between the surfaces. An exponential dependence of the rate on this distance results. Later consideration [109] relies on calculation of Franck-Condon factors. Where small energy gaps are involved, the Franck-Condon factor is dominated by skeletal vibrations, but where the energy gaps are large, these factors are dominated by CH vibrations. Thus deuteration will affect the rate of internal conversion principally only in the latter instance. This would be the case for transitions between the lowest triplet and the ground state for aromatic hydrocarbons.

The final consideration concerns the truly intramolecular process [110,111]. Naphthalene, anthracene, and naphthacene upon excitation into their second excited singlet state exhibit internal conversion to the first excited singlet state under conditions where no collisions occur during the lifetime of the second excited singlet state [110,111,112,113,114]. This clearly indicates that in these cases, interactions of any kind with a solvent are not necessary for internal conversion among the singlet states. In addition for benzene, intersystem crossing between the lowest singlet and triplet occurs at very low vapor pressure [111,115]. Presumably this result would also mean that intersystem crossing, which occurs via internal conversion, can occur without a solvent interaction. All of these results are significant in two ways: first, internal conversion can occur as an intramolecular process between states of the same or different multiplicity in the absence of solvent interaction and, second, the internal conversion process in the absence of solvent interaction can occur for molecules such as benzene, which are relatively small compared to organic molecules as a whole.

The case of naphthalene [111] is important to consider further and can be considered as a specific example for the general considerations. From the fact that the maximum possible intensity of fluorescence from the second excited singlet state is 0.002 of that from the first, the *minimum* value of the rate constant for internal conversion is approximately 2×10^{10} sec^{-1} (for a first-order process). The mechanism for the intramolecular internal conversion has certain parallels to that previously discussed for mechanism 2. A large number of states of the first excited singlet (S_1) lie at the same energy as a given level or state of the second excited singlet (S_2). The large number of states of the first have their origin from vibrations and rotations present in the molecule. If the transition probability of crossing from a given level of S_2 to any given level of the large number of S_1 has a reasonable value, the radiationless transition is expected to be irreversible. Of course the probability is dependent on Franck-Condon factors well as symmetry characteristics. Primarily, the number of levels of S_1 at the same energy of a level of S_2 depends on the number of fundamental vibrational modes, their energies, and the electronic energy difference between S_1 and S_2; that is, these factors will be important in determining the number of different combinations of vibrational quanta which, when added to the electronic energy of S_1, will give an energy level equivalent to any given one in S_2.

Naphthalene has 48 fundamental modes. The electronic energy difference between S_1 and S_2 is approximately 3400 cm^{-1}. With some division of the fundamental modes among vibrations of various energy [111] one can calculate that all of the various combinations that would give an energy of approximately 3400 cm^{-1} results in 1.3×10^5 levels. These would presumably be in a band approximately 415 cm^{-1} wide giving a density of about 300 levels per cm^{-1}. Although some of these may not be able to couple with a given level of S_2 because of symmetry, there certainly would be a large number of states of S_1 near a given one of S_2 so that a radiationless transition would occur to one of them (of S_1).

Further examination of the factors contributing to radiationless transitions between two electronic states of a polyatomic molecule in a solvent matrix indicates that the Franck-Condon factors are of prime importance [116–118]. That is, under the foregoing environmental conditions the radiationless transition can be considered to consist of two steps: (a) a transition between vibronic states (of the two different electronic states) followed by (b) rapid vibrational relaxation within the lower electronic state system (solute-solvent). The former of these, step (a), will depend on vibronic matrix elements as well as Franck-Condon factors and may be rate determining. With such an assumption the rate constant will be [117]

$$k_{AB} = \frac{2\pi\rho_E}{\hbar} J^2 F(E) \tag{94}$$

where ρ_E is the density of states (those resulting from solvent and solute molecular electronic and vibrational states), J is the electronic transition matrix element (from vibronic or solvent interaction), F is the Franck-Condon or vibrational overlap factor, and E is the vibrational energy relative to the zero-point energies of the electronic states of the system (E_A and E_B, for example, where $E_A < E_B$). The Franck-Condon factor is [116–118]

$$F(E) = \sum_p P\left[\sum_n |\langle \gamma'_n(v_n) \mid \gamma_n(0)\rangle|^2\right] \tag{95}$$

where $E = \sum_n v_n \hbar\omega'_n$, v_n is a vibrational quantum number, ω'_n is the vibrational frequency in the final (lower electronic) state, γ'_n and $\gamma_n(0)$ are vibrational wave functions of an oscillator n in the final and initial state (0 level), respectively, and $\sum_p P$ is an operator that permutes phonons among the N oscillators (N normal modes exist). The permutation is subject to an energy conservation condition,

$$E = E_{00} \pm \tfrac{1}{2}\rho_E \tag{96}$$

where E_{00} is the energy of the 0–0 transition between the electronic state systems. Of importance is the fact that (94) shows that the rate constant is proportional to the Franck-Condon factor. Quantitative evaluation of these factors can be made if sufficient spectroscopic data are available or if it is assumed all oscillators are harmonic [117].

For a related group of molecules such as perhydrogenated versus perdeuterated hydrocarbons, it is expected that $F(E)$ will vary. Utilizing observed phosphorescence lifetime and assuming the true lifetime to be 30 sec for all aromatic hydrocarbons, it can be shown that radiationless transitions from the triplet to the ground state are primarily determined by CH stretching vibrational modes (also see later discussion). Thus the rate constant for radiationless transitions can be significantly affected by substitution of deuterium for hydrogen [117,118]. In addition, empirically $\log F(E)$ is a linear function of E/η, where η is $N_H/(N_H + N_C)$, and N_H and N_C represent the number of hydrogen and carbon atoms in the molecule. It is the inclusion of this latter term that is the most significant in determining the linear relationship.

There is additional evidence indicating that the energy of the triplet is important in determining the rate constant for the radiationless process $S_0 \leftsquigarrow T_1$. If it is assumed that the rate of triplet decay is that of the radiationless decay, a plot of $\Delta E_{S_0-T_1}$ versus $\log k_T$ yields a straight line (Figure 30) [119]. Extrapolation of this curve to $\Delta E_{S_0-T_1} = 0$ gives a rate constant of approximately 10^7–10^8. On the other hand, in the region of $\Delta E_{S_0-T_1}$ gaps appropriate for most molecules considered, the rate constant is approximately 10^1–10^2. In addition the observed lifetimes of phosphorescence notably increase with deuterium substitution. There are several significant factors to be deduced from such data. One of these is the fact that there is a high dependence of the rate constant for radiationless transitions $S_0 \leftsquigarrow T_1$ on vibrational overlap or the Franck-Condon factor. Second, since the processes $T_1 \leftsquigarrow S_1$ and $S_0 \leftsquigarrow T_1$ are both intersystem crossings, the data for the rate constant at $\Delta E = 0$ indicates that the process $T_1 \leftsquigarrow S_1$ would be very fast while that from $S_0 \leftsquigarrow T_1$ would be relatively slow. This would occur because of the large energy gap and therefore small Franck-Condon factors in the latter case, but a small energy gap and large Franck-Condon factors in the former case. Finally, as the $\Delta E_{S_0-T_1}$ gap decreases the τ_T should

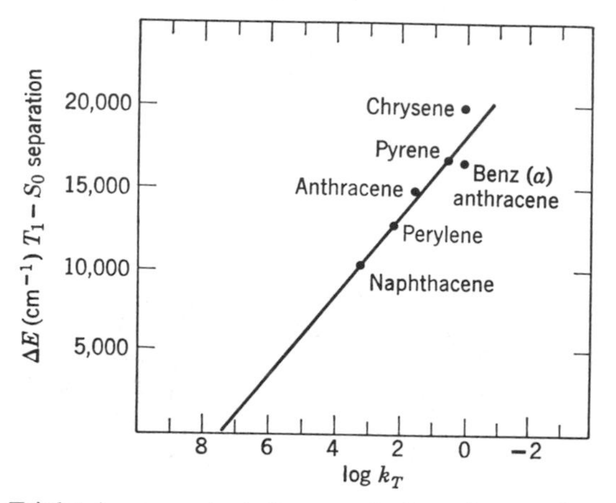

Figure 30 Triplet decay constants for some hydrocarbons as function of $T_1 - S_0$ energy separation [119].

decrease ($>k_T$) because of considerations similar to those just discussed. All of these considerations are consistent with experimental observations. Theoretical considerations [116] indicate that $F(E)$ is dominated by shifts in frequency of distortions (as CH vibrations) of the harmonic oscillators for radiationless transitions where the energy gap E is large. Recall that E is the vibrational energy measured relative to the zero-point energies of the electronic states E_A and E_B, where $E_A < E_B$. These types of oscillators lead to an approximate exponential decrease of $F(E)$ with increasing E. However, $F(E)$ calculated for skeletal modes (displacements) decrease more rapidly than those calculated for distortions. Thus the observation just discussed concerning the relationship of τ_T and ΔE, as well as $F(E)$ and E as well as the marked affect of the substitution of deuterium for hydrogen are capable of being correlated with theory. Therefore, the conclusions [107–109,116–118] regarding the effectiveness of displacement skeletal C—C and distortional CH vibrational modes on radiationless transition probability as a function of the energy gap are in harmony.

In the theories describing radiationless transitions involving the solute in a solid matrix as mechanism 2 [107,109] and the Franck-Condon considerations [116,118], it is assumed that the vibrational relaxation within the lower state is fast and that the transition between vibronic states is rate determining. In addition, except in the case of strong solute-solvent interaction, the rate of the radiationless transition will be independent of the nature of the solvent matrix.

In the case of the lifetime of the triplet state of anthracene and naphthalene, however, there *is* a difference in τ_T depending upon the matrix. The values for anthracene are 0.01 sec in the crystal and 0.1 sec in an EPA glass while for naphthalene they are 0.02 and 2 sec in the crystal and EPA glass, respectively [120]. The implication is that coupling of a vibrationally excited molecule to a solvent matrix is smaller than the corresponding coupling in the pure crystal where the "solvent" is also the solute. It is assumed that intermolecular coupling of vibrations in the pure crystal is via resonance interactions involving transition multipole moments such as quadrapole. This would result in a vibrational exciton bandwidth of the order of magnitude of 1 cm^{-1} and thereby provide an efficient route for vibrational relaxation of the crystal. Thus in solvent matrices versus solute matrices (the pure crystal) resonance energy transfer might be much less likely in the former system. This would occur because the vibrational exciton bandwidth would not necessarily be sufficiently large that the vibronic matrix elements and the Franck-Condon factor mixing the two states are rate determining [120]. The above consideration at least appears to be consistent with the data given earlier for the two aromatic hydrocarbons, naphthalene and anthracene.

The understanding of the radiationless internal conversion process is considerably more advanced than it was just a few years ago. However, certain considerations still remain. Since it is possible to have a mechanism that is truly intramolecular, is it really necessary to involve other mechanisms (such as 1 and 2 given earlier) for a solute in a solvent matrix? It is probable that the intramolecular process and mechanism 1, or mechanism 2, or both, are operating in the case of a solute-solvent matrix system. This is not clear. Further, the rate-determining step is not unequivocal. In addition, the rate-determining step could vary since the vibronic matrix elements mixing the electronic states may vary with the nature of solvent because of changes in coupling of excited vibrational states. Also the vibrational exciton bandwidth may change.

As was implied at the beginning of this section, the solvent viscosity appears to play a role in the radiationless transition probability from the triplet. This is also true for the radiative process (phosphorescence). Such effects were first reported in detail for a palladium metalloporphyrin and *p*-phenylbenzophenone [121]. Later there was an implication that the triplet-singlet transition probability for naphthalene was dependent on the solvent viscosity [122]. Also, both temperature and viscosity exert an influence on the radiationless process of a metalloporphyrin of a different type (magnesium) than that referred to above [123]. It has been established that there is an activation energy of 0.3–0.8 kcal for the radiationless process, $S_0 \leftsquigarrow T_1$, for several aromatic hydrocarbons and their halogen derivatives [124,125, 126].

Based on relative quantum yields and lifetimes of phosphorescence for a palladium metalloporphyrin and *p*-phenylbenzophenone [121,122] it is possible to show that the radiative rate constant changes with viscosity. With certain assumptions [122] it is possible to evaluate the individual radiative and radiationless rate constants over a broad temperature and viscosity range. For each of the two molecules, the radiationless rate constant decreases with increasing viscosity. On the other hand the radiative rate constant can decrease (metalloporphyrin) *or* increase (*p*-phenylbenzophenone) [122]. It should be noted that even if the actual values for the individual rate constants are not correct, the general conclusion concerning the trends noted above is not affected. Thus the viscosity affects the probability of intercombination between the triplet and ground state. It has been suggested [124] that an interaction between the rotational motion of the solute molecules and spin-orbit coupling might exist and be changing thus affecting the probability of intercombinations. It would be expected, that this would cause any change in the radiative rate constant to be in the same direction for all solutes. However, such is not the case [122]. Although it is not possible to promulgate a definitive mechanism, at least several factors seem of importance: (a) the solvent molecular relaxation time, (b) steric influence of the solvent molecules, and (c) coupling between electronic and vibrational states of the solute and lattice modes of the solvent molecules.

Chapter 10 Solvent Effects on Emission Spectra

A. GENERAL SOLVENT EFFECTS

The first point to be considered, an important one, is the relationship between the energy of the 0–0 band in absorption and emission. It will be recalled that this is the vibrational band of lowest energy in absorption and of highest energy in emission. The discussion does not deal with cases in which the electronic transition is symmetry forbidden, in which case the 0–0 band would be missing in absorption. Even in this case the 0–0 band can occur in solution although it will be absent for the pure solute in the vapor state. Ideally, it would be expected that the absorption and emission bands would coincide since the energy change is identical. Figure 27 gives some examples of near coincidence and noncoincidence of these frequencies.

Several approaches concerning this point are appropriate [127,128]. If the solvent and solute are such that one is polar and one is nonpolar or if both are nonpolar, the frequency shifts in absorption and emission are predicted to be the same. The quantitative expressions for the shifts in absorption and emission are equivalent [127,128]. Therefore the 0–0 bands should coincide or very nearly so. This results from considerations similar to those given earlier. The combinations noted above would encompass those cases noted as: I (nonpolar-nonpolar) as benzene in hexane, II (nonpolar-polar) as naphthalene in chloroform, (IIIA (polar-nonpolar) as acetone in benzene, IIIB (polar-nonpolar) as nitrobenzene in cyclohexane, IVA (polar-polar) as acetone in ethanol, IVB (polar-polar) as nitrobenzene in ethanol. The situation for cases IIIA, IIIB, IVA, and IVB is illustrated in Figure 7.

The important factor in cases IIIA and IIIB is that there is no or very little difference in energy between the Franck-Condon state and the equilibrium state. Thus the state attained by absorption and that from which emission originates are nearly the same. Further, it will be recalled that in these cases the polarization term (induced dipole-induced dipole term) and dipole-induced dipole terms are the principal contributing factors in the solvent-solute interaction. Therefore no orientation strain occurs; hence the Franck-Condon and equilibrium states are nearly the same.

When the solute and solvent are both polar, the situation in fluorescence is more complicated, just as it is for absorption. This combination includes IVA and IVB of Figure 7. Examples are acetone in ethanol and nitrobenzene in ethanol. In the simplest instance, if there is negligible dipole reorientation in the excited state, the frequency shift for fluorescence is the same as that for absorption. Again the quantitative expressions for the shifts are identical. In this case the 0–0 bands in the two processes should be the same or nearly so. This situation could arise when a rigid solvent is employed. In such a system the relaxation times are several orders of magnitude greater than the lifetime of the excited state. If the solvent is fluid, however, relaxation is much more rapid and is generally completed in a time less than the lifetime of the emitting state. When this is true, it can be assumed that complete dipole reorientation occurs. Therefore, absorption will occur to the metastable Franck-Condon state but emission will occur from the equilibrium state. The representation of this difference can be seen in Figure 7 for cases IVA and IVB. The quantitative expression for the $\Delta\bar{\nu}$ in absorption is different from that in emission and the 0–0 bands will not coincide. The final relationship expressing this difference is [128]

$$\bar{\nu}(\text{abs}) - \bar{\nu}(\text{emiss}) = \left(\frac{2}{hc}\right)\left(\frac{(m_{00}^u - m_{ii}^u)^2}{a^3}\right)\left(\frac{D-1}{D+2} - \frac{n_0^2 - 1}{n_0^2 + 1}\right)$$
$$+ \left(\frac{2}{hc}\right)\left\{\frac{(\alpha_0^u - \alpha_i^u)[3(m_{00}^u)^2 - 5(m_{ii}^u)^2 + (2m_{00}^u)(m_{ii}^u)]}{a^6}\right\}$$
$$\times \left(\frac{D-1}{D+2} - \frac{n_0^2 - 1}{n_0^2 + 2}\right)^2 \tag{97}$$

where m_{00}^u and m_{ii}^u represent the dipole moment of the solute molecule in the ground and excited states, respectively, a is an effective cavity radius appropriate to the solvent, D is the static dielectric constant of the solvent, n_0 is the solvent refractive index at zero frequency, and α_0^u and α_i^u are the polarizabilities of the solute molecule in its ground and excited states, respectively.

In actual fact instead of predicting the shift to be expected, one often calculates the excited state dipole moments from the equation using the experimentally determined shift. In many cases the second term can be neglected. This term originates from the induced dipole-dipole interaction (solute-solvent). If the solvent is not highly polar, the induced dipole-dipole terms can be considered to make a negligible contribution to the shift. Thus an estimate of the excited-state dipole can be made from experimental shift data and the known ground-state dipole moment. Other expressions similar to (97) have been used to estimate excited-state dipole moments [129] and other methods have been proposed [130]. Also, the general effect of solvent viscosity [121] was discussed in the preceding chapter.

B. SPECIFIC SOLVENT EFFECTS

Perhaps the most important topic here concerns molecules in which $n \leftrightarrow \pi^*$ transitions occur. It is important to note that in many cases in which the n, π^* singlet state is the lowest excited state, the emission that occurs is phosphorescence. This is by far the most common situation among molecules containing a carbonyl group and in certain aza-aromatics. Therefore, most of the discussion regarding $n \leftrightarrow \pi^*$ transitions is appropriate to phosphorescence. Some compounds of the type noted above, however, do fluoresce and the effect of the solvent is important. More details of particular cases are considered in Chapter 12.

One of the first studies relative to the effect of a solvent on fluorescence involved chlorophyll [131]. Early work [132] has shown that the fluorescence of the chlorophylls in benzene is extremely weak but that the addition of small amounts of hydroxyllic solvents increases the intensity. Changes were also noted in the long wavelength absorption band. More complete studies [131] at low temperature show that the chlorophylls in hydroxyllic solvent systems fluoresce strongly. If dry chlorophylls are dissolved in a dry hydrocarbon solvent, however, the usual fluorescence is absent but a new and intense phosphorescence occurs at longer wavelength [131]. Simultaneous with this change in emission, a new shoulder develops on the long wavelength absorption band of the chlorophylls. This has been interpreted in terms of reversal of the n, π^* and π, π^* singlet states [131]. The phosphorescence emission in a dry hydrocarbon solvent is at higher energy than the one in a hydroxyllic solvent indicating the possibility that some emission occurs from a triplet state higher than the lowest one. It may be possible that mixing of the two triplets is absent or sufficiently small so that the rate constant for emission from a higher triplet is greater than that for relaxation to the lowest triplet. It is also possible that the triplet n, π^* and π, π^* states have interchanged order and the emission in hydrocarbon is from the n, π^* triplet. It has also been proposed that the new long wavelength emission in dry hydrocarbon solvents may be due to dimers [133]. However, it has been shown that at the same and even higher concentrations of nonspecially dried chlorophylls in the same hydrocarbon solvent, no similar emission occurs

[134]. It has also been observed [135] that these solvent effects do not occur when the metal is removed from chlorophyll. Pertinent to this point is the fact that an analog of chlorophyll, a magnesium prophyrin, apparently does show solvent complexing directly to the metal [136]. Finally, the emission in dry hydrocarbon has been assigned as a π, π^* phosphorescence originating from an unsolvated chlorophyll monomer [137]. If indeed this is the case, then the energy of the π, π^* triplet is significantly affected by solvation. Based on the location of the phosphorescences (Chapter 13A) aliphatic hydrocarbon desolvation has to raise the energy of the π, π^* triplet states approximately 2100 cm^{-1}. Furthermore the intersystem crossing, or the fluorescence rate constant, or both, is (are) markedly affected [131]; that is, in dry solvents no fluorescence exists whereas in hydroxyllic solvents fluorescence is dominant. The foregoing considerations do not appear to be consistent with one [137] of the suggested assignments.

Other interesting cases concerning n, π^* states are those of acridine [138] and quinoline [139]. There have been varying opinions concerning whether acridine fluoresces. The problem of the fluorescence of this molecule has largely centered around the nature of the solvent required. It has been noted that water or hydroxyllic solvents are necessary for fluorescence. However, acridine does fluoresce in solvents not containing water or hydroxyllic solvents at low temperature [138]. A combined study of low-temperature absorption and emission further shows that the n, π^* singlet state is not the lowest state [138]. Also of interest is the fact that at low temperatures a tricomplex can be formed with acridine even when it is protonated. Moreover, protonated acridine also shows fluorescence considerably to the red of acridine (the absorption in part is also considerably displaced to the red). Here again the specific effect of solvents, including protonation, is important.

The case of quinoline is interesting (see Chapter 12 for more detail). Although the absorption of quinoline and naphthalene appear to be similar, the emissions may be quite different [139]. In hydrocarbon solvents, quinoline is nonfluorescent but shows some phosphorescence (similar to naphthalene). In hydroxyllic solvents, however, a fluorescence appears that is similar to that of naphthalene while the intensity of phosphorescence is lowered. In this and the previous cases cited it is very important to note the specific roles that solvents play. Solvent effects can be dramatic, and a complete change in the nature of the emission can occur with a change of solvent.

In addition to the specific roles played by the solvents in reversing the nature of the lowest singlet state (thus modifying the nature of the emission), other less specific but important effects are possible. Molecules of particular interest are the nitrogen heterocyclics. The phosphorescence lifetimes of quinoline, isoquinoline, and phenanthridine are shorter in a hydrocarbon solvent than in a solvent containing hydrogen-bonding molecules [140].

Further, there is a change in the degree of out-of-plane polarized phosphorescence. The lifetime and polarization data indicate that there is mixing of the n, π^* and π, π^* triplet states through vibronic interaction. Although proof of a similar interaction between the lowest singlet π, π^* state and a higher n, π^* state is lacking, it is not unreasonable that such an interaction exists. Consequently, if this were so, spin-orbit interaction would increase, crossing to the lowest π, π^* triplet would occur, and fluorescence could be quenched (in a hydrocarbon solvent versus a hydroxyllic solvent). The inverse of this argument would account for the fact that even when the π, π^* singlet remains lowest, fluorescence enhancement can be produced by a hydroxyllic solvent relative to a hydrocarbon one. The difference in effect between the two major solvent systems would arise because of the variation they would produce in the separation between the lowest singlet π, π^* state and the higher singlet n, π^* state. The closer the states are, the greater both the interaction and the spin-orbit coupling with consequent diminished fluorescence (the situation for a hydrocarbon solvent).

Additional cases in which the solvents play outstanding roles in effecting characteristics of the emission are discussed in other sections. They include (a) the effect of perdeutero versus perhydro solvents and (b) light and heavy atom-containing solvents such as methane versus argon and alkyl chlorides versus iodides.

Another area of interest regarding solvents is that concerning the nature of the matrix formed on cooling. Certain pure hydrocarbons and mixtures of these with many other solvents produce clear glasses when cooled to low temperature (77°K). Excitation of the solute results in the usual fluorescence with associated relatively broad vibrational fine structure. However, in the case of certain aliphatic hydrocarbon solvents and aromatic hydrocarbon solutes interesting changes can occur. Heretofore the functional group characteristics of the solvent have played dominant roles in determining the spectral effects. In this case we will find it is certain physical characteristics of the solvent molecule that are important [141]. At low temperature (77°K) the emission spectra of aromatic hydrocarbons such as naphthalene dissolved in aliphatic hydrocarbons such as heptane show a remarkedly sharp, quasi-linear vibrational structure. This effect is discussed later in greater detail.

Chapter 11 Hydrocarbons, Intersystem Crossing and Intramolecular Heavy Atom Effect

A. ALIPHATIC AND OLEFINIC COMPOUNDS

Certain polyacetylene derivatives show fluorescence, phosphorescence, or both, in a rigid glass at low temperature [142]; for example, fluorescence is observed for diethyltetraacetylene and diphenylacetylene. Phosphorescence is observed for these as well as others such as diphenyldiacetylene (no fluorescence) and dihydroxytriacetylene (no fluorescence). The fluorescences originate in the 3000–4350 Å region while the phosphorescences originate in the 4550–5900 Å region. The lifetimes of the latter emissions vary from 0.01 to 5 sec. Solutions of some compounds presumably show fluorescence under high-energy excitation as with γ-rays [143]; for example, bromoform, tetrachloroethylene, and dimethylformamide show weak emission. Some other compounds such as diphenylmethane, tetraisobutylene, and 1,6-diphenyl-1,3,5-hexatriene, show somewhat stronger emission, particularly the last one. In other cases, such as aromatic compounds, a generally good correlation of luminescence intensity from γ excitation and ultraviolet excitation exists [143].

Certain substituted polyenes show emission [144]. These are derived from butadiene and hexatriene parents with primarily phenyl substitution or methyl additions in some cases. Quantum yields of fluorescence are as high as 0.43 and as low as zero. Usually the quantum yields are different in benzene and in heptane, being lower in benzene (with some exceptions). The emission spectra vary in location from the ultraviolet to the visible. The simplest hexatriene derivative, 1,6-diphenylhexatriene, emits in the deep blue. The addition of methyl groups to this compound to form the 5-methyl or the 4,5-dimethyl derivatives causes a marked decrease (5–30 fold) in the quantum yields of fluorescence. In fact, in the dimethyl case fluorescence seems not to occur. Other substituted derivatives such as the 1,5-diphenyl-6-(2-pyridyl)-hexatriene show no emission. As the number of phenyl substitutions increase on the butadiene parent, the spectrum moves into the visible. In addition the tetraphenyl (1,1,4,4) derivative has the highest fluorescence quantum yield. The lifetimes of fluorescence for the 1,1,4,4-tetraphenyl-butadiene and 1,6-diphenylhexatriene derivatives are 2.5 and 7.1 $\times 10^{-9}$ sec, respectively, in heptane (shorter in benzene).

A. FLUORESCENCE OF AROMATIC HYDROCARBONS

Aromatic hydrocarbons (see Figure 13) are fluorescent in solution. The spectral position depends on the longest wavelength of absorption. Benzene emits in the ultraviolet and a progressive movement of the emission to the visible occurs through the linear polyacenes with anthracene emitting in the blue, naphthacene in the green, and pentacene in the red. This is in agreement with the predictions of molecular orbital theory. Especially at low temperatures, there is well-defined vibrational structure. The quantum yields of fluorescence increase with increasing number of rings from a value of 0.26 [86] for benzene to about 0.46 [145] for anthracene (however, later data indicate a lower value for each of these, see later discussion and Table 8) and then diminish to a low value at pentacene. At least in the case of benzene and naphthalene, the quantum yields are somewhat dependent on the concentration as well as on the frequency of the exciting radiation. Table 8 gives some values. It must be emphasized that in obtaining quantum yields it is very important to consider factors such as reabsorption, the nature of the solvent, the concentration, and the exciting frequency. The neglect of correcting for reabsorption can result in an appreciable error—as high as 50 percent if 30–50 percent of the absorption spectrum overlaps the fluorescence spectrum and if the fluorescence quantum yield is high [146]. The nature of the solvent can sometimes have a dramatic effect. For example, 9-methyl-10-chloromethyl-anthracene shows a Φ_F of 0.09 in benzene and 0.73 in ethanol (uncorrected for reabsorption) [87]. In the case of parent hydrocarbons, the effect is not so marked but is nonetheless present [88]. At least in certain cases—for example, anthracene in ethanol and fluorene in ethanol or hexane—the Φ_F is independent of wavelength over quite a wide range of wavelengths [83].

Values of fluorescence lifetimes in solution are somewhat less plentiful than those on phosphorescence but in addition there are some values for microcrystalline hydrocarbons [145]. Some values are shown in Table 8. In

Table 8 Spectral Characteristics of Aromatic Hydrocarbons

Molecule	Wavelength Origin—Å		Measured Lifetimes[a]		Quantum Yields		
	F	*P*	*F*(nsec)	*P*(sec)	*F*	*P*	*IS*
Benzene-h_6	~2700	3390	2000[g], 590[h], 29[f]	7.0(21)[b], (28 ± 4)[c], 16[p], 8[p]	0.26[b], 0.20[c], 0.055[e], 0.07[f], 0.08[f], 0.21[p]	0.26[b], 0.20[c], 0.19[p]	0.24[e]
Benzene-d_6			26.6[f]	26[d]	0.08[f]		
Naphthalene-h_8	~3120	4695	96[f], 500[g], 273[z18]	2.6(11)[b], 2.2[6], 2.1[j], 3.0[k], 2.25[l], 2.25[m], 2.4[n]	0.55[b], 0.51[o], 0.39[p], 0.19[e], 0.23[f]	0.1[b], 0.03, 0.008[p]	0.40[q]
Naphthalene-d_8			96[f]	21.6[r], 22[n], 18.3 ± 3[m], (28)[s]	0.60[o], 0.47[s], 0.20[e], 0.23[f]	0.41[s]	0.53[s], 0.38[q]
Anthracene-h_{10}	~3750	6700	4.9[f], 6.5[t], 4.26(16.7)[u], 3.9[w], 4.2[x], 5.7[y], 6.4[z], 4.0[v]	0.04[z1], 0.04(20)[z2], 0.025[z3]	0.32[z4], 0.36[f], 0.99[z5], 0.27[p]		
Anthracene-d_{10}			4.9[f], 5.7[t]	0.15[z6], 0.12[z2]	0.32[f]		0.53[e]
Phenanthrene-h_{10}		4605	71[t], 79[z1]	3.7[m], 3.3[z7], 3.8[i], 3.8[n], 3.5[l], 4.3[p]	0.15[e], 0.12[s], 0.14[p]	0.20[o], 0.11[p], 0.09[s]	0.88[s], 0.76[q]
Phenanthrene-d_{10}			79[t]	13.2[r], 16.4[n], 15.2[m], (38)[s]	0.10[s]	0.24[s]	0.90[s]
Naphthacene-h_{12}	~4720	~10000	6.4[f]	0.0008[z2]	0.21[f]		
Benz(a)anthracene-h_{12}	~3800	5950	60[t], 6.1[z]	0.3[z7], 1.04[z2], 0.39[z8]	0.20[e], 0.23[z8]		0.55[z9], 0.64[z8]
Benz(a)anthracene-d_{12}			60[t]	1.7[z6], 1.90[z2]			
Benzo(c)phenanthrene-h_{12}	~3700	5000		2.5[z10]			

Chrysene-h_{12}	~3610	4975	50[t], 44.7[f], 17.6[z]	2.5[z7], 2.0[z2]	0.23[e], 0.14[f]		0.70[e], 0.67[q]
Chrysene-d_{12}			50	13(32)[s], 12.3[z2]	0.40[s]	0.25[s]	0.60[s]
Triphenylene-h_{12}	~3430	4200	36.6[f], 48[z18]	15.9(29)[b], 14.8[m], 13.3[k], 16[n], 14.3[r], 17.1[p]	0.06[p], 0.08[f], 0.07[e], 0.04[b], 0.15[s], 0.06[p]	0.42[r], 0.53[b], 0.28[p]	0.85[s] 0.95[q]
Triphenylene-d_{12}			38[f]	23[n], 22.1[m], (36)[s]	0.14[s], 0.11[f]	0.47[s]	0.86[s]
Pyrene-h_{10}	~3740	5880	~400[t], ~750, 46.5[z]	0.2, 0.5[n], 0.63[z2]	0.65[z12], ~1[z11], 0.32[f]		
Pyrene-d_{10}			~400[t]	3.2[n], 3.4[z2]			
Picene							0.36[e]
Perylene-h_{12}	~4420	8000	6.4[f], 7.9[t], 5.02(5.65)[u], 4.9[v], 5.2[x], 4.0[z13]		0.94[f], 0.80[z4]		
Perylene-d_{12}			8.2[t]				
Dibenz(a,h)-anthracene		5465		1.5[z7], 1.4[z14]			0.89[o], 0.98[e]
Coronene		5235		9.4[z7], 7.9[k]	0.3[z16]		
Fluorene	~3030	4210	10[f]	4.9[z7]	0.54[p], 0.70[e], 0.80[f]		0.31[q]
Biphenyl-h_{10}	~2930	4385	16[f]	3.6[z7], 4.2[n], 5.1[p]	0.18[f], 0.23[z17], 0.17	0.25[p]	
Biphenyl-d_{10}			17.6[f]	10.3[n]	0.17[f]		
p-Terphenyl-h_{14}	~3250	4850	0.95[f], 1.3[t], 5.5[z]	2.6[n]	0.93[s], 0.93[f], 1.0[z17]	0.006[s]	0.07[s]
p-Terphenyl-d_{14}			1.49[f], 1.3[t]	5.3[n]	0.91[s], 0.92[f]	0.02[s]	0.09[s]
Acenaphthene-h_{10}	3220		46[f]		0.60[f]		0.47[q]
Acenaphthene-d_{10}				20[z6]			
Fluoranthene		5400		0.85[z15]			
Dibenzo(c,g)-phenanthrene		5050		2.5[z10]			
1,12 Benzperylene		6170					0.59[e]
1,2 Benzcoronene				5.65[z8]	0.27[z8]		0.64[z8]
Benz(a)phyrene			8.6[z]				

See Table refs. page 120.

References to Table 8

[a] Numbers in parenthesis are intrinsic lifetimes. For phosphorescence it is assumed all radiationless or photochemical processes occur from the triplet. The intrinsic lifetime of fluorescence can also be calculated from $\tau_0 = \tau/\Phi_F$.

[b] Reference 86, in EPA at 77°K, ±10 percent.

[c] Reference 203, in EPA at 77°K.

[d] M. Wright, R. Frosch, G. Robinson, *J. Chem. Phys.*, **33**, 937 (1960) in methane at 4°K (also see text).

[e] Reference 182a. Φ_{IS} data at 23°C in hexane and alcohol, Φ_{IS} data in EPA at 77°K from $T \leftarrow T$ absorption data, ±15 percent except ±20 percent for anthracene. Value for Φ_{IS} of benzene quoted as from A. Lamola, PhD Thesis, Calif. Inst. Tech. (1965). Also, see W. R. Dawson and N. W. Windsor *J. Phys. Chem.* **72**, 3251 (1968).

[f] I. Berlman, *Handbook of Fluorescence Spectra of Aromatic Molecules*. Academic Press, N.Y. (1965). In cyclohexane except for a few cases in benzene or alcohol at room temperature. In some cases the intensity of fluorescence was higher in nitrogen-saturated cyclohexane solution than in an aerated one. This would indicate some quenching and therefore some modification of lifetime values.

[g] Dammers de Klerk, *Chem. Weekblad.*, **54**, 281 (1958).

[h] J. Donovan and A. Duncan, *J. Chem. Phys.*, **35**, 1389 (1961) in vapor phase extrapolated to zero pressure.

[i] D. Olness and H. Sponer, *J. Chem. Phys.*, **38**, 1779 (1963) in EPA at 77°K.

[j] Reference 185, based on electron paramagnetic resonance data in durene at 77°K.

[k] Reference 98, based on electron paramagnetic resonance data in mixed alcohols at 77°K.

[l] G. von Foerster, *J. Chem. Phys.*, **40**, 2059 (1964) in EPA at 77°K.

[m] Reference 191, in EPA at 77°K, approximately ±2 percent error.

[n] R. Kellogg and R. Schwenker, *J. Chem. Phys.*, **41**, 2860 (1964) in EPA or plastic at 77°K.

[o] R. Keller and H. Rast quoted in R. Keller and S. Hadley, *J. Chem. Phys.*, **42**, 2382 (1965) presumably at 77°K in hydrocarbon glass.

[p] C. Parker and C. Hatchard, *Analyst*, **87**, 644 (1962) in EPA at 77°K; quantum yields relative to fluorene $\Phi_F = 0.54$ at room temperature. Φ values noted to be rough guide only.

[q] Reference 212, in benzene at room temperature.

[r] J. Brinen and W. Hodgson, Private Communication (1967) from electron paramagnetic resonance data at 77°K in 3-methylpentane. Lifetimes slightly shorter based on phosphorescence data.

[s] Reference 211, in cellulose acetate at 77°K from energy transfer experiments. Phenanthrene-d_{10} and triphenylene-d_{12} could have Φ_P values as high as 0.35 and 0.55, respectively.

[t] Reference 147, in EPA at 77°K.

[u] Reference 148 in benzene at room temperature.

[v] J. Birks and D. Dyson, *Proc. Roy. Soc.* (London), **A275**, 135 (1963).

[w] W. Metcalf quoted in footnote "v" above.

[x] W. Ware, *J. Phys. Chem.*, **66**, 455 (1962).

[y] W. Ware and P. Cunningham, *J. Chem. Phys.*, **43**, 3826 (1965) vapor phase extrapolated to zero pressure.

[z] Reference 146, microcrystalline state.

[z1] G. Jackson and R. Livingstone, *J. Chem. Phys.*, **35**, 2182 (1961) from $T \leftarrow T$ data at −25°C in glycerine. In tetrahydrofuran anthracene is ~0.01 sec.

[z2] Reference 119 in plastic at 77°K except anthracene in EPA at 77°K and naphthacene by $T \leftarrow T$ absorption at 25°C.

[z3] Reference 194, in crystal at room temperature.

[z4] Reference 83, anthracene in heavy paraffin at room temperature, perylene in benzene. Data corrected for reabsorption.

[z5] Reference 190, in the crystal at 25°C.

[z6] Reference 116, in EPA at 77°K.

[z7] Reference 260, in EPA at 77°K, approximately ±5 percent error.

[z8] W. Dawson, Private Communication (1967) Φ_F data at 23°C in polymethylmethacrylate. Φ_{IS} and phosphorescence lifetime data in EPA at 77°K (also see text).

[z9] H. Labbart, *Helv. Chem. Acta*, **47,** 2279 (1964) at 25°C.

[z10] E. Clar and M. Zander, *Chem Ber.*, **89,** 749 (1956).

[z11] Reference 199, extrapolated to 77°K in ethanol from data as a function of temperature (also see text).

[z12] Reference 197, at 23°C.

[z13] W. Ware and P. Cunningham, *J. Chem. Phys.*, **44,** 4364 (1966) in vapor phase from 0.02–0.05 mm.

[z14] D. Craig and E. Ross, *J. Chem. Soc.*, **1954,** 905 in EPA at 77°K.

[z15] G. von Foerster, *Z. Naturforsch*, **18a,** 620 (1963).

[z16] Reference 84, at room temperature.

[z17] E. Bowen, *Advances in Photochemistry*. **1,** 23, Interscience, N.Y. (1963) in benzene at room temperature except *p*-terphenyl in petroleum ether.

[z18] B. Stevens and M. Thomas, *Chem. Phys. Letters*, **1,** 549 (1968) at 77°K.

---

some cases such as pyrene and benz(*a*)pyrene in the microcrystalline state, it is doubtful if the lifetime is inherent to the free molecule. The fluorescence spectra [146] of these compounds show characteristics of a dimer-type complex between excited and unexcited molecules and thus, the emission lifetime in microcrystalline state may possibly be associated with this aggregate. Although peculiarities of the spectra have been pointed out [146] and general molecular interaction has been studied, apparently this specific possibility has not been considered. Studies on the fluorescence spectra of benz(*a*)pyrene, pyrene, coronene, and perylene also suggest that part of the emission may be from the excited dimer. Such dimer complexes are discussed later in this section. The crystalline monomethyl derivatives of benz(*a*)-anthracene have lifetimes of approximately 10 nsec except that the 8- and 11-derivatives show considerably longer lifetimes (~100 nsec). In the latter cases there is again an indication of a fluorescence different from that of the other derivatives which is attributable to an excited dimer. Crystalline monomethyl-substituted benzo(*c*)phenanthrenes have lifetimes in the 20 nsec range.

The fluorescence lifetimes, in EPA at 77°K, of perhydro and perdeutro derivatives of the same parent hydrocarbons are essentially the same [147] (Table 8). This indicates that essentially there is no internal conversion to the ground state from the first excited singlet state. Further, at least in the case of anthracene, the lifetime is the same at room temperature. Since the rate constant for fluorescence emission is not expected to be temperature dependent, this would indicate that the rate constant for intersystem crossing is unlikely to vary with temperature. The rate constant for fluorescence emission, however, does show a very slight temperature dependence [148]. Further, the sum of the rate constants for internal conversion and intersystem crossing shows no temperature dependence. In contrast, for 9,10-dichloranthracene the sum of these same rate constants shows a strong temperature dependence. However, the emission rate constant shows only a slight temperature dependence [148]. More details concerning these points will be given later in this section. Nevertheless the results for anthracene, despite the very small variation of the fluorescence rate constant, quite clearly show that all the rate constants considered above are essentially temperature independent and thus that the quantum yield of fluorescence should be temperature independent. This appears to be true [148].

Just as in the case of the linear polyacenes, the spectral location of the fluorescence of geometric isomers of hydrocarbons depends on the location of the longest wavelength absorption. In the isomeric tetracyclics this depends qualitatively on the number of "bends" in the molecule. Thus linear naphthacene emits in the green beginning at approximately 4750 Å; benz(*a*)anthracene, (one bend) at 3800 Å; chrysene and benzo(*c*)phenanthrene (two bends) at 3600 Å and 3700 Å, respectively; and finally triphenylene, 3450 Å [149]. The general trend is in accord with deductions from molecular orbital theory.

The intensity of fluorescence varies considerably among the tetracyclic isomers. Benz(*a*)anthracene and naphthacene are the strongest emitters and triphenylene is the weakest. In compounds containing larger ring systems, such as five or six rings, it is generally true that the fluorescence is at longer wavelengths although there are some exceptions; for example, a benzochrysene (five rings) apparently emits at shorter wavelengths than does benz(*a*)anthracene (four rings) [150].

Partial hydrogenation of the molecule can cause a significant shift of the emission toward the blue or ultraviolet region of the spectrum. Hydrogenation of linearly condensed hydrocarbons at sites that reduce the number of rings in conjugation may cause visible fluorescence to disappear as both the absorption and the fluorescence are displaced into the ultraviolet region. For example 9,10-dihydroanthracene contains two disconnected benzene rings, and the blue fluorescence of anthracene is replaced by an ultraviolet fluorescence similar to that of benzene [151]. Similarly, the fluorescence of fluorene is limited to the ultraviolet region and in agreement with the structure is similar in position to that of biphenyl; however, the quantum yield of fluorescence of fluorene is considerably higher than that of biphenyl. Similarly, the emission spectrum of pentacene begins at approximately 5800 Å, whereas that of one of the dihydro derivatives begins at approximately 3850 Å.

One of the interesting developments regarding the spectra of aromatic hydrocarbons is the possibility of creating unusually well-resolved fine structure. This is accomplished by cooling solutions in aliphatic hydrocarbons to 77°K or below, at which the mixture becomes a frozen crystalline solution. Depending on the dimensions of the solute and solvent molecules, the resulting fluorescence can be quasilinear in nature, with bands as narrow as 10 cm^{-1}. Detailed studies of this aspect of the emission of aromatic hydrocarbons include those of naphthalene, anthracene, benz(*a*)anthracene, benz(*a*)pyrene, coronene, perylene, pyrene, and others [141,152,153]. Certain aliphatic alcohols can also be used as solvents for the development of this effect. The linelike structure is presumed to arise because an aromatic hydrocarbon replaces an aliphatic hydrocarbon at a lattice site and thereby exists in a state that can be described as an oriented gas molecule. Thus the spectra are of the unperturbed systems corresponding to the free molecules. The ideal situation results when the dimensions of the solute and solvent molecules are comparable so that the solute molecule can occupy a definite position in the host lattice with minimum deformation; for example, the emission spectrum of naphthalene in pentane shows quasi-linear structure, but is more diffuse in hexane and heptane. If the solvent molecule is too small, a similar situation arises. The broader, more diffuse spectra that can result tend to be similar to those obtained in clear, rigid glass solvent systems. There can be a variation in the number of lines and their intensity depending on the solvent and temperature. The variations in part appear to originate from the fact that there are local differences in the crystal field surrounding different molecules. Also, the rate of freezing of the mixture is important. Slow freezing results in the loss of some short wavelength components and weakening of some long wavelength components. This probably results from formation of rotational isomers [153]. Under slow freezing it is more likely that a greater proportion of the molecules will be in the most stable conformation and the spectrum will be simple. Such fine-line spectra can be used to determine structural details of the emitting molecule [153].

One factor that requires consideration when studying the fluorescence of aromatic hydrocarbons is the effect of solute concentration. In the cases of anthracene and phenanthrene, for example, there is considerable separation of the 0–0 bands (absorption versus fluorescence) when the concentration is high (10^{-3} *M*) in cyclohexane. However, when the concentration is low ($<10^{-5}$ *M*), there is near perfect coincidence of the 0–0 bands [154]. It should be noted, however, that this effect does not always occur since in certain solvents, such as ethanol, dioxane, and benzene, a separation at low concentration still may exist (50–110 cm^{-1}). The closest coincidences are obtained with aliphatic hydrocarbon solvents. In some cases there is a still wider separation between the 0–0 bands even when low concentrations are used; for example, in biphenyl the separation is 445 cm^{-1} at 10^{-7} *M*. In *trans-trans*-diphenyl octatetraene the separation is even larger (1675 cm^{-1}) [154]. This is not surprising in view of the fact that the structures in the ground and excited states may be quite different (moreover photoisomerization can occur). Further, solvent interactions with the excited state can be appreciably different from those in the ground state.

Finally, it is worth rementioning the interesting situation in the hydrocarbon azulene (a nonalternate hydrocarbon). Here fluorescence may occur from the second excited singlet state and no phosphorescence apparently exists.

1. Excimers

At high concentrations of hydrocarbons entirely new spectral characteristics can arise [155–158]. In the case of pyrene at approximately 10^{-4} *M* only a structured fluorescence is observed in the ultraviolet-violet region. As the concentration is increased to about 10^{-2} *M*, however, the structured fluorescence becomes very weak and a new, broad, structureless fluorescence occurs in the blue region. The intensity of this new fluorescence is proportional to the square of the pyrene concentration. This transformation can also be seen at an intermediate concentration [155]. The spectral change can be interpreted in terms of the formation of dimers formed by the combination of one excited and one unexcited pyrene molecule [155,156]. In order for emission to occur from this species it must form within the lifetime of the excited state of the pyrene molecule. The fluorescence of the monomer and dimer can be quenched with certain molecules [156].

The formation of excited dimers or *excimers* in this way is by no means universal although it is quite common. The best known case is probably pyrene. The existence of such excimers depends on whether the lifetime of the excited-state molecule is long enough for dimer formation, whether the dimer dissociates rapidly, and whether the intermolecular interaction energy is sufficient to cause quenching of the dimer fluorescence. The lifetime of the monomer pyrene fluorescence in deoxygenated cyclohexane is approximately 2.9×10^{-7} sec in the absence of external quenching. The observed lifetime is near this only at concentrations less than 0.5×10^{-3} *M*. At higher concentrations dimers are formed and the lifetime is 4.0×10^{-8} sec and is independent of concentration [158].

There is also evidence for a similar type emission from crystalline hydrocarbons, as mentioned earlier [159]. In these cases, pyrene in particular shows a crystal lattice in which there are essentially pairs of almost completely overlapping molecules. The fluorescence spectrum of the crystalline pyrene exhibits a broad structureless band in the visible similar to that known for dimers in solution. Other molecules having lattices in which there is very high molecular overlap also show this phenomenon, such as perylene, benz(*a*)-pyrene, and coronene.

It also appears possible to have excimers of mixed character; for example, an excimer of perylene and pyrene can exist [160]. The exact interpretation of the emission and excited-state interactions responsible for stabilization of the excimer vary [160–165]. There is good evidence of intermolecular configuration interaction between the perylene-pyrene charge transfer state and the lowest excited singlet of pyrene [160]. This is also probably true for the mixed crystal system. This type of interaction would be at least as important in like-molecule excimers as pyrene-pyrene and perylene-perylene types. The stabilization energy cannot be accounted for, however, by this interaction [160]. Resonance interaction of states does not seem to be a plausible contributor. Thus the emission would seem to originate from a charge transfer type state.

2. Delayed Fluorescence

In addition to the fluorescence resulting from excimers, there is a delayed fluorescence. This is spectrally similar to the ordinary fluorescence but has a considerably longer lifetime. There are three mechanisms that could produce this result. One is an old observation and has been called α-phosphorescence. This results from thermal excitation of the triplet and reoccupation of the first excited singlet followed by fluorescence.

Delayed fluorescence for most molecules is a result of a biphotonic process; that is, the delayed fluorescence intensity varies as the square of the intensity of the exciting light. This type of fluorescence has been found for many molecules including phenanthrene, anthracene, and pyrene [166], naphthalene, hexahelicene, and triphenylene [165]. For the latter three molecules, delayed fluorescence has been seen in rigid glasses at low temperature (77°K). In addition, delayed fluorescence has been seen in the gas phase and in solids. The mechanism involves excitation to S_1, intersystem crossing to T_1, followed by triplet-triplet annihilation with the resultant formation of one molecule in S_0 and the other in S_1 followed by fluorescence emission (delayed) [166, 167]. In fluid solution it is not difficult to visualize the interaction of two triplets. In solids and rigid media, however, the process of interaction is not obvious. The mechanism could involve resonance interaction of triplets via triplet exciton-triplet exciton annihilation [168,169]. Another mechanism is possible, however, [170]. This can be schematically represented as follows with reference to the schematic energy level diagram of Figure 31.

$$^3L_a + {}^3L_a \rightarrow X \text{ (bimolecular state)} \tag{98}$$

$$X \rightsquigarrow S^- \tag{99}$$

$$S^- \rightarrow {}^1L_a + {}^1A \tag{100}$$

$$^1L_a \rightsquigarrow {}^1L_b \tag{101}$$

$$^1L_b \rightarrow {}^1A + h\nu \tag{102}$$

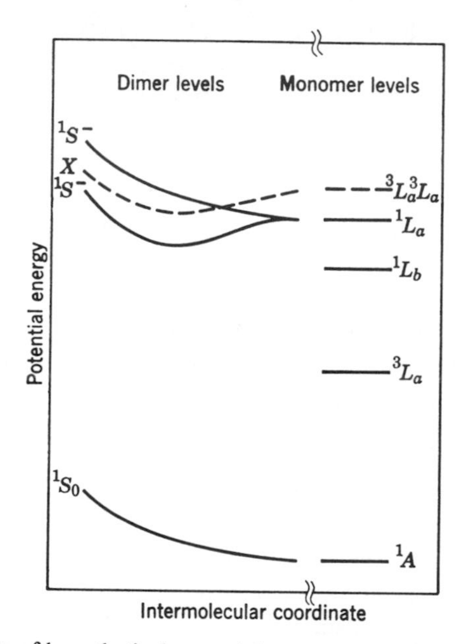

Figure 31 Schematic of hypothetical potential energy curves for two interacting aromatic molecules.
Reprinted by permission from American Institute of Physics, V. G. Krishna, *J. Chem. Phys.*, **46**, 1735 (1967).

In essence the bimolecular state (98) undergoes a radiationless transition to a dissociative state (99) followed by dissociation (100), internal conversion (101), and emission (102). In addition, the S^+ state shown in Figure 31 can be considered as the stable excimer state formed from an excited (^1L_a) and unexcited ^1A molecule from which excimer emission occurs. The nature of the excimer state was discussed earlier.

A third mechanism for delayed fluorescence originates from recombination of ions and trapped electrons to produce excited singlet states. Both one and two photon processes exist [171,172]. In addition, excited singlet states can be slowly formed from solute-solvent charge transfer states followed by fluorescence emission [171]. It is also possible to have delayed phosphorescence originating from a triplet formed by radical-electron recombination.

In Chapter 4-B the effects of pressure on the absorption spectra of various compounds were considered. In addition to affecting the absorption spectrum, the emission spectrum is also modified. The pyrene monomer fluorescence undergoes a red shift slightly greater than the lowest ^1L_b absorption band in polymeric matrices at room temperature [173] at 25 kbar. At high concentrations, the excimer fluorescence intensity decreases considerably with an increased pressure and the shift varies from longer to shorter wavelength at high pressure depending upon the polymer matrix and concentration of pyrene. Structure does appear on the short wavelength side of the excimer emission indicating that a part of the fluorescence arises from a different state or a positional isomer [173]. In the case of chrysene and benz(*a*)-anthracene, there is little or no indication of excimer emission in the crystal. Under high pressure, however, diffuse emission bands arise at longer wavelengths than the usual fluorescence [174]. This indicates excimer emission brought about by molecular or crystalline component reorientation to form transient overlapping molecular pairs. The effect of pressure on phosphorescence spectra is discussed later (Chapter 11-C).

C. PHOSPHORESCENCE, INTERNAL CONVERSION, AND INTERSYSTEM CROSSING IN AROMATIC HYDROCARBONS

Many aromatic hydrocarbons phosphoresce in rigid media (glasses) at low temperature (77°K) (Table 8). Benzene is the simplest member of this series and has received the most intensive study. Although information for hydrocarbons containing more than four rings is scarce, a considerable amount is known for smaller hydrocarbons. Table 8 gives data on the spectral location, lifetime, and quantum yield for some hydrocarbons. As in the case of fluorescence, the phosphorescence wavelength of the linear polyacenes increases as the number of rings increases. This is consistent with expectations from molecular orbital theory. Naphthacene is estimated to have a phosphorescence beginning about 10,000 Å. Also, it is worthwhile to note that the phosphorescence lifetime decreases as the number of rings increases (benzene through anthracene).

In the tricyclics and isomeric tetracyclics hydrocarbons, the origin of the phosphorescence spectra, like that of the fluorescence, decreases in wavelength with an increase in the number of bends of the skeletal structure. This is also consistent with predictions from molecular orbital theory. It will be noted that the phosphorescence lifetimes increase as the number of bends increases (this also represents an increase in the symmetry elements). On the basis of the foregoing relation of the lifetimes and structures it would be expected that benzo(*c*)phenanthrene would have a longer lifetime than benz(*a*)anthracene. This is experimentally verified (Table 8). It is interesting to note that the nonisomeric tetracyclic pyrene has properties very similar to benz(*a*)anthracene except for the $S_2 - T$ separation. It can be seen that the $S_1 - T$ separation is not consistent among the isomers; however, the $S_2 - T$ is amazingly consistent. Moreover, this is true among the monomethyl derivatives of a given individual parent molecule as well as between the monomethyl derivatives of different parent molecules (see later discussion in this section).

The five-ring dibenzanthracene (C_{2h}) has characteristics quite similar to its C_{2h} counterpart in the tetracyclics (chrysene). The six-ring coronene (D_{6h}) shows a resemblance in phosphorescence lifetime to its D_{6h} counterpart benzene. The other characteristics vary, however. Finally, fluorene fits in between benzene and naphthalene which is not expected. Essentially, the benzene portions are isolated but surely some interaction does occur giving the molecule a "benzene-naphthalene hybrid" character. Among the alternate hydrocarbons the general trend in phosphorescence lifetime is to increase with an increase in the symmetry elements (the tricyclics do not fit well).

The lowest triplet state in benzene is ^3L_a or $^3B_{1u}$ [103,107,175]. The assignment for the lowest triplet state of naphthalene is still in question. The phosphorescence has been shown to be polarized parallel to the second singlet-singlet electronic transition (and thus in the plane) as well as perpendicular to the plane of the molecule [176,177]. Even if it were in-plane polarized (and thus presumably $^3B_{3u}$), a clear decision cannot be made without detailed analysis because of vibronic mixing between the two lowest singlet states. It has been noted, however, that electron-spin resonance data on a single crystal of naphthalene in durene [97] permitted assignment of the triplet as ^3L_a or $^3B_{1u}$ (long axis as y and short axis as z). However, the original reference [97] does not report any assignment. In phenanthrene, the phosphorescence is polarized perpendicular to the plane [176,177,179]. The triplet is assigned as ^3L_a or $^3B_1$.

For the tetracyclics there is little information on polarization available. In the case of chrysene, however, it is implied that the phosphorescence is polarized perpendicular to the plane [177,180]. In one case this conclusion is based on the fact that the polarization is negative relative to the first two transitions in absorption. These are quite surely polarized in plane [47]. For all of the isomeric tetracyclics the $S_2 - T$, separation is remarkably constant (10,600–11,200 cm^{-1}) [149]. Moreover, this is true for all the monomethyl derivatives. For example, in the methyl benz(*a*)anthracenes the variation is 10,920–11,220 cm^{-1}. It has been implied that the lowest triplet state has the same symmetry as the second excited singlet state [1,36,47]. There are other indications that triplets in some cases may be polarized in-plane [181]. However, it has been generalized that the phosphorescence is polarized perpendicular to the plane for all polyacenes [177,180]. It is worthwhile to evaluate the significance of the out-of-plane polarization for phosphorescence. The implication is that it is *not* the π, π^* singlet and triplets that mix but rather σ, π^* (or π, σ^*), or Rydberg singlets and π, π^* triplets. Spin-orbit interaction between π, π^* states is extremely small because of vanishing one- and two-center terms for aromatic hydrocarbons [181a,181b]. In addition, the ground state could mix with π, σ triplets [177,180].

It can be seen from Table 8 that the total quantum yield of emission for several hydrocarbons is quite high. Moreover, there is significant variation among the quantum yields of phosphorescence, although unfortunately there is not a large amount of such data available. This fact makes it difficult to generalize on the relationships among the symmetry properties, number of rings, and quantum yields. However, a few comments can be made. Despite the fact that no quantitative data exist for benz(*a*)anthracene, the intensity of phosphorescence for triphenylene is strong while for benz(*a*-) anthracene is weak [149]. Quantitative data support the qualitative data for triphenylene (see Table 8). Further, the qualitative data indicate that intensity of phosphorescence in triphenylene is approximately equal to that of fluorescence in benz(*a*)anthracene. Thus for benz(*a*)anthracene where the phosphorescence is weak, the lifetime is 15–50 times shorter than in the case of triphenylene where it is strong. This could be partly accounted for by radiationless transitions out of the triplet state if it is occupied to a relatively high degree. Very recent data [182] indicate a high degree of occupation, 64 percent. Any remaining difference in lifetime must be accounted for in other ways—for example, by the symmetry of the states and the degree of spin-orbit coupling.

Another point of interest is the effect deuteration has on the spectral characteristics of the hydrocarbons. In Chapter 9-B the significance of deuterium substitution was discussed particularly with reference to radiationless transitions; that is, substitution of deuterium for hydrogen modified the Franck-Condon factors to a sufficient extent that the radiationless transition probability was decreased. This resulted in increased quantum yields and lifetime. Table 8 provides data showing this effect. Particularly, in the case of benzene more information regarding the nature of the couplings can be deduced utilizing deuteration and various types of external solvent perturbations. The lifetime of perdeuterated benzene in different matrices of CH_4, A, Kr, and Xe at 4.2°K is 22, 26, 1, and 0.07 sec, respectively. Ordinary benzene has a lifetime of 16, 16, 1, and 0.07 sec, respectively, in the same matrices [183,184]. The lifetime of perdeuterated naphthalene is also longer than that of ordinary naphthalene [185,186]. Deuterated naphthalene in deuterated durene at 77°K shows a mean lifetime of decay of the paramagnetic resonance signal originating from the triplet state of 16.9 ± 0.7 sec, which is considerably longer than ordinary naphthalene in ordinary durene at 77°K, 2.1 ± 0.1 sec. Ordinary naphthalene in deuterated durene has a lifetime near the 2.1 sec value. A parallel situation also exists for perdeuterated anthracene and ordinary anthracene [183].

It has been suggested that the normal phosphorescence originating from intramolecular perturbation only is highly vibrationally induced [183,187]. The intensity behavior of the 0–0 band of phosphorescence of benzene in the above matrices discussed above adds support to this premise [107,189]. The 0–0 band is strong in Kr and Xe and weak or absent in the matrices of lower molecular weight such as CH_4. In CH_4 and Ar the spectrum is that arising from the inherent intramolecular perturbation. In Kr and Xe the effect of the external perturbation (intermolecular heavy atom effect) predominates over the intramolecular perturbation as can be deduced from both the effect on the lifetimes and intensity behavior of the 0–0 band. Moreover, the effect of deuterium in lengthening the phosphorescence lifetime of several hydrocarbons points to the fact that phosphorescence, in the absence of external perturbation, is vibrationally induced. The presence of a radiationless transition between the triplet and ground states for benzene and naphthalene is consistent with quantum yield data. The combination of quantum yield data, isotopic effect, and solvent perturbation data provides strong evidence for the fact that the normal phosphorescence of hydrocarbons in general is vibrationally induced.

Although a detailed analysis of the implications of the foregoing will not be undertaken, certain deductions are worth mentioning. Assume that the lowest excited singlet to ground state radiationless mode is unimportant. Also, estimates of Franck-Condon factors and spin-orbit coupling terms are uncertain. Nevertheless, by using calculated values for the spin-orbit terms, establishing limits for the Franck-Condon factors, using calculated values for the location of some of the higher triplets, and carrying out a vibrational analysis on the phosphorescence of benzene in the matrices, the principal operating mechanism for the intramolecular perturbation in benzene can be deduced [107]. The results indicate that the two most likely mechanisms involve either indirect interaction of the ${}^1B_{2u}$ state with the triplets via vibronic mixing between the ${}^1B_{2u}$ and ${}^1E_{1u}$ states and spin-orbit mixing of the ${}^1E_{1u}$ and ${}^3E_{1u}$ or vibronic mixing within the triplet manifold of ${}^3E_{1u}$ with ${}^3B_{1u}$. The former mechanism has been proposed to be the more likely one [107]. In addition, based on polarization data and a vibrational analysis involving the e_{2g} progression, a mechanism involving spin-orbit mixing of the ${}^1E_{1u}$ and ${}^1A_{2u}$ states with the ${}^3E_{1u}$ and vibronic mixing of the ${}^3E_{1u}$ and ${}^3B_{1u}$ is satisfactory [103]. The latter mechanism resulting from the polarization and vibrational analysis can account for the mixed polarization and the strongest part of the phosphorescence. This same mechanism has some parallel to the one immediately preceding.

The foregoing considerations actually involve crossing to a triplet state other than the lowest one. It is important, of course, that the upper triplet be properly located energy wise; that is, it should not be significantly higher than the lowest excited singlet state. Direct evidence for the location of the second excited triplet state of benzene has only recently been given [188]. Data on very pure benzene and perdeutero benzene crystals strongly indicate that the second triplet is at 36,560 cm^{-1} for benzene (36,784 cm^{-1} for perdeutero benzene). In addition, oxygen perturbation studies are consistent with this assignment. The lowest triplet in benzene is at 29,674 cm^{-1} and is $^3B^+_{1u}$. The second triplet is assigned as $^3E^+_{1u}$ and is located very close to the lowest excited singlet which is $^1B^-_{2u}$ at 37,805 cm^{-1} [188,189]. Thus the second triplet is located approximately 9/10 up the energy gap between the lowest triplet and singlet excited states.

Before continuing with further consideration of benzene, it will be particularly helpful to consider the case of anthracene. The second triplet state of anthracene is at 26,050 cm^{-1} and is assigned as $^3B^+_{2u}$ [192]. There does not appear to be any internal conversion from the first excited singlet to the ground state for anthracene and certain other aromatic hydrocarbons including phenanthrene, pyrene, benz(*a*)anthracene, chrysene, perylene and *p*-terphenyl [147] however, see discussion later in this section on Φ_{IS} for such compounds. Since the quantum yield of fluorescence of anthracene is about 0.3 [83], intersystem crossing must be competitive with fluorescence. Further, since the intersystem crossing does not appear to be sensitive to deuterium substitution [147,191], it would not seem likely that the crossing is to the lower triplet; that is, deuterium substitution is expected to affect the radiationless transition rate between states separated by large energy gaps but not between those having a small energy difference [107]. The second triplet at 26,050 cm^{-1} would be located just slightly below the lowest excited singlet at 26,700 cm^{-1} [190] or 26,640 cm^{-1} [192]. Of further significance is the fact that the quantum yield of fluorescence for crystalline anthracene is 0.99 $\pm$ 0.03 [190] which is notably higher than the value in solution. Therefore, the probability for intersystem crossing in the crystal has been significantly decreased. This might be accounted for if the lowest excited singlet state of the crystal were lower than in solution and in addition lower than the second triplet. The singlet is at 25,439 cm^{-1} [193] in the crystal and thus is, in fact, slightly lower in energy than the second triplet.

The quantum yield measurements in the anthracene crystal are consistent with the fact that only a very weak phosphorescence exists for anthracene in the crystalline state [194]. The lifetime is approximately 0.025 sec at room temperature. In fact, the emission intensity is so weak that the spectrum has not been recorded. The decay was measured utilizing a time-averaging computer (CAT) to enhance the signal to noise ratio.

Similar considerations to those above can be used to explain the lack of, or the presence of, temperature dependence for the fluorescence rate constant and the quantum yield of triplet occupation for substituted anthracenes [195]. Although more detailed consideration will be given later in this section, the data are consistent with the concept that crossing from the first excited singlet occurs to the second triplet and because of the variable relative location of these states, there will be variable activation energy requirements for intersystem crossing.

A parallel situation appears to exist for pyrene; that is, pyrene with a fluorescence quantum yield of 0.65 at room temperature [196] and negligible internal conversion [197,198] shows a significant dependence of the quantum yield of fluorescence on temperature [199]. Further, the quantum yield of fluorescence at 77°K in ethanol is assumed equal to one. From these results it can be surmised that the process competing with fluorescence is intersystem crossing to a triplet higher than the lowest one. Thus crossing most likely occurs from the lowest singlet ^1L_b at 26,947 cm^{-1} [47] to either the ^3L_b or the ^3B_b estimated [200] to be at 27,100 cm^{-1} and 24,400 cm^{-1}, respectively. Triplet-triplet absorption data locate excited triplets at 28,300 and 29,700 cm^{-1}. These are assigned as $^3B^+_{1g}$ and $^3A^+_{1g}$, respectively [201]. It is assumed that the triplet at 28,300 cm^{-1} is the lowest excited triplet [201]. However, the assumption that $\Phi_F = 1$ at 77°K cannot be rigorously correct since phosphorescence does exist for pyrene $-h_{10}$ and $-d_{10}$ at 77°K (Table 8). However, Φ_F must be essentially one at 77°K; that is, the discrepancy from $\Phi_F = 1$ lies well within the combined errors limits of the fluorescence lifetime measurements [199] and quantum yield measurements of Parker [197,202].

Finally returning to the benzene case, it would appear quite possible for crossing to occur from the lowest excited single state to the second triplet state followed by internal conversion to the lowest triplet state and, subsequently, phosphorescence. Thus it appears reasonable to assume that, mechanistically, crossing from the lowest excited singlet to higher triplets is the preferred path.

Questions of natural lifetime are important. These depend on the known quantum yields and measured lifetimes. The true or natural radiative lifetime would be that expected in the absence of radiationless modes. Based on absolute quantum yields and assuming all radiationless modes originate in the triplet state, a phosphorescence lifetime of 21 sec can be calculated for benzene [86]. Newer quantum yield data give a longer value of 28 $\pm$ 4 sec [203]. Based on deuteration experiments a lower limit of 26 sec exists. Thus the radiative lifetime can be given as 28 $\pm$ 2 sec. Assuming a true radiative phosphorescence lifetime of 26 sec, the difference between this value and the 28 sec lifetime arises because of some singlet to ground state radiationless process ($S_1 \rightsquigarrow S_0$). A quantum yield of about 0.06 (upper limit) can be calculated for this radiationless process [203]. Limits are established on the assumption that for some deuterated hydrocarbons the measured lifetime is the same as the natural lifetime. At least for perdeutero naphthalene, the radiationless transition is not totally quenched by perdeuteration [204]. This means that the measured lifetime cannot be considered the same as the natural lifetime (the former is shorter than the latter). This has been established also by utilizing observed lifetime data and quantum yield data [205]. Intrinsic lifetimes for naphthalene and biphenyl are considerably longer than measured lifetimes for the partially deuterated compounds. Even though these latter results do not rigorously establish the presence of radiationless deactivation because of only partial deuteration, at least it has been established for naphthalene based on the former result [204], in which deuteration appears to be complete. The importance of these findings is that it does not appear reasonable to conclude that the lifetimes of deuterated hydrocarbons are the same or approach their intrinsic lifetime. This conclusion has been purported to be true for those cases where the deuterium effect was expected to be appreciable [206]. Further discussion of benzene with particular reference to internal conversion from the lowest excited singlet state and Φ_{IS} will be considered shortly.

Normally it is assumed that rate of internal conversion between singlet excited states is very much more rapid than that for intersystem crossing from excited singlets to triplets. Early results indicated that hexahelicene and chrysene showed a considerable increase in the ratio of intensity of phosphorescence to fluorescence when excitation occurred to singlet states of higher energy than the first excited state [209]. This has since been shown to be the result of impurities and dimers [208,209,210]. Furthermore the decrease in fluorescence quantum yield of 9,10-dibromanthracene with excitation into upper excited singlets was interpreted in terms of increased intersystem crossing from higher singlets to higher triplets [213]. This phenomenon has been further studied and the decrease in fluorescence is very likely due to photodecomposition [208]. Thus, there does not appear to be any substantiated exception to the assumption that internal conversion among the excited singlet states is sufficiently rapid that intersystem crossing occurs only from the lowest excited singlet.

The foregoing discussions have established several facts relative to (a) state assignments, (b) mechanisms of intersystem crossing, (c) presence of radiationless modes in the lowest excited singlet and triplet and their consequence, and (d) the results of excitation to excited singlet states higher than the first one. Since it is obvious that quantum yields and lifetimes of phosphorescence are affected by the several factors considered, it would be very useful to know the quantum yield of occupation of the triplet state. Table 8 gives data for some aromatic hydrocarbons. Such data also exist for other classes of molecules and are presented in the appropriate chapters. Two methods for determining such information are particularly adaptable for broad use [211,212] In one method, energy transfer from triplet states to singlet states of dyes in plastic films at 77°K is used to measure the quantum efficiencies for fluorescence, phosphorescence and intersystem crossing [211]. The experimental data required are the ratio Φ_P/Φ_F in the absence of the acceptor and the efficiency of phosphorescence, $\Phi_P/k_{IC+}k_P$, where k_{IC} refers to the internal conversion, $S_0 \leftarrow\!\!\sim\!\!\sim T_1$. The ratio $\Phi_P/k_{IC+}k_P$ can be obtained by measuring the change in the ratio of Φ_P/Φ_F for the donor in the absence and presence of the acceptor, assuming the difference arises from energy transfer from the donor triplet to the first excited singlet of the acceptor. It is assumed that internal conversion from the first excited singlet is negligible.

Data relative to the above method appear in Table 8. General agreement with other data is found. Some variation appears in some cases depending upon the acceptor. For example, $\phi_P = 0.24$ and 0.35 for phenanthrene when rhodamine B and rhodamine 6*G*, respectively, are the acceptors. There are indications [211] that a value of 0.10 may be too high for Φ_P of naphthalene-h_8 and thus 0.03 may be more reasonable. Further, based on all the quantum yields and measured lifetimes, the calculated intrinsic lifetimes are longer (Table 8) than previously predicted based on the assumption that the lifetime of deuterated compounds was intrinsic (see later discussion in this section). Finally, there is general agreement for Φ_{IS} values between this method and another that is discussed next although the deviations would appear to be beyond random error (see Table 8).

In the other method, Φ_{IS} for a molecule is determined by measuring the quantum yield of an induced *cis-trans* isomerization process in the acceptor [212]. Acceptor olefins are chosen, where it is known that the triplet states are involved in the isomerization. The donor and acceptor are irradiated in a benzene solution at room temperature. Triplet-to-triplet energy transfer is involved in this method in contrast to the other [211] where triplet-to-singlet transfer occurs. Values of Φ_{IS} for several hydrocarbons are given in Table 8. Other data are available for substituted hydrocarbons, heterocyclics, and ketones and are given in the appropriate chapter.

In several instances the sum of Φ_{IS} and Φ_F does not equal one. This could be interpreted as due to internal conversion from the first excited singlet although, as discussed earlier, quantum yield and lifetime data indicated this was unlikely. It is possible that variation in Φ_F data (Table 8) and errors in the methods can account for at least some of the discrepancy. Despite this possibility it does not seem possible that errors alone can account for all of the discrepancies. For example, at room temperature benzene appears to have a quantum yield of internal conversion from the singlet of approximately 0.7 [182]; naphthalene-h_8 and d_8, approximately 0.4 [182]; benz(*a*)anthracene, 0.1 [182]–0.25 [182a]; and 1,2-benzcoronene, approximately 0.2 [182] (also see Table 8). Other molecules such as fluorene and triphenylene appear to have no internal conversion from the excited singlet. Molecules such as chrysene and phenanthrene appear to have little internal conversion ($\leqslant 0.1$) from the excited singlet. On the basis of low-temperature data (77°K) (see Table 8) naphthalene, phenanthrene, chrysene, triphenylene, *p*-terphenyl-h_{14} and d_{14}, and dibenz(*a*, *h*)anthracene appear to have, essentially, no internal conversion from the excited singlet.

The problem of errors and temperature differences can also be seen as follows. The deviation of quantum yield of intersystem crossing for a molecule such as benz(*a*)anthracene is 0.13 at room temperature. Further, the difference between room and low-temperature (77°K) data is 0.15 for naphthalene. In each case the data are derived by comparison of different methods. Nonetheless, some of the discrepancy appears to be too large to be accounted for by errors alone. Again, for example, in benzene the internal conversion from the singlet is approximately 0.7 at room temperature [213] and a maximum of 0.06 at 77°K [203] utilizing low-temperature (77°K) quantum yield and lifetime data. Utilizing low-temperature fluorescence quantum yield data and room-temperature intersystem crossing data, the quantum yield of internal conversion from the singlet is approximately 0.5 for benzene.

In some part, the problem may well lie in the assumption that the intrinsic lifetime of the triplet is equal to the lifetime of the perdeuterated molecule (see earlier discussion in this section). In addition, the data may be more temperature sensitive than was thought earlier; for example, in the case of benzene the second triplet is below the first excited singlet. The third triplet is estimated to be at 37,990 cm^{-1} and a $^3B_{2u}^-$ state. This would place it barely above the first excited singlet. Therefore, if crossing occurred to this state, the quantum yield of intersystem crossing (and fluorescence) would be temperature dependent. More efforts are required to fully clarify the problems posed at the beginning of this paragraph.

It is worth noting the Φ_{IS} of naphthalene-d_{10} is insignificantly different from that of naphthalene-h_{10}. Based on the lack of an expected deuterium effect if the crossing were to a triplet separated by a considerable energy gap, it could be predicted that crossing occurs to a triplet higher than the lowest one. There does appear to be a very slight temperature dependence for intersystem crossing in naphthalene [148]. The triplet states most likely to be important are the ^3L_a, ^3B_a, and the ^3L_b calculated to be at 22,400, 29,200 and 30,800 cm^{-1}, respectively [200]. The lowest triplet is calculated to be the ^3L_a at 17,700 [214] or 22,400 cm^{-1} [200]. The experimental assignment is still questionable, as noted earlier.

The phosphorescence spectra of aromatic hydrocarbons show a dramatic increase in structure when studied in certain crystalline saturated aliphatic hydrocarbons at low temperature. Fluorene in heptane at 77°K shows fine-line structure for both fluorescence and phosphorescence [215]. The fluorescence spectrum begins at 3015 Å and the phosphorescence at 4069 Å in heptane. In octane, phosphorescence begins with an intense band at 4203 Å and ends near 5800 Å. It is likely that the spectrum originating at 4203 Å in octane is intrinsic and that shorter wavelength lines are caused by impurities. It is interesting to compare results on the isomers benz(*e*)- and benz(*a*)pyrene [216]. The benz(*e*)pyrene has a fluorescence beginning at 3763 Å and extending to approximately 4500 Å and phosphorescence from 5366 Å to 6500 Å in heptane at 77°K. The origin of phosphorescence for benz(*a*)pyrene is at considerably longer wavelength 6820 Å versus 5370 Å than for benz(*e*)pyrene while the fluorescence of the benz(*a*) derivative is only some 250 cm^{-1} lower in energy (compared with 3960 cm^{-1} lower for phosphorescence). In addition, the benz(*e*)pyrene shows a mirror symmetry relationship between absorption and fluorescence which is not present for the benz(*a*)pyrene. Further, the 0–0 band of fluorescence in the benz(*e*) derivative is weak whereas it is the most intense band in the benz(*a*) derivative. Finally, for benz(*e*)pyrene, if the temperature is lowered to 4°K, then not only do the lines further sharpen but some of the lines present at 77°K split up into new, even finer lines.

Several seven-membered ring compounds show fine-line structure in aliphatic hydrocarbons at low temperature [217]. They include a tribenzpyrene and a benzonaphthylpyrene. The phosphorescence spectra of these begin at 6100 Å and 4900 Å, respectively, and the sharpness of the lines is not the same in all solvents. For example, hexane is the best solvent to obtain sharp lines for the tribenzpyrene derivative while octane is the best for the benzonaphthylpyrene derivative. Again it appears quite certain that the phenomenon is a molecular one and not the result of crystallization, etc. (see Section 11-B).

The interpretation of the effect of high pressure on phosphorescence spectra is not clear. The phosphorescence of naphthalene and quinoline in a Lucite polymeric matrix shows an increase in intensity of a factor of 6 and 25 percent, respectively, as the pressure is increased to 20 kbar [218]. Also there is a gradual decrease in the lifetime (by 1.25) over an increased pressure change of 25 kbar. The interpretation of these results is unclear and varies. Possible causes could be (a) an increase in k_{IS} [218], (b) an increase in the radiative rate constant, k_P [219], or (c) an increase in the rate of excitation of the singlet state because of varying transmission characteristics of the cell window and absorption spectral shifts [220]. The phosphorescence of aromatic hydrocarbons shows a small red shift, comparable to that of the fluorescence, in the range of 0–30 kbar in polymethylmethacrylate at room temperature [221]. The phosphorescence lifetime decreases monotonically over the same pressure range for nine of 10 aromatic hydrocarbons (chrysene is the exception). It appears as if both spin-orbit interaction and vibrational overlap factors are increased resulting in increases of both the radiative and non-radiative triplet-to-ground state rate constants [221]. More investigation is required before the nature and magnitude of the factors responsible can be elucidated.

D. SUBSTITUTED AROMATIC HYDROCARBONS AND THE INTRAMOLECULAR HEAVY ATOM EFFECT

Substitutions may alter the fluorescence of a parent molecule by their effect on the location, quantum yield, and lifetime of fluorescence (see

Tables 9 and 10). One of the best studied examples to determine the effects of substitution on spectra is that of benzene. Therefore the following discussion will refer to this as the parent and to other aromatics where appropriate.

Alkyl substitution on both benzene and naphthalene causes small red shifts of the absorption bands and so correspondingly for the fluorescence. The quantum efficiency of the ultraviolet fluorescence of toluene (2700–3200 Å) and the xylenes is higher than that of benzene, whereas that of hexamenthyl benzene is lower. In the case of anthracene, the quantum yield of fluorescence depends quite noticeably on the position of substitution. At 20°C the 9-methyl derivative has Φ_F near 0.4, whereas the 2-methyl derivative has Φ_F near 0.2 [222]. Multiple alkyl substitution causes differences in fluorescence quantum yields that again depend on the relative positions of substitutions [223]. The fluorescence quantum yields of 9-alkyl substituted anthracenes seem quite independent of the chain length (methyl through propyl) [224]. More discussion concerning the interrelation of position of substitution, Φ_F, and temperature dependence of Φ_F will follow shortly.

Halogens have a varying effect in several respects. The monofluoro derivatives of aromatic hydrocarbons in general have their emissions near the same wavelength and their intensities are approximately the same as the parent hydrocarbon. For benzene and naphthalene, the fluorescence is red shifted and the intensity progressively decreases when going to chlorine and bromine as substituents. The iodobenzene compound is not fluorescent [225]. As will be discussed in further detail later in this section, the probability of radiationless transitions to the lowest triplet state (intersystem crossing) increases with the atomic number of the halogen. This competes with the fluorescence process and in effect is a quenching mechanism. In addition, fluorescence can be quenched by photochemical processes that result in rupture of the carbon-halogen bond. Iodobenzene is particularly affected because of the low bond strength. Thus the lower fluorescence is probably due in some part to the competing predissociation process. In the case of anthracene, there is more sensitivity to position of substitution [222,226]; for example, 9,10-dichloroanthracene has a quantum yield of fluorescence near 1 at low temperatures whereas that of the 1,5 derivative is approximately only about 0.1. Moreover, Φ_F of the 9,10 derivative shows steep temperature dependence while that of the 1,5 compound does not [222]. It appears to be a generalization that disubstitution on the 9 and 10 positions notably increases the quantum yield over that of any other disubstituted derivative [222,223,226]. Substitution in the 9 and 10 positions by two halogen atoms or one halogen and an alkyl group results in fluorescence maxima in nearly identical positions [227]. Also, dialkyl or dihalo substituents in the 9- and 10- positions give fluorescence maxima in almost the same region.

Table 9 Spectral Characteristics of Some Aromatic Hydrocarbons[a] and Substituted Derivatives

Compound	Origin of Fluorescence (Å)	Origin of Phosphorescence (A)	Measured τ_P(sec)[b]	Measured τ_F(nsec)	Φ_P	Φ_F
Benzene	~2700	3390	7.0[c]	29[k]	0.26[c], 0.20[d]	0.26[c], 0.20[d], 0.07[k]
Toluene	~2725	3460	8.8, 8[e]	34[k]		0.23[l], 0.17[k]
Durene			5.7[f], 6.40			0.5
Hexamethylbenzene			~5.7[f], 8.55			
Aniline	~3200	3730	4.7	3.9[k]		0.08[k]
Phenol	~2800	3495	2.9	2.1[k]		0.21[m], 0.08[k]
Anisole	~2800	3545	3.0	8.3[k]		0.29[k]
Fluorobenzene	~2700			8[k]		0.24[c], 0.13[k]
Chlorobenzene			0.004		0.06[c]	
Bromobenzene			0.0001		<0.03[c]	
Iodobenzene	Absent	Absent				
o-dichlorobenzene			0.018			
p-dichlorobenzene			0.016			
p-dibromobenzene			0.0003			
1,3,5-trichlorobenzene			0.022			
1,3,5-tribromobenzene			0.00074			
1,2,4,5-tetrachlorobenzene			0.018			
1,2,4,5-tetrabromobenzene			0.00055			
Benzoic acid		3675	2.5			
Hexachlorobenzene		4130				
Naphthalene	~3120	4695[a,g]	2.6, 2.3[g]	96[k]	0.1[f], 0.051[g]	0.55[c], 0.23[k], 0.19[o]
1-Methylnaphthalene	~3150	4760 (4810[g])	2.5, 2.1[g]	67[k]	0.044[g]	0.85[g], 0.25[k]
2-Methylnaphthalene	~3200			59[k]		0.32[k]
1-Aminonaphthalene	~3500	5265	1.5	6.0[k]		0.46[k]
1-Hydroxynaphthalene	~3250	4855[g]	1.9	10.6[k]	0.036[g]	0.76[g], 0.21[k]
2-Carboxynaphthalene		4875	2.5			
1-Nitronaphthalene		5195[g]	0.049		0.083[g]	<0.0001[g]

1-Fluoronaphthalene		4730[g]	1.5		0.056 ± 0.009[g]	0.84 ± 0.08[g] ($\chi = 0.07$)
1-Chlorononaphthalene		4830[g]	0.30, 0.29[g]		0.30 ± 0.06[g]	0.058 ± 0.005[g] ($\chi = 5.2$)[i]
1-Bromonaphthalene		4830[g]	0.018		0.27 ± 0.04[g]	0.0016±0.0005[g] (χ=0.164)[i]
1-Iodonaphthalene		4880[g]	0.0025, 0.002[g]		0.38 ± 0.06[g]	<0.0005[g] (χ = >1000)[i]
2-Fluoronaphthalene		4695				
2-Chloronaphthalene	~3220	4740	0.47	4.2[k]		
2-Bromonaphthalene		4760	0.02			
2-Iodonaphthalene		4760	0.0025			
1-Naphthaldehyde		5060[g]	0.08[g]		0.02[g]	<0.0001[g]
Biphenyl	~2925	4385	3.6	16.0[k]		0.23[n], 0.18[k]
o,o'-Difluorobiphenyl		3890	0.88			
p,p'-Difluorobiphenyl		4350	3.2			
Anthracene	~3750	6700	0.04[j]	4.9[k]		0.36[k], 0.32[p]
2-Methylanthracene			0.021[h]			0.23[q]

[a] For more detailed information on these and other parent aromatic hydrocarbons see Tables 8 and 10.
[b] All data from reference 260 except where noted and in EPA at 77°K, error generally 2-10%.
[c] Reference 86, in EPA at 77°K, ±10%.
[d] Reference 203, in EPA at 77°K.
[e] Y. Konda and H. Sponer, *J. Chem. Phys.*, **28,** 798 (1958), pure toluene at 4°K.
[f] D. Olness and H. Sponer, *J. Chem. Phys.*, **38,** 1799 (1963), in EPA at 77°K.
[g] Reference 225, Quantum yields based on $\Phi_F = 0.55$ for naphthalene. Error in τ_P is 5–10 percent. Solvent is ethanol: ether (2:1) at 77°K.
[h] Reference 195, in Lucite at room temperature from $T \leftarrow T$ absorption data.
[i] $\chi = \Phi_P/\Phi_F$.
[j] Reference 119 in EPA at 77°K.
[k] I. Berlman, *Handbook of Fluorescence Spectra of Aromatic Molecules*. Academic Press, N.Y. (1965), in cyclohexane or ethanol at room temperature. In some cases the intensity of fluorescence in a nitrogen-saturated cyclohexane solution was greater than in an aerated solution.
[l] E. Bowen and A. Williams, *Trans. Far. Soc.*, **35,** 765 (1939), at room temperature
[m] R. Cowgill, *Arch. Biochem. Biophys.*, **100,** 36 (1963). In water at room temperature.
[n] E. Bowen, *Advances in Photochemistry*, **1,** 23, Interscience, N.Y. (1963), in benzene at room temperature.
[o] Reference 182a, in hexane and alcohol at 23°C.
[p] Reference 83, in aliphatic hydrocarbon at 25°C, corrected for reabsorption.
[q] Reference 222, in ethanol at 20°C.

Table 10 Fluorescence Quantum Yields for Some Hydrocarbons and Derivatives

Molecule	Φ^a	$\Phi_0{}^b$
Benzene	0.62[d], 0.20[e], 0.055[y], 0.07[x], 0.63[m]	
Naphthalene	0.55[d], 0.39[n], 0.51[z], 0.19[y], 0.23[x]	
Anthracene	0.24(B)[c], 0.24(P)[c], 0.25(A)[c], 0.32(HC)[c], 0.36(H)[i], 0.30(A)[s], 0.36[x], 0.99[z1]	0.26(B)[c], 0.32(P)[c], 0.27(A)[c], 0.33(HC)[c]
Phenanthrene	0.12[l], 0.14[n], 0.15(P)[k]	
Pyrene	0.65[o], ~1[p]	~1[p]
Chrysene	0.23[y], 0.14[x]	
Perylene	0.94[x], 0.98(B)[v], 0.80(B)[c], 0.79(P)[c], 0.84(A)[c]	0.89(B)[c]
Rubrene	1.0(B)[u]	
Triphenylene	0.04,[d] 0.15[l], 0.06[n], 0.08[x], 0.07[y]	
Coronene	0.3[h]	
Fluorene	0.54[n], 0.70[y], 0.80[x]	
Biphenyl	0.23[v], 0.18[x]	
p-Terphenyl	0.93[l], 1.0(P)[v], 0.93[x]	
Toluene	0.23[t], 0.13[x]	
Durene	0.5[t]	
Fluorobenzene	0.24[d]	
Phenol	0.21[w]	
m-Methylphenol	0.25[w]	
m-Carboxylphenol	0.01[w]	
m-Formylphenol	<0.01[w]	
m-Hydroxymethylphenol	0.25[w]	
1-Fluoronaphthalene	0.84 ± 0.08[j]	
1-Chloronaphthalene	0.058 ± 0.005[j]	
1-Bromonaphthalene	0.0016 ± 0.0005[j]	
1-Iodonaphthalene	<0.0005[j]	
1-Methylnaphthalene	0.85 ± 0.19[j], 0.25[x]	
1-Nitronaphthalene	<0.0001[j]	
1-Naphthaldehyde	<0.0001[j]	
1-Hydroxynaphthalene	0.76[j], 0.21[x]	
2-Aminonaphthalene	0.49(B)[c]	
2-Methylanthracene	0.23[k]	
9-Methylanthracene	0.37[k], 0.35[x], ~1[q], 1.00 ± 0.05[r]	
9-Phenylanthracene	0.49[x], 1.00 ± 0.05[r]	
9-Cyanoanthracene	0.80(B)[c], 0.66(P)[c], 0.74(A)[c], 1[k]	0.88(B)[c]
9-Bromoanthracene	0.05(B)[c], 0.024(P)[c], 0.019(A)[c], 1.00 ± 0.05[r]	0.025(B)[c]
9,10-Dimethylanthracene	0.81(B)[f], 1.00 ± 0.05[r]	0.81(B)[f]
9,10-Diphenylanthracene	1[x], 0.84(B)[c], 0.83(P)[c], 0.81(A)[c], 1.00 ± 0.05[r], 1[k]	0.84(B)[c], 0.81(A)[c]
9,10-Dichloroanthracene	0.55[g], 0.56[k], 0.55[x], ~1[q], 1.00 ± 0.05[r]	0.65[h]
1,5-Dichloroanthracene	0.12[k]	
1-Aminoanthracene	0.61[x]	
1-Aminonaphthalene	0.46[x]	
2-Hydroxynaphthalene	0.32[x]	
2-Methylnaphthalene	0.32[x]	

References for Table 10

[a] Solvents: B is benzene, P is petroleum ether, A is ethanol, AE is ethanol-ether (2:1), HC is aliphatic hydrocarbon, H is hexane and EPA is ether-isopentane-ethanol (5:2:2). See table 8 for additional detailed data on parent hydrocarbons.

[b] Quantum yield data extrapolated to infinite dilution.

[c] Reference 83, at 25°C corrected for reabsorption. Data uncorrected for reabsorption can vary by as much as 20 percent. Consult original reference.

[d] Reference 86, in EPA at 77°K, reabsorption considered negligible; some data slightly concentration dependent.

[e] Reference 203, in EPA at 77°K.

[f] Reference 87, at 25°C

[g] Reference 148, 21°C in ethanol based on 0.28 for anthracene in alcohol at 21°C.

[h] Reference 84, at room temperature.

[i] Reference 148, 21°C in hexane relative to anthracene, see footnote g.

[j] Reference 225, in AE at 77°K, based on $\Phi_F = 0.55$ for naphthalene.

[k] Reference 222, in ethanol at 20°C.

[l] Reference 211, in cellulose acetate film at 77°K, determined from energy transfer experiments.

[m] G. Kistiakowsky and C. Parmenter, *J. Chem. Phys.*, **42,** 2942 (1965), upper limit at very low vapor pressure.

[n] C. Parker and C. Hatchard, *Analyst*, **87,** 644 (1962), in EPA at 77°K.

[o] Reference 197, at 23°C.

[p] Reference 199, assumed at 77°K from lifetime data, see text, p. 132.

[q] Reference 195, estimtaed at 77°K in ethanol from lifetime data.

[r] Reference 208, in EPA at 77°K, all 9- and, 9,10- substituted compounds referenced to 9,10-diphenylanthracene where Φ_F measured as 1.00 ± 0.05. See text concerning the 9,10-dibromoanthracene.

[s] Reference 85, at 25°C.

[t] Reference 85, at 25°C.

[u] E. Bowen and H. Williams, *Trans. Faraday Soc.*, **35,** 765 (1939), at room temperature

[v] E. Bowen, *Advance in Photochemistry*, **1,** 23, Interscience, N.Y. (1963), in benzene at room temperature.

[w] R. Cowgill, *Arch. Biochem. Biophys.*, **100,** 36 (1963), at room temperature in water.

[x] I. Berlman, *Handbook of Fluorescence Spectra of Aromatic Molecules*, Academic Press, N.Y. (1965). In cyclohexane except for a few cases in benzene or alcohol at room temperature. In some cases the intensity of fluorescence was higher in nitrogen-saturated cyclohexane solution than in an aerated one. This would indicate some quenching and therefore some modification of lifetime values.

[y] Reference 182a. Φ_F data at 23°C in hexane and alcohol. Φ_{IS} data in EPA at 77°K from $T \leftarrow T$ absorption data, ±15 percent except ±20 percent for anthracene. Value for Φ_{IS} of benzene quoted as from A. Lamola, PhD Thesis, California Inst. Tech. (1965).

[z] R. Keller and H. Rast quoted in R. Keller and S. Hadley, *J. Chem. Phys.*, **42,** 2382 (1965), presumably at 77°K in hydrocarbon glass.

[z1] Reference 190, in crystal at 25°C.

---

1. Intersystem Crossing in Substituted Anthracenes

As noted above and in earlier discussion of fluorescence and phosphorescence of hydrocarbons, the lifetime and quantum yields of fluorescence of substituted anthracene can show distinctively different temperature dependency. The factor of most importance in determining the nature of the dependence is the position of substitution [222,226,148,195,208]. Energies of activation for fluorescence have been measured and found to vary substantially depending primarily on the position(s) of the substitution and to a lesser degree on the nature of the substituent.

For 9- and 9,10-substituted anthracenes, the fluorescence quantum yields increase with decreasing temperature [222,148,195,208]. Furthermore, the quantum yield of the production of triplets decreases with decreasing temperature [195,208]. The fluorescence lifetimes also undergo a marked change with temperature. The radiative rate constant or lifetime, however, is only very slightly temperature dependent. This means that the sum of the radiationless rate constants k_{IC}^{F} and k_{IS} is undergoing the change and therefore are temperature dependent (anthracene itself does not show this effect) [148,195]. Furthermore, it can be shown from quenching experiments [228], Arrhenius plots [208], and quantum yield of triplet occupation [195] that internal conversion from the first excited singlet is negligible. Therefore, it is apparent the k_{IS} is temperature dependent and that there is an activation energy for the $T_n \leftsquigarrow S$ process where T_n represents some triplet state.

Results for the perdeuterated 9-methyl derivative indicate that the C—H stretching frequency is not an important factor affecting k_{IS} [195]. This would not be expected if the intersystem crossing occurred between states with a large energy gap. Therefore, the triplet state to which crossing occurs, T_n, is higher than the lowest one; that is, $n = 2$ or greater. Several triplet states have been predicted to be near the S_1 level in anthracene [214] and one (probably T_2) has been found [199] to be just below the S_1 level (650 cm^{-1}).

Other triplet states have been found to be nearly at, above, and below the S_1 level [195] for other derivatives of anthracene. The varying temperature dependence of fluorescence for the derivatives could be understood as follows. Derivatives as 9-cyano and 9,10-diphenyl have T_2 sufficiently above S_1 that the large activation energy and unfavorable Franck-Condon factors preclude occupation of T_2 and T_1, respectively. Thus the fluorescence of these compounds should have no temperature dependence and have quantum yields of approximately 1; this is found [222]. For derivatives as 2-methyl and 1,5-dichloro, T_2 is sufficiently below S_1 that the density of states is high at the crossing point. Thus the fluorescence would be temperature independent with a quantum yield of less than 1; this is also found [222]. Finally, if T_2 is nearly at the same energy as S_1, a barrier to intersystem crossing is expected and the fluorescence would be temperature dependent. This is found for the 9- and 9,10-substituted derivatives [222,148,195,208].

Some questions still remain concerning the cause of the relative differences in the location of S_1 and T_2 in different derivatives of anthracene and the nature of any barriers between S_1 and T_2. Table 11 gives the location of S_1, T_1, and T_2 for anthracene and several derivatives. It can be seen that the

Table 11 Singlet and Triplet Energies for Some Derivatives of Anthracene[a]

Compound	S_1(cm^{-1})	T_1(cm^{-1})	T_2(cm^{-1})
Anthracene	26700	14700	26050
2-Methylanthracene	25320	14560	~25600
9-Methylanthracene	25800	14220	~24400
1,5-Dichloroanthracene	25620	14250	~25400
9,10-Dichloroanthracene	24800	14150	~25000[b]

[a] All values except those for anthracene were determined in Lucite at room temperature [195].

[b] This value is quite approximate since only a very weak continuum exists in the 8500–9000 Å region [195].

S_1 level of anthracene is much more affected by substitution than is the T_1 level. It has been suggested that S_1 is lowered sufficiently more than the T_2 in certain cases that it lies below T_2 [208], thus giving an activation energy. However, the 1,5-dichloro derivative which has a relatively large shift of S_1 has only a small fluorescence activation energy [208]. Thus it appears that in the 9- and 9,10-derivatives the Franck-Condon factors are small, tunnelling to the triplet may be slow, and classical crossing may be required. Therefore a relatively large activation energy may be required. However, the foregoing arguments concerning the relative positions of S_1 and T_2 are actually based on the relative sensitivity of S_1 and T_1 to substitution. The data in Table 11 tend to indicate that although T_1 is relatively insensitive to substitution, T_2 is not and is affected significantly and variably; for example, compare S_1, T_1, and T_2 for the 9-methyl and 1,5-dichloro derivatives.

In summary, the data tend to support the idea that the location of the T_2 state relative to the S_1 state is responsible for the varying degree of temperature dependence of the intensity of fluorescence for derivatives of anthracene. This results from the possible presence of barriers to crossing between the S_1 and T_2 states. Some discrepancies, however, still appear to be present such as for 9-methyl (Table 11). Here there is a relatively large activation energy (830 cm^{-1}) and yet T_2 would seem to be sufficiently below S_1 ($\sim$1400 cm^{-1}) that such a value would not be expected (compare with 1,5-dichloro with an activation energy of 170 cm^{-1} and where $T_2 \approx S_1$). Thus the additional consideration of Franck-Condon factors seems to be of implicit importance but their precise role is not clear.

2. Intramolecular Heavy-Atom Effect in Substituted Aromatic Hydrocarbons

The lifetime of phosphorescence of the halobenzenes decreases with increasing atomic number of the halogen (see Table 9). This is in line with expectation based on the discussion in Chapter 8-B concerning the dependence of singlet-triplet transition probability on atomic number. The quantum yield of phosphorescence, however, is not so atomic number dependent. In the fluoro derivative, phosphorescence is absent, it is weak in the chloro case still weaker in the bromo derivative, and absent in iodobenzene. In these cases several factors must be taken into account, including the ratio of the fluorescence and internal conversion rate constants to the intersystem crossing rate constant. Moreover, even if the triplet state were relatively highly occupied, the $T_1 \rightsquigarrow S_0$ internal conversion rate constant could be larger than the radiative one, resulting in a low quantum yield. A detailed understanding of these cases is still lacking. It is interesting that multiple chloro substitution (up to four chlorine atoms) has negligible effect on the lifetime. Moreover, isomeric substitution (*o* and *p*) does not alter the lifetime. Further, multiple bromo substitution seems to have little effect on the lifetime and, in fact, it is possible that the lifetime even may increase somewhat. Here again there is no real understanding of the phenomena.

By comparison an examination of the halonaphthalenes is of value. In the 1-substituted compounds there is a progressive decrease in τ_P as the atomic number of the halogens increases. The quantum yield of phosphorescence increases notably from the fluoro to the chloro compound and then undergoes only minor changes, although that of the iodo compound is the highest (see Table 9). The quantum yields of fluorescence do decrease regularly as the atomic number increases. The total quantum yield of emission of the fluoro compound, however, is more than twice that of any of the others. Further, the decrease in Φ_F from the fluoro to the chloro compound is about 14 fold while the increase in Φ_P is only about five fold. Obviously there is a substantial increase in the rate constant of some radiationless process(es). Between the chloro and the bromo compounds, the change in Φ_P is insignificant but the decrease in Φ_F is about 35-fold, whereas between the bromo and iodo compounds, Φ_P increases again about 30 percent but the Φ_F decreases by a factor of 3 or more. Several general conclusions can be drawn from these results. First, the intersystem crossing rate constant increases notably relative to the fluorescence rate constant when going from the fluoro to the chloro derivative. Second, $\chi = \Phi_P/\Phi_F$ undergoes a rapid increase as the atomic number of the halogen increases (0.07 for fluoro to >1000 for the iodo compound) (Table 9). Third, the rate constant for emission (based on Φ_P) does not, in general, increase as rapidly in magnitude as the intersystem crossing rate constant. As noted earlier, there is an increase in a radiationless mode(s) going from the fluoro to the chloro derivative. These modes are still present and are even more probable when going from the chloro to the bromo derivative since Φ_F decreases dramatically. Proceeding from the bromo to the iodo derivative, the increase in Φ_P is greater than the decrease in Φ_F indicating a decrease in the probability of a radiationless mode(s) of deactivation. The increase in Φ_F between naphthalene and fluoronaphthalene with essentially no change in Φ_P points to a decrease in some rate constant(s) for internal conversion.

In the case of the 2-halo substituted naphthalenes, the lifetimes of phosphorescence are almost the same as in the 1-halo derivatives. Thus it is apparent that the position of substitution is relatively unimportant relative to the effect on the radiative probability. Unfortunately, quantum yield data are not available for further comparison. The assignment of the triplet states in the 2-halo naphthalene cases is still not clear. Earlier indications were that the symmetry of the triplets were the same as that of the ground states [229].

Polarization data on the halonaphthalenes and halophenanthrenes is complicated by the presence of subspectra [230,231,232,233]. In the phosphorescence emission, the 0–0 band is polarized perpendicular to the ^1L_a and ^1L_b transitions, subspectrum I. This is also what is found for naphthalene itself. This subspectrum I consists of only totally symmetric C—C modes. The subspectrum II in the 1-halo compounds, except for the 0–0 band, is polarized parallel to the ^1L_a transition and its magnitude is sensitive to the atomic number of the halogen. In the 2-halo compounds, however, except for the 0–0 band the polarization is essentially depolarized relative to the ^1L_a transition. Furthermore, for the 2-halo derivative the polarization is relatively insensitive to the nature of the halogen but is sensitive to the wavelength of excitation [230,231]. The 1-methyl naphthalene has polarization similar to that of naphthalene indicating the effects described above do not arise from symmetry modification. The halogens by virtue of the presence of subspectrum I enhance the coupling already present in the hydrocarbon. It may be recalled that the latter coupling involves out-of-plane σ, π^* or

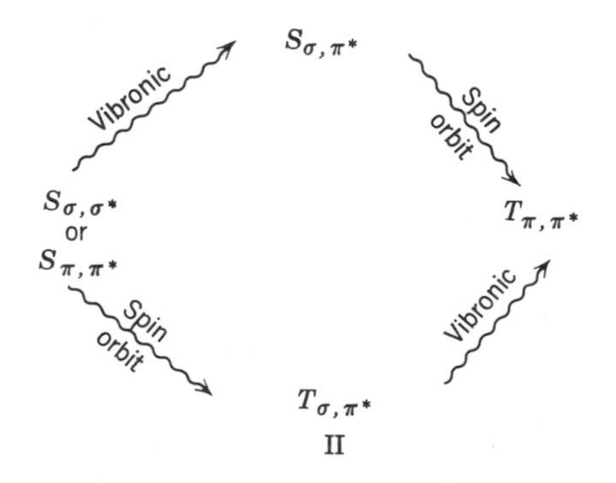

II

Rydberg states. If it is σ, π^*, then the σ orbital involved could be the carbon-halogen bond and a mechanism (I) exists as

$$S_{\sigma,\pi^*} \underset{\substack{\text{orbit}\\ \text{I}}}{\overset{\text{spin}}{\longleftrightarrow}} T_{\pi,\pi^*}$$

Since a subspectrum II exists and is polarized more nearly parallel to ${}^1L_a$, another coupling scheme must exist. The coupling scheme (II) could be [230,231]: If σ, π^* states are involved, the σ orbitals involve the carbon-halogen bond. The σ, σ^* state could be a charge transfer state.

For the 1-halo naphthalenes, as the atomic number of the halogen increases or the number of halogens increases the importance of coupling scheme II increases relative to that of I. For the 2-halo naphthalenes, the relative importance of the two schemes is insensitive to the atomic number of the halogen.

Despite the fact that coupling scheme I results from a first-order perturbation and scheme II from a second-order perturbation, the resulting subspectra are comparable in intensity. This could result from the fact that $S_{\sigma,\pi^*} \leftrightarrow S_0$ transitions are considerably less probable than the S_{σ,σ^*} or $S_{\pi,\pi^*} \leftrightarrow S_0$ transitions. Thus compensating factors operate that tend to equalize the final probability for the appearance of each of the subspectra. Three-dimensional polarization analysis [234] permits evaluation of the relative contribution of the coupling schemes I and II to the intensity of phosphorescence of halonaphthalenes. Coupling scheme I contributes 20–40 percent and II, 60–80 percent. The relative amount of long and short axis polarized emission in subspectrum II is sensitive to the position of the halogen. The variations can be explained if the perturbing transitions involve intramolecular charge transfer states formed by transfer of charge from the halogen to the ring.

Some differences in results are found for the halophenanthrenes [233]. For phenanthrene and two methyl phenanthrenes, the phosphorescence emission is polarized mainly perpendicular to the molecular plane (or out-of-plane). Thus the methyl groups provide no new electronic or vibronic perturbation. The halo derivatives exhibit three general features: (a) the 0–0 band is predominantly polarized out-of-plane and weak, (b) a strong progression exists that is polarized in-plane along the short axis, and (c) there is a strong in-plane polarized progression with no clear polarization orientation. In the case of the weak 0–0 band, the intensity is derived from direct (first-order) coupling between σ, π^* (or σ^*, π) states with the $T_{\pi\pi^*}$ state. The configurations involved are very likely those localized on the halogen. This explanation is similar to the mechanism I described above for the halonaphthalenes. The in-plane short-axis progression apparently derives its intensity from distant spin-orbit coupling between the triplet and singlet π, π^* states of the hydrocarbon [231]; that is, the halogen enhances the stealing from the $\pi^*, \leftarrow \pi$ singlet transitions. The progression that is essentially depolarized probably derives its intensity through second-order interactions involving vibronic and spin-orbit perturbations. This is similar to that proposed for mechanism II above for the halonaphthalenes.

3. General Effects of Various Substitutions on Aromatic Hydrocarbons

The introduction of hydroxyl or methoxyl into the benzene nucleus shifts the fluorescence to longer wavelengths. It is probable that the quantum yields of emission are increased over that of benzene. In aqueous solutions of the hydroxy compounds, the quantum yields may vary markedly with the pH. For example, phenol fluoresces in neutral and acidic solutions but not at pH > 12. Thus the anion appears to be incapable of fluorescence while the neutral molecule is fluorescent. Moreover, anisole fluoresces quite uniformly throughout the entire pH range [235]. This gives added support to the conclusion that it is the phenolate anion that is nonfluorescent.

It appears that in the case of naphthols (the 1- and 2-mono derivatives) the neutral molecules and the anions do fluoresce, at least in aqueous solutions to pH 13.3 [236]. Moreover, in one case (the 2 derivative) the fluorescence intensity of the anion is greater than that of the unionized form. The dihydroxy compounds show fluorescence, as do their singly charged anionic forms in basic aqueous solution [237]. One of the doubly charged anions (the 1,3 derivative) also shows fluorescence [236]. The reasons for the differences among these compounds are not clear. Further, there are conflicting data on the mononaphthols [235,236]. There do not appear to be any fundamental reasons for the differences between the anions of phenol and the naphthols.

The monomethyl-substituted phenols (the cresols) and disubstituted derivatives all show fluorescence in aqueous solution at low concentration (10^{-4} M), which is red shifted from that of phenol. The fluorescence intensity is high from pH = 1–7 and becomes zero at pH = 14. In general, methyl substitution lowers the intensity with respect to that of phenol except for *p*-cresol. The dihydroxy benzenes (catechol, resorcinol, and quinol) fluoresce [235], but like phenol these compounds do not fluoresce at pH = 13. As the pH is lowered they begin to fluoresce weakly at pH 10–11, where the concentration of the uncharged molecular species first becomes appreciable. It thus appears that both the singly charged and the doubly charged anions of the dihydroxybenzenes are nonfluorescent.

The amino, methylamino, and dimethylamino groups cause quite large red shifts of the benzene fluorescence (2700–3100 Å) with generally enhanced intensity. In aqueous solution, the fluorescence maxima of the amino and methyl amino occur in the 3500 Å region with maximum intensity at pH = 7–10 and essentially zero intensity at pH = 1 [235]. This indicates that the cation is nonfluorescent. Similar group substitutions appear to produce intensity enhancement relative to naphthalene with fluorescence from 3300–3600 Å.

The carboxyl group causes a definite red shift of fluorescence. Moreover, it causes a substantial if not complete quenching of the fluorescence of benzene (at least in aqueous solution at all pH values) [235]. Benzoic acid shows a relatively strong blue-violet phosphorescence in ethanol at low temperature (origin at 3670 Å) [238]. The phosphorescence varies in location and intensity depending on the solvent (weakest in petroleum ether). The general spectrum is similar to that of toluene. The phosphorescence spectrum of the esterified compound, methyl benzoate, is similar to that of the acid in all solvents. Thus, since the ester will exist as a monomer, it is quite certain that the spectrum of the acid also represents that of the monomer. The phosphorescence of the ester (in benzene) at 90°K contains strong out-of-plane vibrations indicative of reduced molecular symmetry and molecular deformation compared with the unsubstituted compound. The similar compound, benzamide, has a blue phosphorescence in ethanol and benzene although it is weaker in the latter solvent. Also, the lifetime is shorter in benzene ($\sim$1 sec versus $\sim$5 sec). It is noteworthy that no out-of-plane vibrations are observed in this case. Since the molecular structure of all three of the compounds is expected to be similar, it appears that benzamide is not deformed in benzene at low temperature. The effects resulting from substitution of a carboxyl group can be compensated for by the additional substitution of hydroxyl or amino groups. For example, salicyclic acid, *p*-hydroxybenzoic and *o*- and *p*-aminobenzoic acids all show fluorescence.

In aqueous solution at concentrations $<10^{-4}$ M, the *o*-, *m*-, and *p*-hydroxy benzoic acids appear not to fluoresce as neutral species [235]. The *o*-hydroxy derivative shows fluorescence as the mono and dianion whereas in the other cases only the dianion fluoresces. An increase in concentration could modify these results to some degree although probably not change the essential qualitative order of relative intensity. More complicated acids such as hydroxyphenylacetic acid fluoresce [235]. In general at concentrations of $\sim 5 \times 10^{-5}$ M in aqueous solutions those acids that do not contain the phenol grouping do not fluoresce (for example, phenylacetic acid). In those cases in which the hydroxy group is present, the acids fluoresce only as the mono- or dianion.

For derivatives of naphthalene varying results are obtained. The substitution of methyl in the 1 position enhances the Φ_F but has little effect on Φ_P or τ_P. The spectral characteristics are very similar to those of the 1-fluoro derivative. Other groups in the 1 position have a more marked influence. This is particularly true for an aldehyde group for which Φ_F is strongly decreased, Φ_P decreases somewhat, and τ_P also decreases (see Table 9). In the original discussion of the 1-naphthaldehyde [225] it is inferred that energy is transferred to a triplet with no further explanation. In the aldehyde it is likely that an n, π^* singlet state is lowest and intersystem crossing to a triplet is complete or nearly so. The reason for this is discussed in Chapter 12. However, there must be highly probable radiationless quenching modes associated with the triplet state of the aldehyde. This is manifested by the low quantum yield of phosphorescence and the large decrease, by a factor of about 30, in the lifetime compared to that of naphthalene. This decrease would be expected if the degree of occupation of the first excited singlet state of naphthalene were converted in the aldehyde to a parallel degree of triplet state occupation that disappeared radiationlessly. Consistent with this explanation is the fact that for the 1-hydroxy derivative, in which any $\pi^* \leftarrow n$ transition is not expected to be that of lowest energy, there is no decrease in Φ_F compared with naphthalene. In fact, Φ_F is increased by about 50 percent and only a small decrease in Φ_P and in τ_P occurs. In fact, the total emission quantum yield increases in the naphthol (Table 9). In the aldehydes, of course, some quanta could be lost by a radiationless process originating in the excited singlet state, although this process is probably much less important than that described above. Finally, it appears that the phosphorescence of 1-naphthaldehyde originates from a π, π^* and not from an n, π^* triplet state [239].

The nitro group tends to repress fluorescence rather strongly. Nitrobenzene, the nitronaphthalenes, the nitrophenols, and picric acid do not fluoresce. Nitrobenzene, in addition to having no fluorescence, shows only a very weak phosphorescence. Substitution of a nitro group in the 1 position of naphthalene causes a large decrease in Φ_F, a slight decrease in Φ_P, and a large decrease in τ_P. These facts again indicate the presence of strong deactivational modes, the most important of which is probably between the emitting triplet and ground state. Unfortunately, only lifetime data exist for amino substitution and these indicate a slight decrease in τ_P. The true consequences of amino substitution cannot be evaluated. The decreases in τ_P caused by the hydroxy and amino substitutions in naphthalene are parallel to those in the benzene. Of the 14 possible nitro-substituted 1- and 2-naphthylamines, only three show phosphorescence, whereas the others fluoresce [240]. It is interesting that in several cases the fluorescence maxima are at longer wavelength than some of the phosphorescence maxima. This is unusual and although there may be explanations, it would seem that further substantiation would be worthwhile. A series of dinitro naphthalenes also show phosphorescence (as do the 1- and 2-mono derivatives). In these compounds the origins vary from ~5000 Å to ~6000 Å [240]. The spectra all show structures near the origin except for the 1, 2 and 2, 3 derivatives where steric interactions apparently cause broadening. As with iodo-substituted compounds, the nitro group does not necessarily completely quench the fluorescence if the molecule contains a substituent tending to increase the intensity of fluorescence; for example, 1-nitro-2-naphthylamine is fluorescent as are some dyes containing the nitro group, although the quantum efficiency of fluorescence is low.

The fluorescence of methyl-substituted isomeric tetracyclic aromatic hydrocarbons is generally red shifted from that of the parent molecule [134,241]. This is in agreement with the changes in absorption and in accord with the predictions of molecular orbital theory. In addition, there is considerable variation in intensity among a given methyl-substituted series such as the benz(*a*)anthracenes. For the 12 monomethyl benz(*a*)anthracenes the origin of phosphorescence is displayed to longer wavelength by some 600 cm^{-1} from that of the parent hydrocarbon at 16800 cm^{-1} or 5950 Å [134]. The intensity of fluorescence within the monomethyl derivatives varies about 10-fold, with the 7-methyl derivative being the strongest. Phosphorescence is considerably weaker than the fluorescence in all cases. The separation $S_1 - T_1$ varies from 8900 to 9400 cm^{-1}, whereas for $S_2 - T_1$ it is 10,700–11,400 cm^{-1}. Several dimethyl derivatives show red shifts of the origin from that of the unsubstituted compound, the 7, 12 derivative showing the largest (~1200 cm^{-1}). In the methyl benzo(*c*)phenanthrenes (parent phosphorescence origin at 20,000 cm^{-1} or 5000 Å), the shifts are all to the red but the maximum shift caused by the methyl group is less than that for benzanthracenes [134]. One tetramethyl derivative shows a large red shift (~1700 cm^{-1}). The intensity of fluorescence and phosphorescence is about the same in the parent. The $S_1 - T_1$ separation varies from 6500–7500 cm^{-1} while for the $S_2 - T_1$, it is 10,400–10,900 cm^{-1}. In the 1-methyl derivative, the $S_1 - T_1$ separation is largest and probably results from steric interaction. For the methyl chrysenes (parent phosphorescence origin at 20,100 cm^{-1} or 4975 Å), the shifts are to the red but again they are relatively small. The intensities of fluorescence and phosphorescence are approximately equal in the parent. The $S_1 - T_1$ separation is 7400–7800 cm^{-1}, whereas for the $S_2 - T_1$ it is 10,900–11,200 cm^{-1} (largest value is for the parent in this case.) In the methyl triphenylenes (parent phosphorescence origin at 23,800 cm^{-1} or 4200 Å), the shifts are to the red. The intensity of phosphorescence is considerably greater than fluorescence in the parent and methyl derivatives. The $S_1 - T_1$ separation is 5400–5500 cm^{-1} while it is 10,700–10,800 cm^{-1} for the $S_2 - T_1$. In the methyl pyrenes (parent phosphorescence origin at 17,000 cm^{-1} or 5980 Å), the $S_1 - T_1$ separation is 9700–10,100 cm^{-1} while for the $S_2 - T_1$ it is 12,600–13,000 cm^{-1} (largest value for the parent). It is worth summarizing several facts about the methyl derivatives and the parents [134]. The variation among the separations ($S_1 - T_1$ or $S_2 - T_1$) of the methyl derivatives of one hydrocarbon is minor. More important, among the hydrocarbons and methyl derivatives all of the $S_2 - T_1$ separations are comparable while there is considerable variation for the $S_1 - T_1$ separations. The nonisomeric hydrocarbon, pyrene, has an $S_2 - T_1$ separation considerably larger than for the other hydrocarbons. The non-or small variance in the $S_2 - T_1$ separation for the isomeric hydrocarbons would seem to indicate that the corresponding states are of the same type, that is, ^1L_a and ^3L_a.

Substitution of phenyl rings on hydrocarbons such as anthracene causes red shifts. In addition, phenyl di-substitution in the 9 and 10 positions causes a large increase of the fluorescence quantum yield ($\Phi_F \sim 1$).

The sulfonic acid group, SO_3H, or the sulfonate ion, $SO_3^=$, in general, exerts little influence on the absorption or fluorescence of aromatic hydrocarbons. The group can be introduced if a water soluble fluorescent compound is desired.

The presence of a quinone structure generally results in quenching of fluorescence in aromatic hydrocarbons. Benzoquinone and anthraquinone do not fluoresce in solution, although anthraquinones containing hydroxyl groups capable of forming hydrogen bridges with the carbonyl oxygen are fluorescent (for example, 1,4-dihydroxyanthraquinone).

In benzene and naphthalene the cyano group CN increases fluorescence efficiency without appreciably changing the wavelength. In anthracene the cyano group in the 9 position greatly enhances the fluorescent yield (Table 10).

Styrene, $C_6H_5CH{=}CH_2$, fluoresces in the ultraviolet with greater efficiency than benzene and with displacement to longer wavelengths. Of the stilbenes, 1,2-diphenylethylenes, the *cis* compound shows lower fluorescence intensity than does the *trans*-stilbene. The introduction of amino, hydroxy, and alkoxy substituents into *trans*-stilbene causes displacement of the absorption bands to longer wavelengths, and the corresponding fluorescence extends into the blue region of the spectrum.

Another interesting consideration is that of azobenzene and azonaphthalene and certain of their derivatives. Azobenzene has no luminescence to about 7000 Å. It has been reported [237] that the azo compounds do luminesce when substituents are able to form intra- or intermolecular hydrogen bonds; for example the *o*-hydroxy azo compounds were presumed to fluoresce in normal hydrocarbons at 77°K and as solids in both room and low temperature [237]. However, recently it has been shown that although the *o*-hydroxyazonaphthalenes (or *o*-phenylazonaphthols), I, fluoresce in a hydrocarbon at 93°K (maximum near 575 nm) it is very likely the hydrazone tautomer, II, that is responsible for the emission [273a]

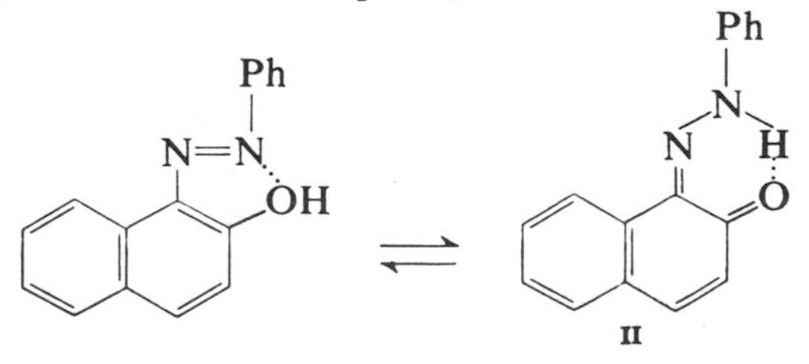

Further, the diphenylhydrazone, III, of *o*-naphthoquinone fluoresces in the same region as II and with a similar shape

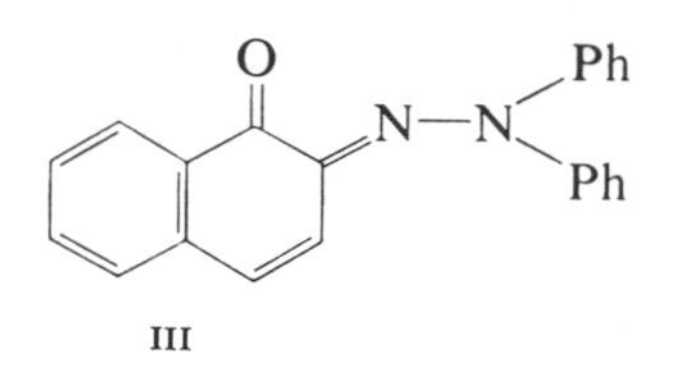

III

The *p*-hydroxyazonaphthalene, IV, also shows a weak fluorescence which is most probably due to the diphenylhydrazone tautomer of *p*-naphthoquinone, V.

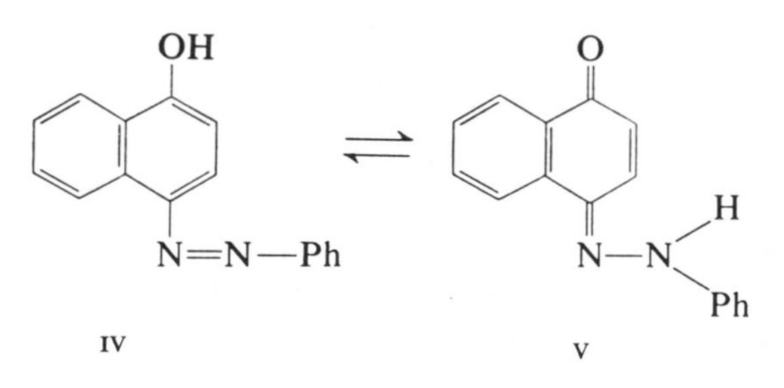

IV V

The fluorescence originally thought to be associated with *o*-hydroxyazobenzene [237] is now believed to have resulted from impurities [237a]. Thus it is apparent that simply the existence of an intramolecular hydrogen bond is an insufficient structural feature to observe fluorescence; that is, some hydrazone tautomer must exist. All of the compounds discussed, except one of the *o*-phenylazonaphthols, show photochemistry involving *cis-trans* isomerization [237a,237b]. In addition, photochemical proton transfers can occur giving hydrazone → azo conversions [237a,237b]. The *o*-aminoazobenzene presumably shows no fluorescence in solution, whereas the *p*-amino derivative shows fluorescence as a solid at room temperature or low temperature [237]. Methylation of the hydroxy [237,237a] or amino groups [237a] causes fluorescence to disappear.

Biphenyl and some derivatives show interesting properties. The sterically hindered *o,o'*-difluoro compound exhibits a rather large blue shift (~3000 cm^{-1}) of its phosphorescent origin as well as a 4-fold decrease in τ_P compared with the unsubstituted compound. On the other hand, the *p,p'*-difluoro derivative resembles biphenyl both in the position of the origin of phosphorescence and in τ_P. The latter result is parallel to the case of naphthalene in which substitution of the same atom had little effect on either the origin of phosphorescence or τ_P although the Φ_F did increase. Further study of such cases would be interesting.

The effect of substituting within the ring to produce heterocyclic compounds depends on the heteroatom. Introduction of nitrogen commonly shifts the fluorescence to the red. The introduction of NH or —O— for CH_2 in fluorene to give carbazole and dibenzofuran, respectively, results in a red shift of the fluorescence. In the case of indene, substitution to give the nitrogen analog, indole, and the oxygen analog, benzofuran, causes a slight blue shift of the fluorescence in the former molecule and fluorescence appears to be absent in the latter molecule [242]. Again, in these cases as additional benzene rings are linearly condensed, the spectrum shifts to the red.

An interesting effect of substitution, from both physical and spectral points of view, occurs for the *p*-polyphenylenes [243]. The members of this series begin with biphenyl and include *p*-terphenyl, etc. As the number of rings increases, a limiting absorption maximum of about 3450 Å is reached, the absorption maximum of biphenyl being at 2500 Å. The maximum of fluorescence also reaches a limiting value. The absorption of the methyl-substituted series also converges but to a shorter absorption limit of approximately 2900 Å (the fluorescence converges as well). If the *p*-polyphenyl compounds are modified by oxygen bridging between adjacent benzene rings, the absorption maxima undergo a significant red shift; for example, *p*-terphenyl has a maximum at 2800 Å while the trioxido *p*-terphenyl has a maximum at 3650 Å. This occurs presumably because internal rotation of the benzene rings is prevented and the *p*-orbital overlap is increased with consequent extension of the total π system for the oxido compounds.

To summarize, the effect of halo substitution on benzene is such that a progressive increase in the atomic number of the halogen results in a progressive decrease in τ_P. Moreover, the Φ_P decreases as well. Multiple halogen substitution has little effect on τ_P relative to the monosubstituted case. Substitution of a methyl group and multiple methyl groups has little effect on τ_P. A similar result occurs for substitution of an amino group. A hydroxyl or carboxyl substitution decreases τ_P by a factor of 2 to 3.

For the naphthalene compounds, an increase in the atomic number of the substituted halogens decreases the τ_P as for the benzenes, but increases Φ_P relative to the parent naphthalene. In the latter respect, the effect of substitution is different from that which occurs in benzene. In addition, except for fluoronaphthalene, halogenation causes a notable decrease in Φ_F. In all cases except fluorine, radiationless modes are introduced. The progressive decrease in τ_P as a function of the atomic number of the halogen is independent of the position of substitution. A methyl group has little effect on τ_P and Φ_P but increases Φ_F, thus reducing the probability for deactivation by radiationless mode(s).

For naphthalene, an aldehyde group effectively and completely quenches fluorescence and decreases τ_P and Φ_P. A dramatic decrease in the total quantum yield of emission shows the introduction of strong radiationless modes that appear to originate principally in the triplet ($S_0 \leftsquigarrow T_1$).

For naphthalene, a hydroxy or amino group causes a small reduction in τ_P. The former substitution also causes a small reduction in Φ_P but a notable (50 percent) increase in Φ_F and the total quantum yield of emission. Thus the probability for a radiationless mode(s) has been decreased. The position of substitution of the hydroxyl has little effect on τ_P.

For naphthalene, a nitro group in effect completely quenches fluorescence, increases Φ_P slightly, and decreases τ_P. The total quantum yield of emission decreases dramatically showing the introduction of strong radiationless modes; further, the principal one appears to originate in the triplet state. A carboxyl group has little or no effect on τ_P. It can be seen that certain general rules appear in both series (benzene and naphthalene), but in many respects there are differences.

The effect of methyl groups on the spectroscopic properties of tetracyclic aromatic hydrocarbons is variable. The strongest influence is noted on the molecule of lowest symmetry, benz(*a*)anthracene. The second singlet state-triplet state separation ($S_2 - T_1$) is remarkably constant for all the parent hydrocarbons and all of the methyl derivatives of the isomeric tetracyclic hydrocarbons. This is not true for the first singlet-triplet ($S_1 - T_1$) separation.

Chapter 12 Molecules with n, π^* States and the Nature of Couplings

Fluorescence commonly does not occur in compounds that have an n, π^* singlet state as the lowest excited singlet state. The reason for this is that almost complete intersystem crossing takes place from the lowest n, π^* singlet state to a triplet, from which phosphorescence occurs. Examples are benzophenone, acetophenone, and quinoline (hydrocarbon solvents). It is known that some molecules, such as formaldehyde, biacetyl, acetone, pyrimidine, and pyrazine, fluoresce weakly even when the n, π^* singlet state is lowest.

Usually when n, π^* states can exist, this state is that of lowest energy in the singlet series. Thus it is usually to be expected that such molecules will not fluoresce but will phosphoresce. This is generally, but not always, true. In particular, most aromatic N-heterocyclics have been thought to be solely phosphorescent. It is now known, however, that many of them exhibit a very weak fluorescence in addition to a relatively strong phosphorescence. In the case of molecules containing a carbonyl group most do, in fact, exhibit only a phosphorescence and the exceptions are few.

Some general criteria can be used for assignment of the nature of the triplet state as π, π^* or n, π^*. The assignment is based on several factors including (a) lifetime and polarization [32,243a,244,245], (b) electron spin resonance [245], (c) triplet-singlet splitting [247], (d) vibrational structure [32,243a, 248], and (e) the effect of heavy atom solvents on the $\pi^* \leftarrow n$ absorption [249]. Table 12 presents some criteria for triplet state assignment. Several other remarks are pertinent. For the lifetime criterion values of >1 sec and <0.01 sec are quite good bases for assignments as $T_{\pi, \pi}$ and T_{n, π^*} respectively. The in-between time values probably do not provide a basis for assignment. Relative to the vibrational structure, it is very common that the

Table 12 Criteria for Assignment of Triplet States as T_{n,π^*} or T_{π,π^*}

Property	T_{π,π^*}	T_{n,π^*}
Lifetime (sec) of phosphorescence	>1	$<10^{-1} - 10^{-2}$
Polarization 0–0 band of phosphorescence	Predominately out-of-plane	Predominately in-plane
Vibrational structure of phosphorescence	Variable	Prominent CO, NO or C-N-C progressions
Triplet-singlet split	$>3000\ cm^{-1}$	$\lesssim 2500\ cm^{-1}$
Intensity $T \leftarrow S$ absorption	$f \sim 1 - 30 \times 10^{-9}$	$f \sim 1 - 7 \times 10^{-7}$
External heavy atom effect on		
(a) Intensity $T \leftarrow S_0$ absorptions	(a) $\sim$twofold increase	Very small
(b) Lifetime of phosphorescence	(b) Decrease	Very small
EPR of triplet	ZFS parameter, D^*, positive and small	Insufficient data

emission spectra from an n, π^* state of aromatic carbonyls, nitro and azine molecules, will contain C=O stretching, N=O stretching, and C—N—C bending-mode vibrations as dominant structural features. In the case of the epr criterion, data on a significant number of molecules of such quite different structures as biphenyl [246], phenyl *s*-triazines [246], phenoxazine [250], and aromatic hydrocarbons [251] having T_{π,π^*} lowest indicate that the zero-field splitting (ZFS) parameter, D^*, is small and positive.

A. CARBONYL COMPOUNDS

It is *generally* valid to assume that *unsubstituted* aromatic carbonyl-containing molecules show solely phosphorescence. Table 13 gives some spectral properties for a large number of carbonyl compounds. In addition, certain substituted ones also show solely phosphorescence; however, others show both fluorescence and phosphorescence. A few unsubstituted carbonyl compounds show phosphorescence, or fluorescence, or both. More concerning the last two points is given later in this section.

Important factors in the discussion are (a) the assignment of the lowest triplet and singlet states—that is, whether they are n, π^* or π, π^* states—and (b) the consequences this has on the type of the emission that will result and the properties associated with the emission such as lifetime and polarization

In formaldehyde, the lowest triplet state is $^3A_2$ of n, π^* type [243a]. The salient consideration is that of the nature of the singlet state that is mixed with this triplet by a spin-orbit perturbation to give allowed character to the $^3A_2 \leftrightarrow {}^1A_1$ transition. Matrix elements of the type $\langle {}^3\psi_{A_2} H_{SO} {}^1\psi_K \rangle$, in which ${}^1\psi_K$ is some excited singlet, are responsible for mixing the singlets and triplets.

Table 13 Emission Spectral Properties of Carbonyl Compounds

Molecule[a]	Origin A	τ_P (sec)	Φ_P	Φ_{IS}	Type Triplet Lowest
Acetone-h_6[b]	4550 (max)	4×10^{-4}, 6×10^{-4}[c]	0.03 ± 0.01	1.0 ± 0.1	n, π^*
Acetone-d_6[b]		1×10^{-3}	0.08 ± 0.02		n, π^*
Methyl ethyl ketone[c]		0.00085			
Diethyl ketone[c]		0.00126			
Diisopropyl ketone[c]		0.0037			
Di-*t*-butyl ketone[c]		0.0086			
Benzophenone	4125 (~4500 max)	0.0077, 0 006[c]	0.75[c], 0.71[o]	1.00[d]	n, π^*
C_6H_5–C_6H_4–CHO[a]	4750				
C_6H_5–CO–C_6H_4–C_6H_4–OCH_3[a]	4850	0.48 ± 0.02	0.50 ± 0.10		
C_6H_5–C_6H_4–CO–C_6H_4–OCH_3[a]	4660	0.28 ± 0.02	0.65 ± 0.06		
C_6H_5–C_6H_4–CO–C_6H_5[a]	4710	0.30 ± 0.02, 0.32 ± 0.01[e]	0.47 ± 0.05		
$C_{10}H_7$–CO–C_6H_5[a]	4970	0.74 ± 0.03	0.25 ± 0.03		
C_6H_5–CO–$C_{10}H_6$–CO–C_6H_5[a]	5020	0.56 ± 0.02	0.29 ± 0.03		

Table 13 (continued)

Molecule[a]	Origin A	τ_P (sec)	Φ_P	Φ_{IS}	Type Triplet Lowest
Naphthalene[a]	4700	2.3 ± 0.1	0.05 ± 0.005		π, π^*
1-Naphthaldehyde[a]	5070	0.08 ± 0.01	0.03 ± 0.01		
2-Naphthaldehyde[a]	4810	0.35 ± 0.02[a], 0.25[l]	0.03 ± 0.01		π, π^*[f]
2-Acetonaphthone[a]	4820	0.97 ± 0.03, 0.95[c]	0.05 ± 0.01	0.84[d]	π, π^*
2-Chloro-1-naphthaldehyde[a]	5190		0.06 ± 0.02		
p,p′-bis(dimethylamino)benzophenone	~4680	0.27 ± 0.02[e]		1.00[d]	π, π^*[g]
p-Aminobenzophenone	~4700 (max)	0.40[e]			π, π^*[g]
p-Dimethylaminobenzophenone		0.35 ± 0.03[e]			π, π^*[g]
p,p-Dimethoxybenzophenone	~4050	0.065 ± 0.007			n, π^*[f]
p-Hydroxybenzophenone	4230				n, π^*[f]
p-Methylbenzophenone[h]	4110	4.5	1.04		n, π[k]
p-Bromobenzophenone[h]	4150	3.6	0.97		n, π^*
o-Chlorobenzophenone[h]	4090	2.8	1.00		n, π^*
o-Bromobenzophenone[h]	4090	2.0	0.92		n, π^*
p,p′-Dibromobenzophenone[h]	4170	3.8	1.00		n, π^*
p,p -Dichlorobenzophenone[h]	4150	4.9	0.96		n, π^*
Acetophenone	3870	0.008[c], 0.0023[i] 0.004[j]	0.63[c], 0.74[j]	1.00[d]	n, π^*[f,j]
p-Methylacetophenone[j]	3920	0.084	0.61		π, π^*
3,5-Dimethylacetophenone[j]	4020	0.11	0.51		π, π^*
p-Methoxyacetophenone[j]	4000	0.25	0.35		π, π^*[f,j]
m-Methylacetophenone[j]	3940	0.074	0.64		π, π^*
3,4-Dimethylacetophenone[j]	4000	0.17	0.56		π, π^*
3,4,5-Trimethylacetophenone[j]	4065	0.20	0.46		π, π^*
m-Methoxyacetophenone[j]	3950	0.25	0.35		π, π^*
p-Bromoacetophenone[h]	4020	0.0066			π, π^*[h]
p-Hydroxyacetophenone[h]	4000				π, π^*[h]
2,4 -Dibromoacetophenone[h]	4130				π, π^*[h]
Butyrophenone[k]	3830	0.002			n, π^*[f,k]
p-Chlorobutyrophenone[k]	3950	0.014			n, π^*[f,k]
p-Methylbutyrophenone[k]	3880	0.009			n, π^*[f,k]

Table 13 (continued)

p-Hydroxybutyrophenone[k]	4030	0.084		π, π^{*}[f,k]
p-Aminobutyrophenone[k]	4430	0.084		π, π^{*}
p-Methoxybutyrophenone[k]	3900	0.051		π, π^{*}
o-Methoxybutyrophenone[k]	3900	0.039		
Benzaldehyde	3970[i]	0.0015[i], 0.0023[l]	0.56[l]	n, π^{*}[f]
p-Methylbenzaldehyde[l]	4013	0.043		
p-Methoxybenzaldehyde[l]	4030	0.15		
o-Hydroxybenzaldehyde[i]	4065			
Ethylphenylketone[i]		0.0037		
Carbazole[i]	4085	7.25		
Benzoin	3900[i]	2.4[i], 0.018[o]		
Biacetyl	~5080	0.00225[c]	0.23[m]	
Anthraquinone	~4550[n]	4.0[n], 0.004[o]	0.41[o]	
2-Methylanthraquinone[n]	~4550	4.4		

[a] All data between —CHO and 2-chloro-1-naphthaldehyde from reference 263 except where noted and in ethanol ether (2:1) at 77°K.

[b] Reference 255. Phosphorescence in ether-isopropanol at 77°K and fluorescence at room temperature in hexane. Quantum yields of fluorescence of $-h_6$ and d_6 are 0.01 ± 0.003 and the lifetime is 2.5×10^{-8} sec. The rate constant for intersystem crossing is 4×10^{7} sec^{-1}.

[c] Reference 260, in EPA at 77°K except benzophenone data (±10%) from reference 85.

[d] Reference 212, in benzene at room temperature.

[e] Reference 264, at 77°K.

[f] Reference 249.

[g] References 264 and 265, based on the lack of photopinacolization.

[h] Reference 245, in ether-toluene (4:1) at 77°K. Quantum yields are relative to benzophenone as 1.00 and have an error of ±10 percent, τ_P data ±15 percent. Actual quantum yield of benzophenone is 0.75.

[i] A. Terenin and V. Ermolaev, *Trans. Faraday Soc.*, **57**, 1042 (1956), in ethanol-ether (2:1) at 77°K.

[j] N. Yang, D. McClure, S. Murov, J. Hauser, and R. Dusenbery, *J. Amer. Chem. Soc.*, **89**, 5466 (1967), EPA at 77°K.

[k] E. Baum, J. Wan and J. Pitts, *J. Amer. Chem. Soc.*, **88**, 2652 (1966). All data in ethanol-methanol (4:1) at −190°C except data for *p*-chlorobutyrophenone in 2-pentene.

[l] D. Murov, Ph.D. Dissertation, Univ. of Chicago, 1967, in EPA at 77°K.

[m] Reference 256, in EPA at 77°K.

[n] W. Neely and H. Dearman, *J. Chem. Phys.*, **44**, 1302 (1966), in EPA at 77°K.

[o] C. A. Parker and C. S. Hatchard, *Analyst*, **87**, 664 (1962) in EPA at 77°K; quantum yields relating to fluorene, $\Phi_F = 0.54$ at room temperature. Φ values noted to be rough guide only.

Further, the intensity of the $^3A_2 \leftarrow {}^1A_1$ depends on the foregoing matrix elements, the intensity of the singlet transition $^1K \leftarrow {}^1A_1$ and the energy of the ^1K and $^3A_2$ states. This consideration neglects mixing of $^1A_1$ with excited triplets and interaction of the triplet with all other excited singlet states $^1\psi_L$, $^1\psi_M$, etc. If one singlet state is particularly effective in perturbing the triplet, however, the sum of the others contribute only in a minor way to the intensity of the $^3A_2 \leftarrow {}^1A_1$ transition.

Excited singlet states can mix with the $^3A_2$ through the various components of the $H_{SO}(\sigma_l)$ as [243a]

$$
^3A_2 \begin{cases} \nearrow \text{with } ^1A_1(\pi^*, \pi) \text{ by } \sigma_z \\ \rightarrow \text{with } ^1B_1(\pi^*, \sigma) \text{ by } \sigma_x \\ \searrow \text{with } ^1B_2(\sigma^*, n) \text{ by } \sigma_y \end{cases}
$$

The appropriate matrix elements can be evaluated from the appropriate potential terms and atomic orbital functions in an LCAO-MO calculation [243a]. One center terms make the largest contribution. From this the following results:

$$
\begin{aligned}
\langle \psi_{3_{A_2}} | H_{SO} | \psi_{1_{A_1}} \rangle^2 &= 2.2 \times 10^{-5} \text{ eV} \\
\langle \psi_{3_{A_2}} | H_{SO} | \psi_{1_{B_2}} \rangle^2 &= 1.1 \times 10^{-6} \text{ eV} \\
\langle \psi_{3_{A_2}} | H_{SO} | \psi_{1_{B_1}} \rangle^2 &= 1.9 \times 10^{-5} \text{ eV}
\end{aligned} \tag{103}
$$

After further consideration of the intensities of the appropriate spin-allowed transitions, such as $^1A_1 \leftarrow {}^1A_1$, and the energy of the transitions, the singlet state most likely mixing is the $^1A_1$; that is, the contribution of the $^1A_1 \leftarrow {}^1A_1$ transition to the intensity of the $^3A_2 \leftarrow {}^1A_1$ is 10–100 times greater than either of the other two possibilities see above [243a]. This conclusion implies that the phosphorescence should have a 0–0 band and the vibronic bands should be polarized parallel to the C=O axis. Further, the lifetime should be $\sim 10^{-2}$ sec and the oscillator strength equal to 1.5×10^{-7} for the $T_{n,\pi^*} \leftarrow S$ transition.

A very important aspect of the foregoing is the implication relative to other carbonyls of C_{2v} or perturbed C_{2v} symmetry such as acetophenone; that is, the oscillator strength of the $T_{n,\pi^*} \leftarrow S$ transitions should be $\sim 1.5 \times 10^{-7}$, the intrinsic lifetime of the $S \leftarrow T_{n,\pi^*}$ phosphorescence should be $\sim 10^{-2}$ sec, and the principal vibronic bands of the $S \leftarrow T_{n,\pi^*}$ phosphorescence should be polarized parallel to the C=O bond axis.

Analysis of the $T \leftarrow S$ vapor absorption spectrum of 1,4-benzoquinone [252] shows that there is a strong 0–0 band of parallel (to the C=O axis) type. This requires the triplet to be A_u of the $^3U(n, \pi^*)$ type. In addition, in a significant number of other molecules, such as benzophenone, benzaldehyde, acetophenone, propiophenone and others [32], the principal vibronic bands are polarized parallel to the C=O bond axis (or in-plane) in conformity with the theory outlined above. Thus the primary mixing singlet is $^1A_1(\pi, \pi^*)$ and intensity is stolen from the $^1A_1(\pi, \pi^*) \leftarrow {}^1A_1$. Furthermore, the lifetimes of the phosphorescence for the previously mentioned compounds plus others such as *p*-methybenzophenone are consistent with that predicted; see above paragraph. Finally, the oscillator strengths of $T_{n,\pi^*} \leftarrow S$ transitions are 1–7 × 10^{-7} for a large group of molecules [245,249]; see above paragraph.

For those cases in which a π, π^* triplet is expected to be lowest, several differences are expected. The bands of the principal vibronic progression in phosphorescence are expected to be out-of-plane polarized [32,177], the singlet-triplet split is expected to be relatively large [247], the phosphorescence lifetime relatively long (10–1000 times longer than that from a T_{n,π^*} state [122,244]), the $T_{\pi,\pi} \leftarrow S$ absorption 10–100 times weaker than $T_{n,\pi^*} \leftarrow S$ [249], and the vibrational progression is not expected to show a dominant C=O frequency spacing. The criterion regarding lifetime is based on several factors. Spin-orbit interaction between π, π^* states for aromatic hydrocarbons is vanishingly small due to vanishing one- and two-center terms [181a,181b]. Thus the generally strongly allowed, energetically close singlet transitions ($\pi^* \leftarrow \pi$) do not mix, thereby giving no strongly radiative character to the π, π^* triplet. Furthermore spin-orbit perturbation is thought to involve n, π^* and σ, π^* states. Neither $\pi^* \leftarrow n$ nor $\pi^* \leftarrow \sigma$ are strongly allowed and thus, no strongly radiative character is introduced into the π, π^* triplet. Further, in aromatic molecules mixing between the π, π^* triplet and the ground state singlet is not allowed in first order and is allowed only in second order to vibronic levels of the ground state [107]. Thus overall, T_{π,π^*} phosphorescences are expected to be relatively long lived. Experiment seems to be in agreement with this proposal thus far. This would indicate that the carbonyl-containing group is acting as a π-electron perturbation (of C_{2v} nature) on the aromatic ring. It should be noted here that spin-orbit mixing of n, π^* and σ, π^* states with the lowest T_{π,π^*} state should give phosphorescence that is polarized out-of-plane (or perpendicular to the plane). It will be recalled that phosphorescences from a T_{n,π^*} state should be in-plane polarized because of spin-orbit mixing with π, π^* states.

Such theoretical and experimental considerations as given above and their agreement is the principal source of the first four criteria in Table 12.

The last criterion in Table 12 relative to optical spectroscopy concerns the effects of heavy atoms on $T \leftrightarrow S$ transitions. Although more detailed consideration of $T \leftarrow S$ transitions and perturbations of these will be given later (Chapter 14), certain aspects are of importance here. On the basis of all the foregoing discussion concerning spin-orbit coupling, it would be expected that the $T_{n,\pi^*} \leftarrow S$ transition probability would be 100–1000 times greater than that for the $T_{\pi,\pi^*} \leftarrow S$. Experimental data are in general agreement with this although it appears as if the lower limit may be as low as 10 [249]. Theoretical considerations of the effect of an internal heavy atom on the intensity of a $T \leftarrow S$ transition vary [244,253]. Experimentally, it appears that heavy atoms do not have a significant affect on $T_{n,\pi^*} \leftarrow S$ transitions

[245]; that is, there is not more than a factor of 2–3 increase in the transition probability and this may well be because of a substitution effect rather than a heavy atom effect. This same conclusion is evident for the $S \leftarrow T_{n,\pi^*}$ emission based on lifetime data [224]. In fact, the factor is very probably less than 2–3 based on the intrinsic lifetime data and more nearly like 1.5 as an upper limit. Thus it would appear that there is little effect, maximum of a factor of 1.5, of substituted heavy atoms on the $T_{n,\pi^*} \leftarrow S$ transition probability.

External heavy atom perturbations cause very little or no enhancement of the $T_{n,\pi^*} \leftarrow S$ transition probability [245,249]. There does appear to be an increase in the transition probability of the $T_{\pi,\pi^*} \leftarrow S$ transition by a factor of approximately 2 [249]. It would be expected that the lifetime of phosphorescence from a T_{π,π^*} would be decreased, whereas that from a T_{n,π^*} would be relatively insignificantly changed by an external heavy atom perturbation. More experimental information relative to external heavy atom perturbed T_{π,π^*} phosphorescence lifetimes needs to be available before the criterion can be considered as firmly established and limits drawn.

Certain nonaromatic ketones such as acetone and biacetyl as well as certain substituted aromatic ketones and quinones such as *p*-aminobenzophenone fluoresce as well as phosphoresce. Certain substituted anthraquinones appear to only fluoresce such as the amino, di-amino and dihydroxy derivatives while others such as anthraquinone itself and the monomethyl and chloro derivatives only phosphoresce [254].

Acetone shows both a very weak fluorescence and a phosphorescence (see Table 13) [255]. In view of the fact that Φ_{IS} is essentially unity there is considerable internal conversion from the triplet to the ground state. Deuteration reduces this loss but there is still a large amount of internal conversion. The lowest singlet state is n, π^*, the lifetime is short and the quantum yield of fluorescence is unaffected by deuteration (Table 13).

Another case of interest is biacetyl, since it shows both fluorescence and phosphorescence in the gas phase, in dilute solution and in a rigid glass. In these phases the Φ_F is 0.0023, 0.01, and >0.005, respectively [256]; the Φ_P is 0.145, 0.007, and 0.23, respectively [256,257]; the lifetime of phosphorescence is 0.0018 [258], 0.001 [259], and 0.00225 [260] sec, respectively. The ratio of $\Phi_P:\Phi_F$ undergoes large changes, being 64 [261] in the gas phase, 7.5 in benzene solution, and quite high in the rigid medium [256]. Other values of Φ_F in different fluid solvents are about 0.01. The natural lifetime of phosphorescence calculated, assuming that intersystem crossing is virtually complete, is approximately 0.01 sec in all three phases. There is considerable interest in biacetyl in terms of energy transfer or sensitization phenomena (see Chapter 18).

An interesting affect is noticeable when the alkyl chain on one side of a diketone is increased. In a series of 2,3-aliphatic diketones beginning with biacetyl and ending with 2,3-octanedione, the intensity of phosphorescence decreases as the alkyl chain length increases [262]. The relative intensities of the fluorescences, however, are approximately the same and weak. The phosphorescences all have a maximum in the 5240–5340 Å region, whereas the fluorescences occur in the 4660–4760 Å region. At 77°K the phosphorescences are very intense and the fluorescences are essentially absent. The phosphorescences can be sensitized by benzophenone.

Returning to aromatic carbonyls, those from benzophenone through 2-chloro-1-naphthaldehyde in Table 13 show no fluorescence but do phosphoresce [283]. The phosphorescence lifetimes are quite different from that of benzophenone. In all cases the phosphorescence origin of the ketones is significantly red shifted compared to that of benzophenone (as are the aldehydes also). In fact the emission spectra resemble those of other derivatives of biphenyl or naphthalene (see Table 13). For example, the emission of 2-acetonaphthone looks very similar to that of 1-chloronaphthalene in both origin and shape [263]. Further, the phosphorescence spectrum does not contain the 1600–1700 cm^{-1} carbonyl vibration as does that of benzophenone. Instead, an aromatic ring-frequency vibration of approximately 1400 cm^{-1} exists. The results quite clearly indicate that for the compounds considered (see above), excitation results in occupation of a singlet state followed by intersystem crossing and emission from a π, π^* triplet state instead of an n, π^* triplet state as for benzophenone.

Acetophenone exhibits solely phosphorescence originating from an n, π^* triplet state. Several different substituted derivatives, however, such as *p*-bromo and *p*-methoxy have phosphorescence originating from a π, π^* triplet state [249] (see Table 13). Some derivatives of benzophenone such as *p*-bromo-, *p,p′*-dibromo-, and *p,p′*-dimethoxy show phosphorescence that also originates from an n, π^* triplet [249] (see Table 13). Other derivatives as *p*-phenyl- and *p*-aminobenzophenone show both a fluorescence and a phosphorescence [264,265]. The lifetime of phosphorescence for these and some other derivatives is between 0.2 and 0.4 sec compared with benzophenone as 0.006 sec (see Table 13). On the basis of spectroscopic data [266] and chemical evidence [265] (see later discussion and Chapter 18) it seems quite certain that the phosphorescence originate from π, π^* triplet states.

Several other considerations are worthy of examination. The *p*-hydroxy benzophenone has the n, π^* triplet state lowest while *p*-hydroxyacetophenone has the π, π^* triplet state lowest—both parent unsubstituted compounds have the n, π^* triplet state lowest. A parallel situation exists for the *p,p′*-dibromo-, and *p,p′*-dimethoxy-benzophenones and a dibromo- and *p*-methoxy acetophenone, respectively (see Table 13). In all cases spectra [249] indicate the lowest singlet state is of n, π^* character. Spin-orbit coupling between singlet states of different configurations—for example, between S_{n,π^*} and T_{π,π^*} states—is expected to be approximately 100 times greater than that between singlet and triplets of the same configuration [122,267]. Thus it would seem that the pathway of intersystem crossing for the benzophenones would be

$$S_{n,\pi^*} \rightsquigarrow T_{\pi,\pi^*} \rightsquigarrow T_{n,\pi^*} \rightarrow h\nu \qquad \text{(I)}$$

whereas for the acetophenones it would be

$$S_{n,\pi^*} \rightsquigarrow T_{\pi,\pi^*} \rightarrow h\nu \qquad \text{(II)}$$

For the substituted benzophenones that show both fluorescence and phosphorescence and the π, π^* triplet is lowest, different considerations may be necessary. If the S_{n,π^*} is lowest, then the mechanism could be similar to that in (II) above. If, on the other hand, the S_{π,π^*} is lowest, which seems to be the case based on the presence of fluorescence, then the pathway could be

$$S_{\pi,\pi^*} \rightsquigarrow T_{\pi,\pi^*} \rightarrow h\nu \qquad \text{(III)}$$

or

$$S_{\pi,\pi^*} \rightsquigarrow T_{n,\pi^*} \rightsquigarrow T_{\pi,\pi^*} \rightarrow h\nu \qquad \text{(IV)}$$

It is expected that purely conjugative substitution will blue-shift $\pi^* \leftarrow n$ singlet transitions and red-shift $\pi^* \leftarrow \pi$ singlet transitions. A + I inductive effect should raise the energy of the n orbital and the π^* and could cause a red or blue shift depending on the relative changes. In the case of the phenyl-substituted acetophenones only the p-bromo is slightly red shifted (200 cm^{-1}), whereas the p-hydroxy and p-methoxy derivatives are blue shifted 1100 cm^{-1} and 600 cm^{-1}, respectively [249]. For the $\pi^* \leftarrow n$ triplet transitions the red shift is less (no shift for the p-bromo case) and the blue shifts are greater. The $\pi^* \leftarrow \pi$ triplet transition cannot be seen in acetophenone. Compared with p-bromo, however, the $\pi^* \leftarrow \pi$ triplet transitions of other p derivatives are all red shifted whereas the $\pi^* \leftarrow n$ triplet transitions are all blue shifted [249]. Unfortunately shifts of the $\pi^* \leftarrow \pi$ singlet transition for the same substitutions are not known. Nonetheless it appears that the T_{n,π^*} state is more affected than the S_{n,π^*} state and, moreover, raised in energy while the energy of the T_{π,π^*} state is lowered. This could then cause a reversal of the T_{n,π^*} and T_{π,π^*} states of substituted derivatives relative to acetophenone which is consistent with experiment [249]. Thus the T_{n,π^*} state seems more affected by substitution than does the S_{n,π^*} state. There is not sufficient data to evaluate the effect of individual substituents or the consequence of their relative position of substitution.

In substituted benzophenones that fluoresce and phosphoresce (π, π^* triplet lowest) the substituents, amino, dimethyl amino, and phenyl, all strongly conjugate with the benzene ring. In these cases it is then likely that the lowest singlet is no longer n, π^* but is π, π^*. This assignment would be consistent with the presence of a relatively intense fluorescence for the compounds. Furthermore, the T_{π,π^*} state would be expected to be of lowest energy since the general tendency resulting from conjugative substitution (particularly a strong one) would be for the T_{n,π^*} to be increased in energy while the T_{π,π^*} would be decreased. Thus mechanism III or IV would be the most likely ones operating.

There are a few exceptions to the fact that unsubstituted aromatic carbonyl compounds phosphoresce but many of these show a variation depending upon the nature of the solvent. For example, pyrene-3-aldehyde does not fluoresce at room temperature in heptane or diethyl ether, fluoresces with very low intensity in acetonitrile, but fluoresces with moderate intensity in alcohols and acetic acid [268]. Accompanying this increase in intensity is a red shift of the fluorescence and loss of vibrational structure. The appearance or nonappearance of fluorescence in a compound such as pyrene-3-aldehyde is likely to depend on the relative location of the n, π^* singlet state and the lowest π, π^* singlet state. If the n, π^* singlet state is lower than the π, π^* singlet state, intersystem crossing from the n, π^* singlet to a triplet state will be probable with consequent quenching of fluorescence.

It will be recalled that the solvent can play an important role in determining the nature of the emission. Because the n, π^* singlet and π, π^* singlet states are close together, they may be reversed in energy relative to one another by a change in solvent. This interchange affects the probability of intersystem crossing to a triplet state and thus the relative intensity of fluorescence and phosphorescence. In cases such as benzophenone or benzaldehyde, it is likely that no hydrogen bonding (hydroxlic) solvent can increase the energy of the n, π^* singlet state to a value above that of the π, π^* singlet and thus intersystem crossing to a triplet is very probable in all solvents with consequent emission of phosphorescence solely.

One reason why intersystem crossing is so likely when the n, π^* singlet is lowest compared to that when the π, π^* singlet is lowest is the following. The intensity of all $\pi^* \leftarrow n$ singlet transitions is very weak (e.g., in benzaldehyde $\varepsilon_{max} \sim 25$). Therefore, in accord with (67), the emission lifetimes are expected to be quite long. Thus on excitation to higher π, π^* singlet excited states (having high transition probabilities), internal conversion occurs to the lowest n, π^* singlet state. With such a low rate constant for fluorescence the intersystem rate constant can be and usually is significantly greater than the fluorescence rate constant. The triplet state is highly or completely occupied, little or no fluorescence occurs, and phosphorescence is the sole emission. The quantum yield can be high or low depending on competing internal conversion modes to the ground state and photochemical reactions (see Table 13). In other cases, however, the intensity of the $\pi^* \leftarrow n$ transition in a ketone may be much more comparable to that of a $\pi^* \leftarrow \pi$ transition in an aromatic hydrocarbon. Nonetheless a ketone such as benzophenone will show only phosphorescence while the aromatic hydrocarbon, such as benzene, will have about equal intensity of fluorescence and phosphorescence.

A parallel situation can exist between an N-heterocyclic and an aromatic hydrocarbon. Thus a low radiative character (fluorescence) for n, π^* singlet states is certainly not the only answer, see below.

As noted earlier, intersystem radiationless transitions between states of different configuration such as S_{n,π^*} and T_{π,π^*} are approximately 10^2 faster than those between states of the same configuration. This results because, to first order, spin-orbit coupling between states of the same configuration is forbidden [267]. Thus, for example, if there is a T_{π,π^*} state between the S_{n,π^*} and lowest T_{n,π^*} states, intersystem crossing will be fast and sufficiently competitive with fluorescence to quench it. The consideration in the preceding paragraph as well as the one given here provides a further basis for one of the pathways, I, given earlier. The proposal in this paragraph has experimental support based on the presence of and relative location of the different triplet states [249].

For pyrene-3-aldehyde, solvents such as alcohols could induce fluorescence because they lower the energy of the π, π^* state below that of the n, π^* state. On the other hand solvents like chloroacetic acid could raise the n, π^* state above the π, π^* state, thereby also inducing fluorescence [268]. Apparently phosphorescence occurs at low temperatures ($<160°K$) in a hydrocarbon solvent and is assigned as originating from the lowest π, π^* triplet state [268]. Without further evidence this assignment is questionable. Presumably the phosphorescence intensity of pyrene-3-aldehyde would also vary with the solvent but this is not known [268].

Other aldehydes as naphthalene-1-aldehyde and phenanthrene-9-aldehyde show no fluorescence in hydrocarbon or hydroxyllic solvents. However, naphthalene-2-aldehyde, anthracene-9-aldehyde and pyrene-3-aldehyde are fluorescent in ethanol but nonfluorescent in heptane. Pyrene-4-aldehyde and tetracene-9-aldehyde are fluorescent in both ethanol and heptane [268].

Fluorenone shows a fluorescence at both room and low temperature [264]. No phosphorescence inherent to the molecule apparently exists. However, a triplet state occupation (equivalent to Φ_{IS}) of 0.93 has been reported for fluorenone [212]. In light of the absence of phosphorescence there must be essentially 100 percent internal conversion from the lowest triplet to the ground state. The fluorescence intensity and wavelength vary as the solvent is changed. The reason for these latter effects is not known. The lowest singlet state is π, π^* [269].

Aromatic acids such as benzoic, phthalic, and gallic [$C_6H_2(OH)_3COOH$] acids emit phosphorescence beginning at 4051, 4600, 4295 Å, respectively [270].

Quinones such as camphorquinone show phosphorescence from the T_{n,π^*} state in the visible spectral region in a hydrocarbon glass at 77°K with a lifetime of 7.1×10^{-3} sec. Here there is a question as to the specific T_{n,π^*}

state assignment because of the presence of two orbitals containing n electrons. This situation arises also in certain other cases such as biacetyl and in diazaheterocyclics compounds.

As mentioned earlier, photochemical reactivity has been used as a criterion for the assignment of the lowest energy state [264,271]. In certain substituted benzophenones, in contrast to benzophenone itself, photoreduction to the pinacol cannot be observed (Chapter 18). This could arise if the lowest triplet state of the photochemically unreactive compounds were of π, π^* character rather than of n, π^* character since the expected charge distribution (di-radical type) would not exist and the property of the oxygen site to abstract hydrogen would be lost. Also, certain ketone and aldehyde sensitizers do not cause racemization of optically active alcohols. It appears that the photo-chemically unreactive group of ketones and aldehydes may have lowest triplet states of π, π^* character. For more detailed discussion of the molecules involved see Chapter 18.

Several summary statements can be made. Generally if a carbonyl-containing molecule has an S_{n,π^*} state lowest, it will not fluoresce but will phosphoresce. The phosphorescence can originate from either T_{π,π^*} or a T_{n,π^*} state. The nature of the state can be determined from the polarization characteristics of the phosphorescence. Generally the lifetime of phosphores-cence from a T_{n,π^*} state is considerably shorter than that from a T_{π,π^*} state. If an S_{π,π^*} state is lowest, both fluorescence and phosphorescence occur.

Considerations pertinent to fluorescence activation and intersystem crossing are (a) to first order, spin-orbit coupling between states of the same configuration is forbidden and (b) spin-orbit coupling is extremely small between π, π^* states because of vanishing one- and two-center terms. Furthermore, it is possible that vibronic coupling between n, π^* and π, π^* states can alter the rate of intersystem crossing and thus modify the relative intensities of fluorescence and phosphorescence. More concerning the effect of vibronic coupling will be given in the discussion of N-heterocyclics.

Substitution has a variable effect on the nature of the emitting triplet state This apparently depends principally on the relative energy difference between the T_{n,π^*} or T_{π,π^*} states in the parent molecule.

Solvents can affect the emission properties. This results from a change in the nature of or energy relationship between the lowest S_{n,π^*} and S_{π,π^*} states.

Table 12 gives a summary of characteristics associated with molecules having either a T_{n,π^*} or T_{π,π^*} state lowest.

B. HETEROCYCLICS

Table 14 gives spectral data for N-heterocyclics. It has been generally considered that N-heterocyclics would only phosphoresce. Generally this

Table 14 Spectral Properties of *N*-Heterocyclics

Molecule	Emission[a] (Å)		Lifetime (sec)		Quantum Yield	
	Fluorescence	Phosphorescence	Fluorescence	Phosphorescence	Fluorescence	Phosphorescence
Pyridazine[b]	3756 (n, π^*)	Not observed	2.6×10^{-9}		0.01	0.0
Pyrimidine[b]	3261 (n, π^*)	3534 (n, π^*)	1.5×10^{-9}	0.01	0.0058	0.14
Pyrazine[b]	3266 (n, π^*)	3767 (r, π^*)		0.02	0.0006	0.30
sym-Triazine[c]		3787 (n, π^*)		0.4		vvw
2-Phenyl-sym-triazine[d]		4081 (π, π^*)		1.4		
2,4-Diphenyl-sym-triazine[d]		4098 (π, π^*)		1.2		
2,4,6-Triphenyl-sym-triazine[d]		4065 (π, π^*)		1.1		
2-Phenyl-4,6-dimethyl-sym-triazine[d]		4048 (π, π^*)		1.9		
2,4,6-Trimethyl-sym-triazine		~4000 (n, π^*)		0.65		
Quinoline		4570[e] (π, π^*)		1.3[e], 1.3 (0.9)[f]		0.19
Isoquinoline		(π, π^*)		0.9 (0.7)[f]		
Isoquinoline-d_7		(π, π^*)		4.5 (3.3)[f]		
Quinoxaline[k]		4650 (π, π^*)		0.25		
Phenanthridine		(π, π^*)		1.2 (0.9)[f]		
9,10-Diazaphenanthrene	4329[g] (n, π^*)	~5800[h] (π, π^*)				
1,10-Diazaphenanthrene[i]		4502 (π, π^*)		1.6		
9,10-Diazanthracene[j]		6420 (π, π^*)		0.02		
5,6-Benzoquinoline				2.9 (2.6)[f]		

[a] 0–0 origins unless noted otherwise. States in parenthesis represent the assignment of the emitting state.
[b] Reference 274, intersystem crossing yields are pyridazine, 0.2; pyrimidine, 0.12; pyrazine, 0.33 and determined at 298°K. All other data at 77°K in hydrocarbon solvent. Phosphorescence lifetime for pyrazine from reference 260.
[c] Emission and intensity data from reference 285, lifetime from J. Paris, R. Hirt, and R. Schmitt, *J. Chem. Phys.*, **34**, 1851 (1961).
[d] Reference 295, in 3-methylpentane, 77°K.
[e] Reference 282, in hydroxyllic solvent, 77°K.
[f] Reference 283, 284 in EPA at 77°K. Values in parenthesis are for a hydrocarbon solvent at 77°K.
[g] Reference 293, for the crystal at 4°K.
[h] Reference 292, ~77°K.
[i] Reference 294, in EPA at 77°K, Triplet assigned as ^3L_a.
[j] R. Harrell, PhD Dissertation, Fla. State Univ., 1959.
[k] J. Vincent and A. Maki, *J. Chem. Phys.*, **39**, 3088 (1963), emission data in EPA or durene, lifetime in durene, 77°K.

has been true with only a couple of exceptions. Recently [272,273,274], however, it has been found that still more N-heterocyclic molecules exhibit fluorescence. Pyridine exhibits no emission. Oxygen perturbation studies show that a singlet-triplet transition exists at approximately 3350 Å [275]. Further, solvent studies indicate that the triplet state is of π, π^* type rather than n, π^* type.

The molecules sym-tetrazine and dimethyl-sym-tetrazine have lowest excited singlet states of the n, π^* type. However, the emission from these are fluorescences with lifetimes estimated to be less than 10^{-5} sec [276]. Also, 9,10-diazaphenanthrene is known to fluoresce [277]. Pyrazine (1,4-diazine) and pyrimidine (1,3-diazine) were considered as prime examples of molecules with an n, π^* singlet state as the lowest state and showing only phosphorescence. All these molecules, however, do exhibit fluorescence and except for pyridazine they also phosphoresce [272,273]. The observed fluorescence lifetime in the case of pyrimidine is $\sim 10^{-9}$ sec. Pyridazine (1,2-diazine) also shows a fluorescence with a short lifetime of 2.6×10^{-9} sec [272]. These are very short compared to that estimated for sym-tetrazine. In all cases fluorescence occurs in hydrocarbon solvents and other solvents. It is also worth nothing that deuteration causes a notable increase of the quantum yield of fluorescence for pyridazine but not pyrimidine or pyrazine [274].

In addition to the above considerations polarization data, coupled with theoretical calculations and experimental results, have improved the understanding of the reason for increased spin-orbit coupling in these types of molecules. From symmetry considerations mixing of n, π^* singlet and triplet states is forbidden [139,278]. Further, phosphorescence from and absorption to the T_{n,π^*} for both pyrazine and pyrimidine [93,278] should be in-plane polarized. Experimental polarization data confirm this [278,279]. Also, theoretical considerations predict that the degree of spin-orbit coupling between n, π^* singlet and n, π^* triplet states is approximately 100–1000 times less than that between n, π^* singlet and π, π^* triplet states [139]. If an n, π^* singlet state is lowest, the above considerations plus the lack of good overlap between the vibrational wave functions of the n, π^* singlet and n, π^* triplet states indicate that the intersystem crossing may well proceed to a π, π^* triplet state. More concerning these and other cases will be given after consideration of some theoretical concepts regarding spin-orbit coupling between states of the same and different configuration.

Spin-orbit coupling between states of the same configuration S_{n,π^*} and T_{n,π^*} states, depends on matrix elements of the type

$$\langle {}^1\psi_{n,\pi^*} | H_{SO} | {}^3\psi_{n,\pi^*} \rangle \tag{104}$$

where the ψ's are the zero-order antisymmetrized state wave functions. The H_{SO} has components as previously mentioned that can be expressed in terms of the derivative of the potential or the potential itself [93,243,280,181a]. Further, for mono-aza-heterocyclics and approximating H_{SO} as the sum of one-electron operators [281,139], (104) becomes

$$\langle n| H_{SO} |n\rangle + \langle \pi_i| H_{SO} |\pi_i\rangle \tag{105}$$

Substituting for H_{SO} in (105), the integrals vanish for any component of H_{SO} [139].

If the n, π^* states to be coupled are of different configurations, other type integrals arise. These are of type $(\pi_i H_{SO} \pi_j)$ for a mono azine and $(n_+ H_{SO} n_-)$ for a diazine where n_+ and n_- arise as linear combinations of the two nonbonding orbitals on nitrogen; for example, $n_+ = \frac{1}{2}(n_1 + n_2)$. Integrals of the type $(\pi_i H_{SO} \pi_j)$ are vanishingly small since the one- and two-center terms vanish [181a]. The other integral $(n_+ H_{SO} n_-)$ also vanishes upon substituting for H_{SO} [139].

Similar conclusions can be reached group theoretically as well. In order for the integral (104) to have a finite value, it must transform as or give the totally symmetric representation. The components H_{SO} transform as rotations, R_x, R_y, R_z. Thus the direct product of ${}^1\psi_{n,\pi^*}$ and ${}^3\psi_{n,\pi^*}$ must transform as one of the H_{SO} components; that is, as R_x, R_y, or R_z. Further, generally for mono and diazines, the symmetry is not high and degeneracies do not arise. Thus, since the spacial part of the ${}^1\psi$ and ${}^3\psi$ states are the same, the direct product of ${}^1\psi \times {}^3\psi$ will be totally symmetric. The H_{SO} components also must then be totally symmetric and generally this does not occur for those point groups to which planar mono- and diazines belong. Similarly, group theoretical consideration can be applied to integrals of the type $(n_+| H_{SO} |n_-)$ [139].

Spin-orbit coupling between n, π^* and π, π^* states can be allowed. For example, in pyrazine, D_{2h}, the ${}^1B_{1u}(\pi, \pi^*)$ state can mix with the lowest ${}^3B_{2u}(n, \pi^*)$ state through the σ_x component of H_{SO}. For this case the appropriate integral or matrix elements contain one center term. Consequently it would be expected that the phosphorescence would be polarized in-plane and parallel to the N—N axis (not out-of-plane because of no n, π^* state mixings). The phosphorescence is polarized this way [278]. Group theoretical treatment of other monocyclic mono- and diaza aromatics shows that phosphorescence should be polarized in plane because of (a) the nature of the H_{SO} component required, (b) the symmetry of the mixing singlet and, (c) the symmetry of the excited triplet state (which cannot perturb the ground state [278]).

Some general important conclusions that result from the above considerations are as follows. When the S_{n,π^*} state is lowest, the singlet states mixing with a lowest T_{n,π^*} state will be of π, π^* type, thus causing the phosphorescence to be polarized in the molecular plane. Further, mixing of S_{n,π^*} into T_{π,π^*} is of the order of 10^3 times more probable than mixing of

$S_{n,\pi*}$ into $T_{n,\pi*}$ [139]. Thus when a $S_{n,\pi*}$ state is lowest, the increased spin-orbit coupling that results will likely result in intersystem crossing to a $T_{\pi,\pi*}$ state. In those cases where $S_{\pi,\pi*}$ and $T_{\pi,\pi*}$ states are lowest, the intersystem crossing is likely to proceed via $S_{\pi,\pi*}$ to $T_{n,\pi*}$ followed by internal conversion to and emission from the $T_{\pi,\pi*}$ [139]. In addition, since the singlets mixing with the $T_{\pi,\pi*}$ would be of the n, π^* and σ, π^* type (recall that $S_{\pi,\pi*}$ and $T_{\pi,\pi*}$ mixing is vanishingly small), the phosphorescence would be polarized out-of-plane. It should be emphasized that the above considerations do not explicitly include a consideration of indirect mixing via vibronic coupling (see later discussion and Chapter 9-A).

In the case of quinoline only phosphorescence occurs in a hydrocarbon solvent, while both fluorescence and phosphorescence are present when the solvent is hydroxyllic [139,282]. The quantum yield of phosphorescence is 0.19 in alcohol-ether hydroxyllic solvents [282]. Further, based on calculated rate constants for internal conversion and fluorescence it has been qualitatively suggested that the n electrons play the important role in spin-orbit coupling. In the hydroxyllic solvent the π, π^* singlet state is lowest. Further, the phosphorescence of naphthalene is similar regarding both intensity and lifetime to that of quinoline. The fluorescence rate constant for quinoline, however, is expected to be about ten times greater than that for naphthalene. Thus some mechanism must be responsible for enhancing the intersystem rate constant for quinoline. It is postulated that the presence of an n, π^* triplet state between the lowest π, π^* singlet and triplet states is responsible for the enhancement [139]. Quinoxaline exhibits a phosphorescence in a hydroxyllic glass. The phosphorescence is highly out-of-plane polarized. The emitting state is assigned as a π, π^* triplet with which an n, π^* singlet, or a σ, π^* singlet, or both states have mixed.

There is clear indication that there is a vibronic coupling between the $S_{n,\pi*}$ and $S_{\pi,\pi*}$ states and between the $T_{n,\pi*}$ and $T_{\pi,\pi*}$ states [283,284]. For molecules containing only H, C, or N and neglecting spin-vibronic contributions (see Chapter 9-A) the transition moment between the ground state and lowest triplet, M_{ST}, can be written

$$M_{ST}(\text{2nd order}) = \sum_{l,n} ({}^3E_1^0 - {}^1E_l^0)^{-1}({}^3E_1^0 - {}^1E_n^0)^{-1}\langle {}^1\phi_l^0 | H_v | {}^1\phi_n^0\rangle\langle {}^1\phi_n^0 | H_{SO} | {}^3\phi_1^0\rangle M_{0l} + \sum_{l,n} ({}^3E_1^0 - {}^1E_l^0)^{-1}({}^3E_1^0 - {}^3E_n^0)^{-1}\langle {}^1\phi_l^0 | H_{SO} | {}^3\phi_n^0\rangle\langle {}^3\phi_n^0 | H_v | {}^3\phi_1^0\rangle M_{0l} \quad (106)$$

where ${}^3E_1^0$ and ${}^3\phi_1^0$ refer to the pure or zero-order triplet energy and corresponding wave function, ϕ_n^0 and E_n^0 represent, respectively, the wave function and zero-order energies of the intermediate singlet and triplet states which are vibronically or spin-orbitally coupled to ${}^1\phi_l$ or ${}^3\phi_1^0$, ${}^1\phi_l$ is any of

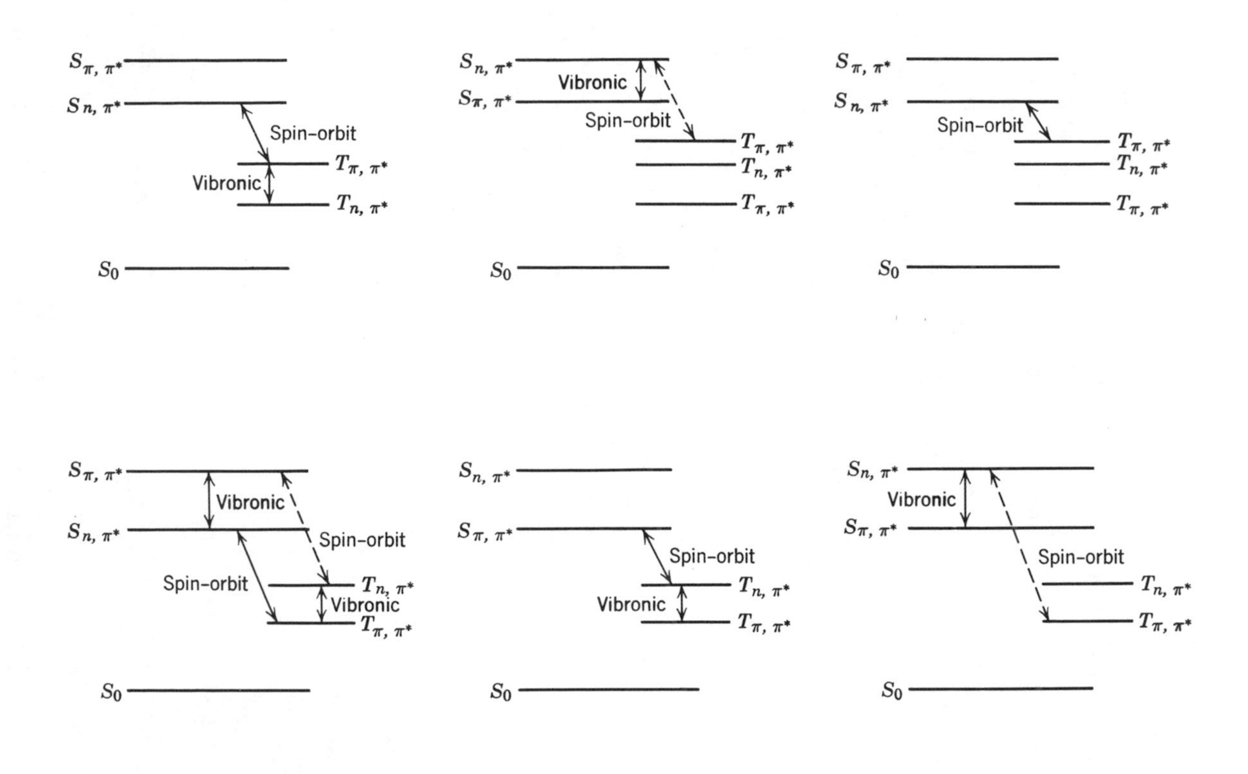

Figure 32 Pathways for intersystem crossing in *N*-heterocyclics.

the zero-order singlet states, M_{0l} is the transition moment of the perturbing singlet transition ${}^1\phi_l^0 \leftarrow {}^1\phi_0^0$, and H_v and H_{SO} are the matrix elements arising from vibronic and spin-orbit perturbations.

It should be noted that the polarization of the phosphorescence is determined by that of the perturbing transition ${}^1\phi_l^0 \leftarrow {}^1\phi_0^0(M_{0l})$ or state. It will be recalled that the first term in (106) is responsible for vibronic coupling between the perturbing singlet, ${}^1\phi_l^0$ with some intermediate singlet, ${}^1\phi_n^0$, which is coupled to the lowest triplet, ${}^3\phi_1^0$, by a spin-orbit perturbation. The second term couples (spin-orbitally) the perturbing singlet, ${}^1\phi_l^0$, to an intermediate triplet, ${}^3\phi_n^0$ which is vibronically coupled to the lowest triplet, ${}^3\phi_1^0$. If indirect or second-order mixing of the types noted above is involved, then some in-plane polarization should be noticeable except for the 0–0 band. It should be recalled that through first-order mixing, general out-of-plane polarization is expected for the π, π^* phosphorescence because of the nature of the perturbing states involved—that is, n, π^* and σ, π^*. Possible pathways for intersystem crossing in *N*-heterocyclics is shown in Figure 32.

The foregoing considerations appear to be of particular importance in those cases in which a change in solvent causes fluorescence quenching and, further, where there has been no interchange of the lowest S_{n,π^*} and S_{π,π^*} states or where the S_{π,π^*} singlet is lowest; for example, consider the cases of the heterocyclics, quinoline and phenanthridine. The degree of polarization of the 0–0 band of phosphorescence (π, π^*) is nearly independent of the nature of the solvent while the remaining bands are more in-plane polarized in hydrocarbon versus hydroxyllic solvents [283,284]. Correspondingly, the lifetime is also sensitive to the nature of the solvent (see Table 14). On the other hand, for the aromatic hydrocarbon carbocyclic analogues, naphthalene and phenanthrene, the lifetime and polarization are not sensitive to the nature of the same solvents. As a matter of fact, the solvent independence of the polarization of the 0–0 phosphorescence band for the above heterocyclics is a good indication of the absence of first-order spin-orbit coupling between the S_{n,π^*} and T_{π,π^*} states. The sensitivity of the lifetime and polarization to solvent change indicates second-order spin-orbit coupling involving vibronic interaction. In addition, it is likely that the vibronic coupling occurs between the T_{n,π^*} and T_{π,π^*} because of the energy term in the denominator of (106).

There is also the consideration of the effect of solvents on intersystem crossing. Even though the lowest singlet state may be S_{π,π^*}, as for isoquinoline and phenanthridine (lowest triplets are π, π^*), fluorescence is quenched in a hydrocarbon solvent relative to a hydroxyllic solvent [283,284]. This may be explained in the following way. Vibronic coupling occurs between the lower S_{π,π^*} and higher S_{n,π^*} state. Spin-orbit coupling occurs to a nearby T_{π,π^*} and internal conversion through a T_{n,π^*} to the lowest T_{π,π^*} state from which emission occurs [284]. This second-order coupling (vibronic, spin-orbit) would be more important in hydrocarbon solvents where S_{π,π^*} and S_{n,π^*} states would be energetically closer. Isoquinoline, as an example, shows a substantial difference in the relative intensities of fluorescence and phosphorescence between a hydroxyllic and hydrocarbon solvent. The fluorescence is significantly lower in the hydrocarbon solvent (phosphorescence is more intense) [283,284].

For the polycyclic monazines it appears that there are two main mechanisms operating to quench fluorescence when changing from a hydroxyllic to hydrocarbon solvent. One of these is the interchange of the lowest excited singlet state—that is, from a S_{π,π^*} state in a hydroxyllic solvent to a S_{n,π^*} state in a hydrocarbon solvent. The other (S_{π,π^*} state lowest) is via vibronic and spin-orbit coupling in which in a hydrocarbon solvent the S_{n,π^*} and S_{π,π^*} (vibronic between singlets) or T_{n,π^*} and T_{π,π^*} (vibronic between triplets) states are energetically less separated, mixing is greater, and intersystem crossing is therefore increased. The resulting pathways and mechanisms would be as shown in Figure 32. It should also be mentioned that despite the fact that state interchange may occur, as may be the situation for quinoline, vibronic coupling is still evident [283,284].

The situation in the monocyclic diazines appears to be quite complicated. These had been considered to be solely phosphorescent or to show no emission at all (pyridazine) [139,285,274] (see Table 14). Pyrazine and dimethyl pyrazine both show phosphorescence at 3850 Å and 3800 Å, respectively, in a hydroxyllic solvent [286]. In addition, both show singlet-triplet absorption; however, the unsubstituted compound shows better resolved fine structure both in absorption and emission. The principal point of interest regarding these molecules is the particular nature of the triplet state. The triplet could be the analog of the benzene triplet, but displaced by the nitrogen perturbation, or it could be of the n, π^* type. The intensity of the transition plus the presence of a 0–0 band in absorption indicate that it is not a symmetry-forbidden transition. Theory predicts the n, π^* triplet to be near the experimentally determined location. Further, since it is predicted that on conjugative substitution a π, π^* triplet state will red shift, whereas an n, π^* triplet state will blue shift [174], the data for the dimethyl derivative supports the assignment of the triplet as n, π^* ($^3B_{2u}$). The vibrational fine structure indicates that the σ orbitals of nitrogen rehydridize upon $n \rightarrow \pi^*$ excitation.

An important consideration in the case of the azines is the nature of solvent interaction and its effect on the absorption and emission spectra. An interesting case in point is pyrimidine, although the following discussion is not restricted to that molecule; for example, pyrazine, also acts similarly [288]. Pyrimidine exhibits a phosphorescence in both hydrocarbon and hydroxyllic glasses (3540 Å) which is assigned as originating from an n, π^* triplet state ($^3B_1$). Moreover, the hydrochloride salt has a phosphorescence in a hydroxyllic glass virtually identical with that of the free base, whose detailed emission spectra is essentially independent of the nature of the solvent. Of further importance is the fact that the emission spectrum of pyrimidine in a hydroxyllic glass does not bear a mirror-image relationship to the absorption spectrum associated with the $\pi^* \leftarrow n$ transition in a hydroxylic solvent glass but to the absorption spectrum in a hydrocarbon glass. Thus the solute-solvent interaction in the excited state must be similar to that in a hydrocarbon glass. This supports the view that hydrogen bonding is essentially absent in the excited state. There is no question but that the pyrimidine hydrochloride has a strong hydrogen bond in the ground state. The emission spectrum, however, is similar to the absorption spectrum of the pyrimidine base in a hydrocarbon solvent and not to that of the hydrochloride. These facts definitely show that a hydrogen bond is either absent or very weak in the n, π^* triplet state for both the base and its hydrochloride. Furthermore, the spectral evidence just referred to indicates that the hydrogen bond is absent in the n, π^* singlet state. In the light of these facts, the shift expected in absorption on going from a hydrocarbon to a hydroxyllic solvent depends on both the stabilization energy of the ground state (resulting from the solvent interaction with the nonbonding electrons) and the Franck-Condon destabilization energy in the excited state. These interactions will result in an overall blue shift. On the other hand since the emission data show that the hydrogen bond is weak or absent in both the excited state and the ground state (after excitation in rigid media); the spectral shift in emission will depend only on the Franck-Condon destabilization energy in the ground state. This will result in a small red shift. It is likely that these results are general for $n \rightarrow \pi^*$ transitions in N-heterocyclic compounds [288].

The monocyclic diazines, however, do show weak fluorescence [273,274] (see Table 14) which requires some special interpretation. Several factors are important. Because of two nonbonding orbitals, two $\pi^* \leftarrow n$ can be expected. These will be allowed or forbidden depending on whether there is a node in the π^* orbital between the nitrogen atoms. Experiment indicates that the lowest excited singlet is the forbidden n, π^* state, S^F_{n,π^*}, and the second singlet is the allowed one, S^A_{n,π^*} [274,289,290,291]. Further, fluorescence apparently originates from the higher S^A_{n,π^*} [274] and phosphorescence from the allowed n, π^* triplet state T^A_{n,π^*} [274]. Figure 33 gives a summary of the pathways responsible for the emission results [274]. General comments can be made. The weak fluorescence is apparently the result of rapid internal conversion between the emitting S^A_{n,π^*} and S^F_{n,π^*} rather than an unusually rapid intersystem crossing to a triplet. The lack of phosphorescence for pyridazine appears to be due to an unusually fast internal conversion from the emitting triplet to the ground state. This may be explained by an additional pathway present in this molecule not present in the others (see Figure 33).

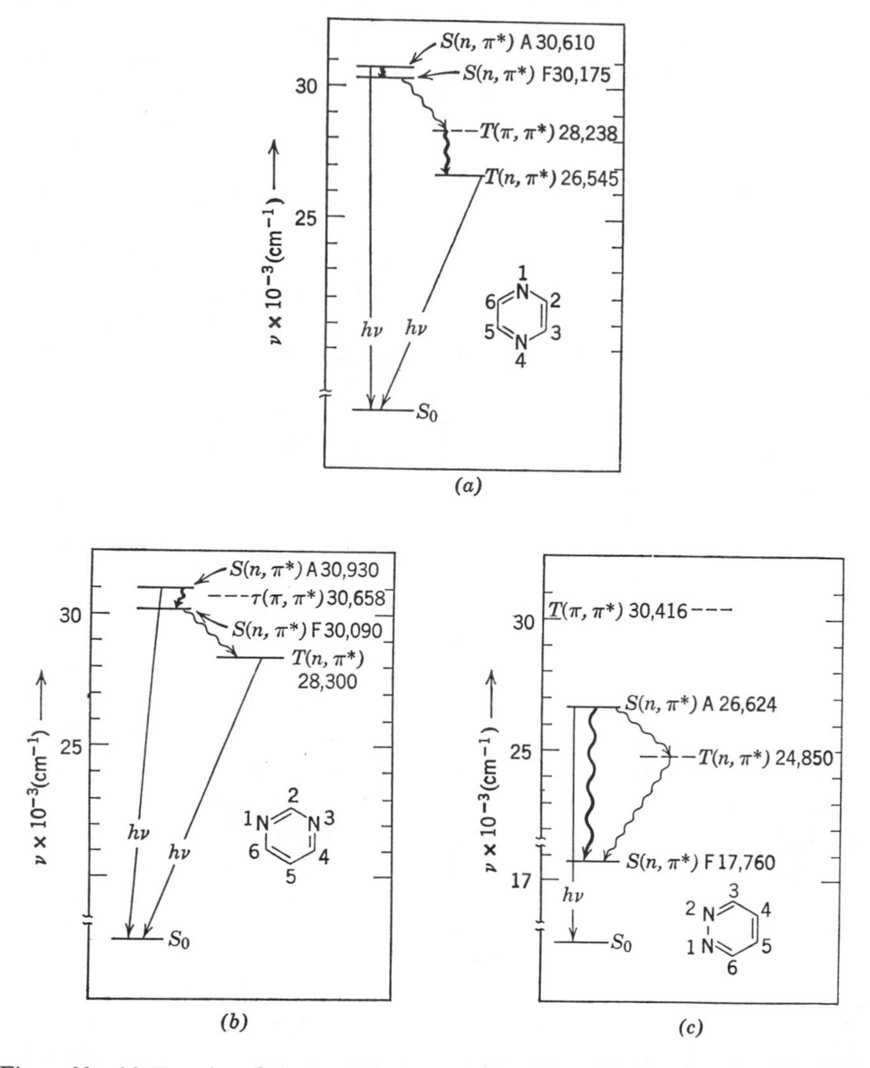

Figure 33 (*a*) Energies of the low-lying states of pyrazine; (*b*) energies of the low-lying states of pyrimidine; (*c*) energies of the low-lying states of pyridazine. State energies which have been experimentally determined are indicated by solid lines. Dashed lines denote calculated values. Fast radiative ⟶ and radiationless ⇝ processes are denoted by dense lines.

Reprinted by permission from American Institute of Physics, B. J. Cohen and L. Goodman, *J. Chem. Phys.*, **46**, 714 (1967).

The idea of extensive internal conversion to the ground state is supported by the fact that the quantum yield of fluorescence for pyridazine is increased by a factor of three by deuteration whereas no change occurs for the other monocyclic diazines.

Increased internal conversion from the lowest excited singlet to the ground state also occurs for monoazines. In the polycyclic monoazines, acridine and isoquinoline, the quantum yields of fluorescence and intersystem crossing are increased approximately 30 percent by deuteration [209]. This is not true for the hydrocarbon analogs.

A point of some importance is that concerning the probability of intersystem crossing between $S_{n,\pi*}$ and $T_{n,\pi*}$ verses $S_{n,\pi*}$ and $T_{\pi,\pi*}$. It will be recalled from earlier discussion that intercombination between the $S_{n,\pi*}$ and $T_{n,\pi*}$ is expected to be approximately 100–1000 times smaller than between $S_{n,\pi*}$ and $T_{\pi,\pi*}$. A combination of quantum yield and rate data for substituted diazines indicates that either the $T_{n,\pi*} \leftsquigarrow S_{n,\pi*}$ is only somewhat less likely than the $T_{\pi,\pi*} \leftsquigarrow S_{n,\pi*}$ or that second-order considerations determine the rate of intersystem crossing [274]. Furthermore it appears that it is the overlap factors between the singlet and triplet states which are the dominant factors in determining the intersystem crossing rate. Earlier discussion in this section quite clearly points up the fact that second-order processes can occur. It is difficult to assess the magnitude of the second-order processes in the diazines. It appears, however, that more attention should be given to this latter possibility, the significance of overlap factors, as well as a reconsideration of the degree of prohibition of first-order spin-orbit coupling between singlet and triplet states of n, π^* type. It also appears that detailed knowledge of the location of allowed and forbidden n, π^* states as well as the location of π, π^* states is important since fluorescence and intersystem crossing, in particular, can show marked sensitivity to deuteration. Furthermore, differences within and between mono- and diazines are sufficiently great that general rules regarding the spectral behavior of azines as a class do not seem justified.

Polycyclicdiazines also show emission. The 9,10-diazaphenanthrene shows a relatively strong fluorescence [292,293] and weak phosphorescence [292]. The 1,10-diazaphenanthrene exhibits a phosphorescence [294] and its fluorescence properties are apparently not known. In the 9,10-diaza case the lowest singlet state is of n, π^* type [293]. The lowest triplet is assigned as of π, π^* type [293]. Once again we have an example of a molecule that fluoresces strongly despite the fact the lowest excited singlet state is of n, π^* type. Several considerations are of importance here. It has been proposed that the reason for the small degree of intersystem crossing was the fact that the $T_{\pi,\pi*}$ was above the lowest singlet state [139]. Consequently since $T_{n,\pi*} \leftsquigarrow S_{n,\pi*}$, is presumably 100–1000 times less probable than $T_{\pi,\pi*} \leftsquigarrow S_{n,\pi*}$, fluorescence from the S_{n,π^*} state would be highly competitive with intersystem crossing to T_{n,π^*}. The T_{π,π^*} state, however, is not above the lowest singlet state [293]. Thus the foregoing possibility would not constitute a valid explanation of the results. If indeed the variation in overlap factors is of as dominant importance as implied in the earlier discussion of the diazines, then the probability of intersystem crossing between the lowest S_{n,π^*} and T_{π,π^*} states might be low (there is a quite large energy gap between them—about 4500 cm^{-1}). This decreased occupation of the triplet would be amplified if it is true that a relatively strong prohibition factor (~1000) exists for the crossing rate between the S_{n,π^*} and the next higher triplet of T_{n,π^*} type.

For the 1,10-diaza case the triplet state is also π, π^* [294] with a long lifetime of 1.6 sec (phenanthrene is ~3.4 sec). The lowest excited singlet state is apparently of π, π^* type also [294]. The mono and diprotonated molecule also show phosphorescence with a red shift for the former and a slight blue shift for the latter relative to the unprotonated species.

Sym-triazine as well as various phenyl and phenyl-methyl substituted triazines show phosphorescence [295] (see Table 14). The parent sym-triazine has a T_{n,π^*} state lowest while all the remainder have a T_{π,π^*} state lowest. A comparison between heterocyclic and carbocyclic analogs such as 2,4-diphenyl-sym-triazine and *m*-terphenyl shows that lifetimes of the hydrocarbons are longer than the heterocyclics. Also the emitting triplets for the hydrocarbon are of π, π^* type. The lowest triplet state of triphenyl-sym-triazine is π, π^*, whereas for the trimethyl-sym-triazine it is n, π^* (the nature of the lowest triplet of the latter molecule is the same as that of the unsubstituted parent).

Certain aromatic N-heterocyclic compounds have already been discussed. In the following discussion additional compounds and more complex cases will be considered. Solutions of the heterocyclic molecules, pyridine, pyrrol, furan, thiophene, and benzofuran, do not fluoresce. Also these do not phosphoresce. Table 15 gives phosphorescence data for some heterocyclics. However, benzopyrrol, indole (I), does fluoresce in a buffered water solution [296]. The quantum yield in neutral aqueous solution is 0.40 [146].

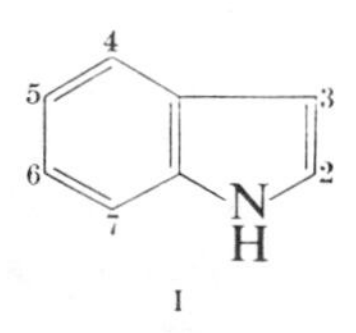

I

It is possible to make many monoaza derivatives of indole. The complete series of 2, 3, 4, 5, 6, and 7 monoaza indoles show fluorescence in buffered water solutions at concentrations of 10^{-5}–10^{-6} M [296]. The cation of each of these derivatives is also fluorescent. The fluorescence maxima are in the

Table 15 Phosphorescence Spectral Properties of Some Heterocyclic Compounds

Molecule	Origin (Å)	Lifetimes (sec)
Fluorene[d]	4080[a]	4.9[b], 6.9[c]
Dibenzofuran[a]	4080	5
Carbazole[a]	4070	10, 7.6[c]
Indene[d]	4050	1
Thianaphthene[a]	4160	0.5
Indole[a]	4040	5, 6.3[c]
9,10-Dihydroacridine		3.5

[a] R. Heckman, *J. Mol. Spec.*, **2**, 27 (1958), in EPA at 77°K.
[b] Reference 260, in EPA a 77°K.
[c] V. Ermolaev, *Optics and Spec.*, **11**, 266 (1961), in ethanol-ether (2 : 1) at 77°K.
[d] Not heterocyclics, for comparison purposes

ultraviolet (3100–3900 Å) except for the 4 and 5 derivatives, which fluoresce at about 4100 Å. The aza substitution causes a relative red shift of the fluorescence maximum, except for the 2 and 3 derivatives. Among these compounds there is quite a variation in the intensity of fluorescence in spite of the fact that the absorption intensities of at least some of the compounds are comparable and that the spectral location of the fluorescence maxima is the same. It is further interesting to note that in the monoaza derivatives, substitution into the pyrrole portion shifts the maximum toward shorter wavelengths, whereas substitution in the benzene portion causes a shift to longer wavelengths. In ethanol the fluorescence maximum for indole is at a slightly shorter wavelength than that of the carbocyclic analog indene (2800 Å vs 2900 Å). Also, all of the neutral monoaza indoles show fluorescence in 95 percent ethanol and all maxima occur in the ultraviolet. Usually, compared to indole, the fluorescence maxima of the aza derivatives are displaced to longer wavelengths; however, they are displaced to shorter wavelengths for the 2- and 3-aza substituted compounds.

The diaza indoles all have fluorescence maxima in the ultraviolet, whether the solvent is water or ethanol. This is also true of the cations. The 2,3- and 3,5-diazaindoles have maxima located very near that of the parent indole while the spectrum of the 3,4 derivative is displaced to the shorter wavelengths. Among the diaza compounds there is no apparent regularity of shift if both nitrogen atoms are in the pyrrole portion or if one is in each portion. The pyrrole-substituted monoaza indoles and the 2,4-diazaindole have a fluorescence more intense than that of the parent indole while that of the others is weaker.

One triazaindole (the 3,4,7 derivative), purine, has a fluorescence maximum [~3700 Å in aqueous solution, 3500 Å in alcohol] at longer wavelengths than indole, the diaza indole compounds mentioned above, and the monoaza pyrroles where the substitution is in the pyrrole ring. The cation shows no fluorescence. The intensity appears to be weaker than that of all the other cases except for the 2,3-diazaindole. In conclusion, the substitution of one nitrogen in the pyrrole portion causes a marked blue shift but the substitution of two nitrogens restores the maximum to a position near that of the parent indole. One nitrogen substituted in the benzene portion always causes a red shift. The effect of substituting one nitrogen in each portion depends on the position of substitution. The substitution of two nitrogens in the benzene portion is capable of more than compensating for the blue shift produced by substitution of one nitrogen in the pyrrole portion.

Acridine fluoresces in pure nonhydroxyllic and hydroxyllic solvents beginning at about 3900 Å [138]. Substitution into the 9 position of acridine with alkyl groups, olefinic groups, aromatic groups, or combinations, causes little change on the absorption spectra. However, when NH_2, OH, or $N(CH_3)_2$ groups are substituted on to the benzene portion of a styryl substituent on acridine, there are notable changes in absorption with accompanying changes in the fluorescence polarization curve [297]. Substitution in the 9 position by OH to produce acridone gives a compound that fluoresces in the region from 4000–5000 Å. Apparently little association occurs in dioxane or alcohol up to a high concentration although some fluorescence quenching does occur. An increase in polarity of the solvent causes a red shift as does transformation to the cationic form [298].

Proflavine or 3,6-diaminoacridine cation (II) has a strong fluorescence at

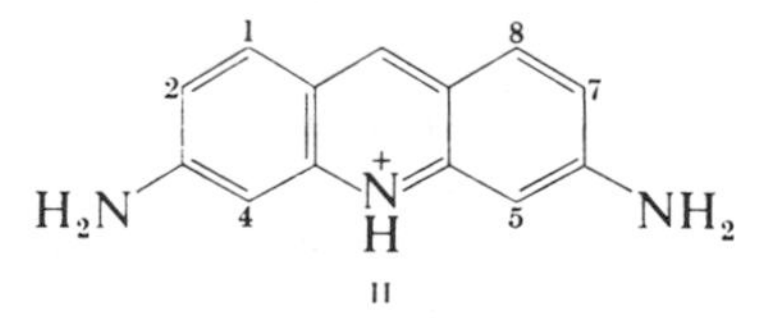

II

~4750 Å at low temperature (−180°C) and a broader maximum at ~4900 Å at room temperature in 95 percent ethyl alcohol [299]. Substitution in the 2,7 positions by fluorine, chlorine, and bromine causes a progressive red shift of the maximum. At the same time the quantum yield of phosphorescence increases.

Coumarin (III) has a fluorescence maximum at 3515 Å in 95 percent ethanol

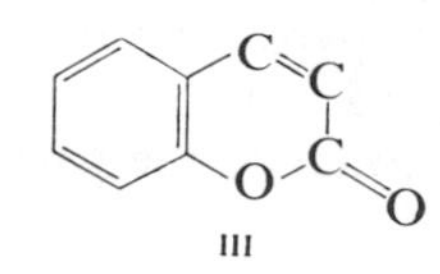

III

[300]. Methyl substitution causes a red shift of the maximum, and also both hydroxyl and methoxyl substituents cause relatively large red shifts, hydroxy producing the large effect. In addition, the shifts also depend on the position of substitution. Electron-withdrawing groups in the 3 position and electron-donating groups in the 7 position enhance the fluorescence intensity.

The fluorescence of the hydroxy quinolines depends on the position of substitution and the pH of the aqueous solution [235]. The 5- and 8-hydroxy derivatives do not fluoresce and the 6-hydroxy quinoline fluoresces weakly. The 2-hydroxy derivative shows uniform fluorescence independent of the pH if excitation occurs in the 3250 Å region but decreases at high pH (~14) if excited at ~2800 Å. The other hydroxy derivatives all fluoresce but the intensity depends on the pH and in some cases, on the wavelength of the exciting light.

Finally, a heterocyclic polyene, triazacyclooctatetriene, emits a yellow fluorescence [301].

C. COMPOUNDS OF BIOLOGICAL INTEREST

There has been a vast increase in the information and interest in molecules of biological interest. These include amino acids, various heterocyclic compounds, and proteins. It is not possible to consider all cases but some representative ones are discussed (see Table 16).

Three aromatic amino acids, phenylalanine (IV), tyrosine (V), and tryptophan (VI), show broad, structureless fluorescence in the ultraviolet with maxima at approximately 2800, 3000, and 3500 Å, respectively, in neutral water solutions at room temperature (Table 16).

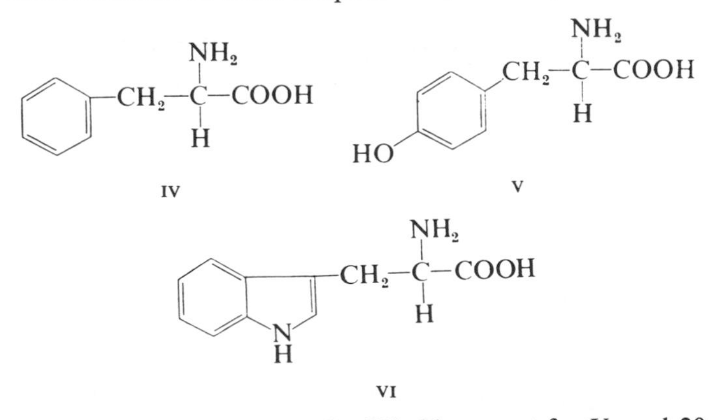

The quantum yields are 4 percent for IV, 21 percent for V, and 20 percent for VI [302]. It is probable that energy transfer can occur from IV to V to VI and from all the aromatic amino acids to the heme groups of hemeproteins [302].

Table 16 Luminescence and pK_a Properties of Some Molecules of Biological Interest

Compound	Fluorescence (max Å)	Fluorescence Φ	Fluorescence (onset Å)	Phosphorescence (max Å)	Phosphorescence Φ	Phosphorescence τ sec	pK_a Properties[p] p$K_a^{S_0}$	pK_a Properties[p] Δp$K_a^{S_1}$	pK_a Properties[p] Δp$K_a^{T_1}$
Thymidylic acid[a]	3180[e]	0.10[e]	~3800	4500		0.33	10	+2	+5.7
Adenylic acid	2980[f]	0.004[f]	3725[a]	4025[f]	0.004	2.45[f]			
Protonated adenylic acid			3650	4250		1.75	4.5	+3.3	−5.3
DNA	3560[g]		3800[a]	4500[a]	0.002[g]	0.30[a]			
Indole[b]	3350	0.42						−7.5	
Benzimidazole[c]	2900	0.67						+6.3	
Indazole[c]	3100	0.46						+5.0	
Benzotriazole[c]	3500	0.02						+2.4	
8-Aza-adenine[c]	3300	0.01						−3.4	
2-Aminopyrimidine	3470	0.15		3870		0.22			
6-Amino-dimethylpyrimidine	3040	0.11		4100		0.03			
Cytosine[d]	3120	0.06		none[h]					
Cytidine[d]	3150	0.03		none[h]					
Thymine[d]	3160			none[h]					
Thymidine[d]	3180			none[h]					
Uracil[d]	3150	0.008		none[h]					
Uridine[d]	3180	0.001		none[h]					
N_1-Methylcytosine[d]	3200	0.02		none[h]					
N_1-Methylthymine	3300	0.02		none[h]					
N_1-Methyluracil[d]	3280	0.01		none[h]					
Adenine[d]	2940	0.06		3840	0.035	2.2			
Adenosine[d]	3150	0.003		4010	0.008	2.5, 2.9[i]			
Guanine[d]	3200	0.06		4090	0.07	1.4			
Guanosine[d]	3250	0.02		4030	0.03	1.3	1.45		

Guanylic acid[d]	3300	0.03		4050	0.04	1.25
Insulin		0.037[j]				
Ribonuclease		0.014[j]				
Phenylalanine[k]	2800	0.04				
Tyrosine[k]	3000	0.21				
Tryptophan[k]	3500	0.20				
Purine				3570[l]		1.6[l], 1.6[i]
Riboflavin[m]	5310	0.023				
All-*trans* retinal[n]	5100					
Rhodopsin[o]	5750					

[a] Reference 327, phosphorescence in ethylene glycol-water mixture (1 : 1) at 77°K and pH = 7.
[b] Reference 328, indole in dimethylsulfoxide.
[c] Reference 328, in H_2O, room temperature, pH = 7.
[d] Reference 328, glycol-water (1:1), 77°K, pH = 7.
[e] M. Gueron and J. Esisinger quoted in reference 328.
[f] Reference 328, emission shows vibrational structure, 0–0 band identifiable at shorter wavelength.
[g] Reference 326, ethylene glycol-water (1:1), pH = 7, 77°K.
[h] Reference 328, phosphorescence is observed in basic solution.
[i] C. Helene, R. Santus and P. Douzou, *Photochem. and Photobiology*, **5**, 127 (1966), in ethanol at 77°K.
[j] Reference 304.
[k] Reference 320, at room temperature in neutral water solution. Also see data table 17.
[l] Reference 337, in EPA at 77°K, wavelength is for 0–0 band.
[m] Reference 339, Φ_F varys significantly with solvent, see text.
[n] Reference 340, at 77°K in 3-methylpentane and EPA.
[o] Reference 341, at 77°K and 3°C in glycerol-water (1:1).
[p] $pK_a^{S_0}$ refers to the pK_a in the ground state singlet, $\Delta pK_a^{S_1}$ refers to the change in pK_a upon excitation to the first excited singlet state relative to the ground state and $\Delta pK_a^{T_1}$ refers to the change in the pK_a upon excitation to the lowest excited triplet state relative to the ground state. For more detail on pK_a determination in excited states, see Chapter 17.

The fluorescence intensity of the amino acids varies with the pH [303,304]. The fluorescence of tyrosine is at a plateau maximum between pH 4 and 9. The fluorescence of tryptophan is maximum at a pH between 10 and 11. Particularly in the case of tyrosine, there is a very rapid fall off of fluorescence intensity above and below the pH values noted. In the case of tryptophan, the fluorescence intensity decreases rapidly above and below 10.5 but levels off on the low pH side until a pH of 3 is reached, then it decreases dramatically again. This is where the $-NH_2$ group becomes $-NH_3^+$. Several compounds related to tyrosine, such as phenol and tyrosine methyl ester, show a dependency of fluorescence intensity on the pH. Also, some derivatives of tryptophan act like tryptophan. In the case of tyrosine, removal of the side chain, acetylation, or removal of the carboxyl has little affect on the quantum yield. However, glycylation of the amino group or esterification of the carboxyl decreases the quantum yield to one third or one quarter of the value of tyrosine. In the case of tryptophan where the side chain is removed (to produce indole), or when the side chain is short and has no carboxyl group, the fluorescence quantum yield is increased. Glycylation or esterification, however, produces a decrease in quantum yield parallel to that found for tyrosine [303]. Fluorescence quenching of tryptophan by bases seems to occur via a mechanism involving transfer of hydrogen from nitrogen to the hydroxyl ion [303]. This conclusion is supported by the *lack* of quenching in *N*-methylindole in base up to 5 *N*. However, indole derivatives show no fluorescence in 1 *N* sodium hydroxide. The proton transfer presumably occurs in the first excited state to produce an anion that is deactivated principally or totally by radiationless modes. In addition, tyrosine and phenol are nonfluorescent as the anions (pH >10). However, substitution of the hydroxyl to give an ether, for example, delays quenching until the pH exceeds 14.

The shape or position of the room-temperature fluorescence spectrum of tyrosine does not depend on pH whereas for tryptophan there is a shift of about 50 Å to the red at pH above 8. The intensity of the fluorescence of both compounds decreases with increasing temperature. Some proteins containing tryptophan show a notable decrease in the intensity of the fluorescence characteristic of tryptophan as the temperature is increased.

The principal importance of the fluorescence of tryptophan and tyrosine is in the analysis of proteins. Proteins are subjected to hydrolysis and the content of these amino acids is determined by their fluorescence spectra. High analytical accuracy is achieved, particularly for tryptophan. Many metabolites of these amino acids have been investigated using fluorescence techniques with excellent results.

Tryptophan has two sets of emissions each of which is composed of a short-lived and long-lived component [305]. Excitation of tryptophan in frozen aqueous solution at 3500 Å gives a blue short-lived emission ($\sim$4500 Å) and a green long-lived emission (~5000 Å). Excitation at 2800 Å gives a fluorescence in the ultraviolet (~3500 Å) and a very weak long-lived emission in the green (~5000 Å). These results are unusual. When excited in a frozen glucose or methanol solution at 2800 Å, the ultraviolet fluorescence still exists but now a blue phosphorescence replaces the weak green one. Excitation at 3500 Å still gives a short-lived blue emission and a long-lived green one. Certain compounds are good quenchers. The amino acid, cysteine, quenches the ultraviolet fluorescence with little enhancement of the long-lived blue emission. Both trinitrobenzene and dinitrophenol quench all of the tryptophan emission. Riboflavin and folic acid are also effective. These latter four results probably occur because of energy transfer to the triplet state of the electron acceptor molecules since the emission observed is essentially that of the acceptor molecules themselves [305].

All three of the aromatic amino acids show fluorescence and phosphorescence at 77°K in EPA (0.01 *M* HCl) and 0.5 percent glucose solution (see Table 17). The greatest shift of fluorescence between room and low temperature occurs for tryptophan. In ethylene glycol-water solutions (1 : 1) or 0.5 percent glucose solutions at 77°K, tryptophan has a resolved phosphorescence with an onset near 3900 Å. Tyrosine under the same conditions has an unresolved phosphorescence with an onset near 3500 Å [305]. Also tyrosine has a total quantum yield very near 1 at 77°K in a variety of solvents [307].

Other amino acids have very little or no fluorescence. In several cases, however, the amino acids have been reacted with another molecule to give products showing high fluorescence [308]. Among such reactants are histamine, histidine, and arginine.

The fluorescence properties of proteins and protein complexes have been actively investigated. Many proteins show fluorescence as a broad structureless band in the ultraviolet with maxima generally in the 3000–3500 Å region [309,310]. The proteins include myoglobin, pepsin, ovalbumin, hemoglobin, fibringogin, insulin, and others. The quantum yield of fluorescence of a group of 21 proteins in water varies from 0 percent (1 case) to 15 percent. For those proteins soluble in an organic solvent such as propane 1,2-diol, the quantum yields increase relative to water as a solvent except in the one case where the fluorescence was originally absent. Cysteine or the peptide linkage in general make no contribution to protein fluorescence.

One interesting consideration is that of energy transfer between aromatic amino acids contained in proteins [311–317]. Earliest indications were that tyrosine fluorescence was detectable only in the absence of tryptophan moieties. In some proteins only tryptophan fluorescence is observed regardless of the energy of the exciting light and even if the number of phenylalanine and tyrosine moieties far outnumber those of tryptophan. There are now indications that tyrosine phosphorescence [315,316] and possibly

Table 17 Emission Characteristics of Aromatic Amino Acids[a]

Molecule	Fluorescence (max Å)		Phosphorescence (max Å)		F/P[c]		τ_P (sec)[d]	
	EPA (0.01 MHCl)	0.5 (percent) glucose[b]	EPA (0.01 MHCl)	0.5 (percent) glucose[b]	EPA (0.01 MHCl)	0.5 (percent) glucose[b]	EPA (0.01 MHCl)	0.5 (percent) glucose[b]
Phenylalanine	2850	2900	3800	3850	0.7	1.7	5.6, 6.4[e]	5.5
Tyrosine	3050	3000	3950	3950	0.8	1.1	2.5, 1.9[e]	2.6, 3[f]
Tryptophan	3250	3300	4900	4350	3.7	3.9	6.3, 4.1[e]	5.7

[a] All data from J. Nag-Chauduri and L. Augenstein, *Biopolymers Symp.*, **1,** 440 (1964), at 77°K except where noted.
[b] 0.5 percent glucose in water.
[c] Ratio of the intensity of fluorescence to phosphorescence.
[d] Maximum deviation ±10 percent, from reference in footnote a except where noted.
[e] T. Shiga and Z. Pretti, *Photochem. and Photobiol.*, **3,** 223 (1964), at 77°K in 6 *N* HCl, given as half-lives from epr data.
[f] J. Moling, K. Rosenheck, M. Weissbluth, *Photochem. and Photobiol.*, **4,** 241 (1965), at 77°K, given as half-life. Lifetime given as 2 sec for ionized molecule.

fluorescence [313,314] exist in a number of proteins containing tryptophan. In one protein RNase, it has been postulated that some tyrosine residues are electronically perturbed whereas others are not [316]. Further, because the phosphorescence of tyrosine moieties in proteins containing only tyrosine (as RNase or insulin) is different from tyrosine under other conditions, it appears apparent that this amino acid is perturbed from interactions with neighboring polypeptide groups [317]. In the case of α-chymotrypsin and its amino acid mixture (2.5 percent tyrosine + 5 percent tryptophan) the phosphorescence is essentially only that of tryptophan. However, in amino acid mixture of liperidase (6 percent tyrosine + 0.4 percent tryptophan) the phosphorescence contains a large percentage of emission typical of tyrosine. Lysozme which has 11 percent tryptophan, shows a relative percent of tyrosine phosphorescence typical of its tyrosine content. In this case and many others, the percent phosphorescence due to tyrosine is wavelength dependent. Generally excitation at 240 mμ results in a greater percentage phosphorescence typical of tyrosine than that at 280 mμ [317].

The results from lysozme indicate that one tyrosine moiety may be relatively spatially isolated from the nearest tryptophan. This factor alone does not appear to be the reason for a relatively large percentage of tyrosine phosphorescence. In general it appears that quenching of excited tyrosines, or of tryptophans, or both, and energy transfer from some excited tyrosine to tryptophans result in variable percentage contributions to the phosphorescence and fluorescence by each of these in different proteins. Specific environmental conditions could cause specific perturbations and thus vary from protein to protein.

Specific examination of tyrosine indicates that the ratio of intensity of phosphorescence to fluorescence varies according to the wavelength of excitation (225 and 275 mμ), the solvent, the pH, and the ionic strength [318]. In part, this has been interpreted on the basis of increased intersystem crossing resulting from excitation into higher excited states [317]. However, much if not all of the reference data from which this idea evolved is now known to be anomalous because of photochemistry, impurities, and dimer formation of the hydrocarbons and *not* because of notable changes in the intersystem crossing efficiency resulting from higher energy excitation.

If a prosthetic group is attached to a protein, it is generally found that the fluorescence is considerably weaker than for the protein without the prosthetic group. These include such species as myoglobin, hemoglobin, and carbon monoxide complexes with myoglobin [313,320]. At least in some cases it is known that the energy is utilized to produce photochemical changes. When an enzyme is complexed with a coenzyme, the intensity of fluorescence of the complex is usually greater than that of the coenzyme itself; however, it is generally lower than for the protein itself (at least in dehydrogenases) [32].

Another area receiving wide attention is the chemical binding of a dye to a protein to give a complex that emits fluorescence in the visible region [319, 322–325]. Probably the most commonly used dye is 1-dimethylamino-naphthalene-5-sulfonyl chloride, which binds to the protein by reacting with the amino groups to produce sulfonamides. Normally, excitation is accomplished via the protein absorption band followed by energy transfer to the dye and emission occurs in the visible region. This technique, with modification of the dye, has been used in the fluorescent antibody technique [308]. Other dyes such as protoporphyrin have also been used to give a complex with a protein that fluoresces red [308].

In the case of DNA, the triplet state is known and appears to reside at the thymine residue [326,327]. It is not unequivocal whether the triplet is associated with neutral thymine or the ionized thymine anion. However, based on epr data, phosphorescence spectra, decay times, and the effect of deuterium substitution, the evidence is in favor of association with the neutral species [327].

DNA has a broad phosphorescence with onset at ~3800 Å (max ~4500 Å) and a lifetime of 0.30 sec [327]. The phosphorescence of thymidylic acid (TMP) onsets, maximizes, and has a lifetime essentially the same as that of DNA. The phosphorescence of ionized TMP onsets and maximizes at shorter wavelength than the unionized TMP and has a lifetime of 0.50 sec. Protonated adenylic acid (at the ring nitrogen adjacent to the amino group), $AMPH^+$, has a phosphorescence onset near 3650 Å, maximum at ~4250 Å, with a lifetime of 1.75 sec. This emission occurs only upon sensitization [327] of $AMPH^+$ (for example, by acetone) which clearly indicates that electronic excitation is lost from the excited singlet state upon direct excitation. The TMP has an increase in pK_a of 2.0 and 5.7 in the excited singlet and triplet states, respectively, while $AMPH^+$ shows an increase of 3.3 and a decrease of 5.3 pK units in the singlet and triplet states, respectively (Table 16). The latter change is unusually large for a molecule in the triplet state, even for a protonated species (see Chapter 17).

In addition to the above biomolecules, the fluorescence, phosphorescence, and change in the pK_a value is known for a considerable number of other purines, pyrimidines, nucleosides, and nucleotides [320,327–336] (see Table 16).

Natural-occurring pyrimidines such as uracil, thymine, and cytosine fluoresce over a widely varying range of pH; however, they show phosphorescence only at high pH where a proton is lost [328]. Methyl pyrimidines in the lactim form at pH = 7, fluoresce at room temperature, and phosphoresce at 77°K. This is not the case for natural-occurring pyrimidines and indicates that the lactam form is the dominate species for the natural pyrimidines.

The purines, adenine and guanine, fluoresce and phosphoresce at pH = 7 and higher at 77°K. In acid solution there is no emission from adenylic acid, the intensity of phosphorescence is strongly reduced for adenosine, and the emissions of adenine are only slightly changed. Cytidlic acid and cytidine have no or at best a weak phosphorescence [328].

Polyadenylic acid (poly A) in a single-stranded arrangement has spectroscopic properties similar to those of adenylic acid itself; that is, at 77°K both exhibit a poorly resolved fluorescence and a resolved phosphorescence ($\tau = 2.4$ sec). Each of these emissions is red shifted in poly A compared with the monomer [330]. In double-stranded poly A, however, the emissions are completely quenched when one half of the bases are protonated. This and the results of heterogeneous polymers [330] indicate that quenching occurs at the excited singlet level and result from hydrogen bonding with accompanying excited-state proton transfer. The other possibility of singlet-singlet energy transfer between protonated and unprotoned bases seems less likely [330].

Purine exhibits a strongly resolved phosphorescence at 77°K (Table 16) which presumably originates in a π, π^* triplet. The fluorescence intensity varies with the polarity of the solvent [337]. In 2-aminopurine the quantum yield of fluorescence (max ~3600 Å) is near unity and no phosphorescence is observed in neutral solutions at 77°K [338]. In the 2,6-diamino derivative the quantum yield of fluorescence is decreased and phosphorescence is observed. In the diamino compound, the fluorescence maximizes at 3200 Å, the phosphorescence at 4030 Å, and the latter emission has a lifetime of 2 sec at 77°K and pH of 3 [338].

Riboflavin has a poorly resolved fluorescence with a maximum at ~5300 Å and quantum yield of 0.023 in water at room temperature. The wavelength and quantum yield varies with the solvent. The emission is at its shortest wavelength (max ~5200 Å) and highest quantum yield (0.40) in a 98 percent dioxane-2 percent water mixture [339]. *All trans* retinal, a long-chain polyene aldehyde, exhibits a fluorescence at 77°K with a maximum at 5100 Å [340]. The emission is also present at room temperature but it is considerably weaker. The 11-*cis* isomer apparently has no or a very weak fluorescence. The visual pigment rhodopsin appears to show a weak fluorescence at 77°K and 3°C with a maximum at 5750 Å [341].

Many other molecules such as vitamin A (a polyene), pyridine nucleotides, other vitamin condensate products, carbohydrate condensate products, steriods in acid media, etc., show fluorescence.

Chapter 13 Molecular Complexes

A. Metal Complexes

A great variety of metal complexes, including the porphyrins, chlorophylls, hydroxyquinolines, phthalocyanines, and others, fluoresce and phosphoresce.

When considering the kind of emission to be expected from a metal complex, several salient factors must be considered. The broad considerations are (a) the nature of the ligand, (b) the nature of the metal, (c) the nature of the perturbation—that is, does the metal perturb the ligand levels or does the ligand perturb the metal levels—(d) intramolecular energy transfer processes, (e) solvent interactions, (f) temperature. In the first area, obviously important are the emission characteristics of the uncomplexed parent molecule. With regard to the metal, important points are its position in the periodic table—categories include the transition elements, the rare earths, and the remainder. In addition, the diamagnetic or paramagnetic nature of the metal ion and its oxidation state are of great importance, as is the third factor, since the nature of the excited states and their intercombination are largely determined by the nature of the perturbation. The possibility of intramolecular energy transfer can notably alter the nature of the emission. This is particularly true for rare-earth chelates as will be discussed later. Solvent interactions arise but exert a dominant effect only in particular instances. As usual, the solvent may shift the position of the emission and modify its appearance. The temperature is particularly important because of its effect on the viscosity of the solvent. This, in turn, affects the collisional deactivation of excited states and thus *both* the intensity and nature of the emission. Also, the temperature affects the fine structure character of the emission. The importance of these factors and others that will be developed in affecting the emission spectra cannot be overemphasized. Many investigations, particularly earlier ones, have produced incorrect results or further effort was thwarted because of a lack of understanding of certain general basic premises. Many of the areas of discussion can be examined utilizing a restricted group of compounds but the results are generally applicable to the broader consideration of various types of complexes.

Probably the metallo complexes that have been most completely investigated are the porphyrins [342,350] and porphyrinlike compounds, including the chlorophylls [131,132,346,351–356] and phthalocyanines [346,357]. Table 18 presents the spectral results for 22 metal porphyrins as well as two different parent molecules for comparison [347,350]. These include closed-shell diamagnetic ions, open-shell diamagnetic and and open-shell paramagnetic transition-ion metallo porphyrins. The discussion will be divided into considerations of the non-transition metal ion and transition metal ion derivatives because of the significant differences in the nature of the ions involved. From the table, one striking fact is found: *no transition metal porphyrin fluoresces, whatever the oxidation state of the metal may be*. This fact alone is very useful, although earlier work had indicated contrary results in a few cases. This knowledge is of great value in the analysis of a single component or a system of components containing metal complexes by a fluorescence technique. The fluorescence intensity of metal complexes varies greatly with the metal ion, generally decreasing with increasing atomic number.

Also, within one period—that is, a group of diamagnetic nontransition metal ions—it might be expected that the complexes would exhibit reasonably smooth trends in the intensity (quantum yield) of phosphorescence, or the lifetime, or both. The reason for this being that with parallel electronic configurations the spin-orbit coupling would be expected to increase progressively with the atomic number of the perturbing metal ion. A glance at Table 18 will show that this is not true and in some cases the deviation from expectation is great. The following discussion will thus be illuminating as to the qualifications that must accompany generalized statements concerning the enhancement of spin-orbit coupling with respect to the parameters upon which it normally depends.

Unfortunately, the relation between fluorescence and the atomic number of the metal in a porphyrin complex is not completely clear-cut; for example, among the metallo complexes of Groups IIA, IIB, and IVA diamagnetic divalent metals (Be, Mg, Ca, Sr, Ba; Zn, Cd, Hg; Sn and Pb) two are nonfluorescent (Hg and Pb) which is something probably not expected. Second, although within the IIA and IIB series the general expectation of decreasing fluorescence intensity with increasing atomic number is followed, the fluorescence of the strontium ($Z = 38$) and barium ($Z = 56$) complexes seems unduly weak. In the IIB and IVA groups, lead and mercury ions have unusually large ionic and covalent radii. The location and shape of the absorption

Table 18 Luminescence and Lifetime Data for Various Metalloporphyrins [348,349]

Compound	Atomic Numbers of Meta Atom	Fluorescence		Phosphorescence		Singlet-Triplet Separation[b] (cm^{-1})	Triplet State Lifetime (msec)
		Position of First Band (Å)	Relative Intensity[a,c]	Position of First Band (Å)	Relative Intensity[a,c]		
Etioporphyrin II		6236[d]	vs	8060[d]	vw		
Mesoporphytin IX dimethyl ester		6175	vs	7515	vw	2871	14
Be mesoporphyrin IX dimethyl ester complex	4	6065	vs	Not observed			12
Mg ethioporphyrin II	12	5815	s	7150	w	3247	160
Ca mesoporphyrin IX dimethyl ester	20	5870	s	7200	m	3313	63
Zn mesoporphyrin IX dimethyl ester	30	5715	m	7010	m	3126	83
Sr mesoporphyrin IX dimethyl ester	38	5890	vw	7365	m	3482	7
Cd mesoporphyrin IX dimethyl ester	48	5820	w	7260	m	3223	7
Sn mesoporphyrin IX dimethyl ester	50	5715	w	7020	s	3087	21
Ba mesoporphyrin IX dimethyl ester	56	5960	vw	7495	s	3201	8
Hg mesoporphyrin IX dimethyl ester	80	Absent		7845	w	4578	≤ 0.5
Pb mesoporphyrin IX dimethyl ester	82	Absent		7910	vw	4655	≤ 0.5

Mn(+2)mesoporphyrin IX dimethyl ester	25	Absent	7520	vw	3816	<0.5
Mn(+3)acetate mesoporphyrin IX dimethyl ester	25	Absent	6580		~2000	
Fe(+2)mesoporphyrin IX dimethyl ester	26	Absent	6805	vw	3647	<0.5
Fe(+3)acetate mesoporphyrin IX dimethyl ester	26	Absent	6815	vw	~2400	
Co(+2)mesoporphyrin IX dimethyl ester	27	Absent	6720	w	3294	<0.5
Co(+3)chloride mesoporphyrin IX dimethyl ester	27	Absent	6815	w	~2900	
VO(+2)mesoporphyrin IX dimethyl ester	23	Absent	7095	w	~3380	<0.5
Cu(+2)mesoporphyrin IX dimethyl ester	29	Absent	6820	vs	3228	0.10
Ni(+2)mesoporphyrin IX dimethyl ester	28	Absent	6810	vw	3464	11
Pd(+2)mesoporphyrin IX dimethyl ester	46	Absent	6580	vs	3099	2.0
Pt(+2)mesoporphyrin IX dimethyl ester	78	Absent	6380	vs	2931	0.14

[a] These intensities are assigned relative to the parent whose fluorescent intensity is considered very strong (vs); and s (strong), m (medium), w (weak), vw (very weak).

[b] Calculated as the frequency difference between the absorption band at lowest wavelength and the phosphorescence band of shortest wavelength. The frequency shift between the lowest wavelength absorption band at room temperature and the shortest wavelength fluorescence band at low temperature is at most 200 cm^{-1} and usually 100 cm^{-1} or less.

[c] No fluorescence exists for any of the transition metal complexes considered.

[d] Wavelength of first strong band, reference 346.

spectra of the lead and mercury complexes are different from those of the other compounds, particularly in the case of lead. Also, the energy separation between the lowest singlet and triplet is unusually large (Table 18). All these factors point to an unusual bonding scheme; that is, the metals are very probably not coplanar with the rest of the porphyrin ring structure.

The concept of increased probability of radiationless transition as a result of geometrical distortion is supported by a general consideration of the data on the luminescence of the complexes of porphyrins with the metals of Group IIB (Zn, Cd, Hg) and IVA (Sn, Pb). As is shown in Table 18, the fluorescence intensity of the cadmium derivative is much less than that of the zinc complex, although the phosphorescence intensity of the two complexes is about the same. For the mercury compound, however, in which the metal atom has a very large ionic and covalent radii, fluorescence is absent and phosphorescence is weaker than for the cadmium complex. It is clear that with increasing size of the metal, the rate constant increases for one or more modes of radiationless deactivation to the ground state. Absorption data [349–350] indicate strongly that there is a distortion from planarity in the complexes containing large metal atoms as the cause of this increase. Comparison of the emission properties of the tin and lead complexes points to the same conclusion. The large lead (II) quenches the fluorescence, which is of weak intensity in the tin (II) complex, and also greatly reduces the intensity and the lifetime of the phosphorescence. Weak phosphorescence would not normally be expected since such a high atomic number center should induce a high probability of crossing to the triplet with consequent strong phosphorescence. The distortion factor plus the high atomic number provide quenching mechanisms for the fluorescence via intersystem crossing to the triplet state and internal conversion to the ground state from both the singlet and triplet states. This is also strongly supported by the lack of fluorescence and the unusually weak phosphorescence for both the lead and mercury (II) complexes. The absorption data provide additional support for the nonplanar nature of the lead and mercury (II) complexes.

With regard to the anomalies exhibited by strontium and barium porphyrin complexes, the decrease in the fluorescence intensity in passing from the calcium to the strontium derivative is abnormally large while the intensity of phosphorescence is approximately the same. The barium complex shows little further change in the intensity of fluorescence despite an atomic number increase of 18 relative to strontium (Table 18) while the intensity of phosphorescence increases. The ionic and covalent radii of barium are large and, in fact, larger than those of lead and mercury ions. Despite this, the barium complex fluoresces and, moreover, also phosphoresces strongly. The absorption spectrum of the barium complex appears normal relative to the normally planar metallo derivatives—for example, magnesium—while those of lead and mercury are abnormal (in spite of the larger size of the barium ion). Because of the low electronegativity of barium, the complex is probably essentially ionic; that is, the barium acts essentially as an external perturbation, but is powerful because of its proximity. Thus little molecular distortion is to be expected. Accordingly fluorescence is highly, though not completely, quenched and, equally important, there is not a high degree of internal conversion to the ground state since a strong phosphorescence exists. The strontium complex appears to have a total quantum yield which is lower than that for the barium complex and this probably results from geometric distortion with consequent increased total internal conversion to the ground state. Even in the zinc-cadmium comparison, the overall quantum yield appears to decrease with increasing ionic size. Here again, molecular distortion resulting from the size of the cadmium ion may play a role.

There are several salient points emerging from these facts. Diamagnetic metal complexes can, in general, both fluoresce and phosphoresce. There are general trends regarding the relationship of atomic number and intensity of fluorescence (higher atomic number centers quench the fluorescence more). The data, however, quite clearly establish the fact that geometrical distortion increases the probability of radiationless transitions to the ground state. It also appears that this occurs principally between the triplet and ground states [349]. In addition to distortion, it is necessary to consider the effect of metal-ion electronegativity and electronic configuration. Any of these (but particularly metal size and its effect on geometrical distortion) may counteract and even reverse spectroscopic trends expected from an increase in atomic number. Because of the complexity of the various effects, it is obvious that it is difficult to predict accurately a result. If the considerations discussed above are taken into account, however, it should be possible to project what might be expected in other metallo complexes containing similar such ions.

Table 18 also shows the emission behavior of metallo porphyrins containing ions of transition elements. Again, none of these show fluorescence. The oxidation state of these ions is of no consequence regarding the presence of fluorescence. Finally, even if the metal ion (and the complex) is diamagnetic, no fluorescence is found. Early work [345] indicated fluorescence for both the nickel and silver (II) compounds but this probably resulted from impurities (the parent prophyrin). The main difference between these cases and the earlier ones considered lies in the nature of the perturbation. In the transition metal complexes, the ligand field splits the *d* levels. Those of appropriate symmetry can combine with ligand orbitals ultimately to produce additional states. Hence additional couplings can occur and thereby promote a high degree of intersystem crossing to the triplet states [348]. Moreover, the presence of new states between the usual porphyrin states and the ground state may increase the probability of internal conversion.

In passing from the zinc to the nickel complex (with a decrease in atomic number), the moderately intense fluorescence of the zinc complex is completely quenched and a sharp decrease occurs in the intensity of phosphorescence. A moderate decrease in lifetime also occurs. Clearly in the nickel complex, the probability of radiationless transitions to the ground state is greatly increased; nevertheless, there is little change in the ionic or covalent radius and the electronegativity [348]. The outstanding difference between the zinc and nickel (II) ions is the presence of unfilled *d* orbitals in the nickel. The presence in the nickel complex of an intermediate ligand field triplet state between the ground state and the normal π-triplet state of porphyrin and geometrical distortion appear to be the factors responsible for the increased probability of a radiationless transition to the ground state [348, 358]. This transition appears to occur from the normal π-triplet state of the porphyrin via the ligand field triplet [348]. Internal conversion within the singlet manifold also appears likely.

The spectral changes occurring among the complexes of the nickel subgroup are also interesting. None of the complexes fluoresce. The very weak intensity of phosphorescence of the nickel compound is greatly increased in the palladium compound but then stays nearly the same for the platinum derivative despite an increase of the atomic number by 32 (see Table 18). The lifetime progressively decreases but a much larger change occurs between the palladium and platinum compounds than between the nickel and palladium complexes even though the atomic number change is the same as that between the nickel and palladium complexes. It is obvious that in this sequence the rate constant for the radiationless mode present in the nickel compound has been markedly decreased. Consideration of all pertinent information leads to the conclusion that it is probable that intersystem crossing to the triplet state is complete in all cases (via indirect coupling through new states). Because of the variable location of an intermediate ligand field triplet, however, the probability for radiationless transitions varies in the series being strongest in the nickel compound [348,358]. Also the molecular geometry of the nickel complex is probably an important consideration. Furthermore, despite essentially complete intersystem crossing the rate constant for emission from the triplet to the ground state singlet appears to depend strongly upon the atomic number. Thus two different factors may be involved in the mixing of the functions that determine the occupation of the triplet state and those that determine the lifetime [348]. There is a larger change in lifetime between the palladium and platinum compounds than between the nickel and palladium complexes. This is very likely the result of a decrease in the probability for radiationless transitions between the nickel and palladium derivatives (as reflected by an increase in the quantum yield of phosphorescence) which offsets the expected decrease in lifetime associated with an increase in the atomic number of the perturbing metal atom.

We shall now turn to complexes of porphyrins with transition metal ions containing varying numbers of *d* electrons. In all cases fluorescence is absent, the phosphorescence intensity is weak, and the phosphorescence lifetime is short (see Table 18). It is very likely that intersystem crossing is complete in these cases and for reasons similar to those given for the nickel subgroup [348]. Also, the low quantum yield of phosphorescence results from the presence of intermediate ligand field states with which is associated strong radiationless deactivating modes (except for the copper complex). The paramagnetic nature of the ion and the radiationless modes are responsible for the short lifetimes. The silver (II) complex represents a unique and unusual case since neither fluorescence nor phosphorescence exists. The +2 oxidation state for silver is not common so that either a ligand field state or charge transfer state must be located at a position such that there is a very high probability for radiationless transitions [348]. It may be possible that an emission does occur from the silver complex, but beyond the region of normal investigation (~10,000 Å).

In conclusion, it can be seen that even in the foregoing where essentially complete spin-orbit coupling exists, the expected strong phosphorescence does not appear because other factors dominate and cause significant deviation from expectation. Moreover, complexes containing diamagnetic metal ions of almost identical atomic number can show very different emission spectral characteristics because of strongly modifying factors such as the presence of ligand field levels. Even among the complexes containing diamagnetic ions from the same group of the periodic system, significant deviation of the emission characteristics from those expected can arise. Further, triplet state lifetimes among metal complexes containing transition metals of the same periodic group may depend on a significantly different factor than that responsible for occupation of the state. Finally, there is *no* fluorescence for any of the transition metal complexes. The varying behavior (particularly that different from expectation) exhibited by the metal complexes forewarn us of the naivity of making *a priori* predictions of emission spectral characteristics from inadequate premises.

Other metal complexes that show phosphorescence are those of the chlorophylls and their derivatives (see Figure 19 for structural formulas) [346,359]. The principal parent actually could be considered to be pheophytin and the others, derivatives of it. Pheophytin, chlorophylls *a* and *b*, the ethyl chlorophyllides, and pheophorbides all emit fluorescence in the red region (approximately 6600–7000 Å). It will be recalled that the chlorophylls have a typical strong absorption in the red and blue regions. The absorption in the red is more intense than the usual porphyrin absorption in the green-yellow region. The modification of the region of absorption and thereby the region of fluorescence occurs principally because of the addition of two hydrogen atoms to one of the porphyrin pyrrole rings and the addition of the pentacyclic ring between one pyrrole ring and a methine carbon. Chlorophylls *a* and *b* have attracted attention because of their participation in photosynthesis. These are already metallo derivatives (Mg). Quantum yields of fluorescence of both chlorophylls vary with the solvent and are from 0.32 in methanol to 0.35 in pyridine for chlorophyll *a* and from 0.084 in methanol to 0.16 in ethyl ether for chlorophyll *b* [360]. The quantum yields *in vivo* are considerably lower. The best studied case is chlorophyll *a* for which the values *in vivo* range from 0.015 to 0.027 [360]. The lifetimes of fluorescence of chlorophyll *a* in different solvents at room temperature have been found to be 5.1–7.8 $\times$ 10^{-9} sec. For chlorophyll *a* the lifetime *in vivo* is approximately one quarter that *in vitro*. Calculations have been made of quantum yields based on the formula $\Phi = \tau/\tau_0$ and compared to the measured ones [361]. The comparison was favorable.

Both chlorophylls *a* and *b* show a weak phosphorescence in a rigid glass at 77°K at 8850 Å and 8750 Å, respectively [359]. In addition, both these chlorophylls show a different, stronger phosphorescence when dry and in a dry hydrocarbon solvent at low temperatures [131]—7750 and 7330 Å for chlorophylls *a* and *b*, respectively.

The copper derivatives of pheophytin *b*, pheophorbide *a*, and chlorophyllide *a* are not fluorescent [131,146] while the cadmium derivative of the chlorophyllide *a* is fluorescent [134]. The zinc chlorophyllide derivative also shows fluorescence [134]. Ethyl chlorophyllide shows a weak phosphorescence. The copper ethyl chlorophyllide *a* and copper pheophytin *b* both show intense phosphorescence at 8635 Å and 8740 Å, respectively [359]. In these latter cases it should be noted that the emissions are considerably stronger than in the chlorophylls (which are magnesium derivatives). This intensity difference is expected because of the strong spin-orbit coupling caused by the paramagnetic copper ion. Here again, it appears as if transition metal derivatives show no fluorescence. Transition metal derivatives do show phosphorescence, which is strongest in the copper case [131,134,346].

The phthalocyanines are closely structurally related to the porphyrins (Figure 19). Earlier work on these compounds was limited to the magnesium derivative [357]. The parent fluoresces in the deep red (6920 Å) as does the magnesium derivative (6705 Å) [346]. The zinc derivative also has a red fluorescence (6730 Å) [346,347], and also the tin derivative fluoresces [134]. However, the nickel(II), cobalt(II), and copper(II) derivatives show no emission to approximately 10,000 Å [134]. Here again it is interesting that the complexes of the transition metal ions show no fluorescence. Phosphorescence has not been found for any of the metallo derivatives or for the parent itself. The decay of the triplet of the parent has been measured using a laser technique in which the recovery of the singlet absorption is measured as decay occurs from the triplet back to the ground state [362]. There is a marked variation of the phosphorescence lifetime with temperature. The lifetime range is from 10^{-6} sec at room temperature to 150×10^{6} sec at 77° K. The fact that no phosphorescence exists and that there is a strong dependence of the lifetime upon temperature suggest that a highly efficient internal conversion pathway exists between the triplet and ground state.

Rare-earth chelates of phthalocyanines show only molecular fluorescence [363] with varying intensity. No molecular phosphorescence or rare-earth line emission occur. More concerning this problem will be given when rare-earth chelates are considered in greater detail.

Several of the metal derivatives of 8-hydroxyquinoline show emission [364,365]. The discussion of the rare-earth complexes will be postponed until later. The aluminum(III), gallium(III), iridium(III), zinc(II), cadmium(II), magnesium(II), lead(II), and thallium(III) complexes fluoresce at room temperature in solution. However, the divalent paramagnetic copper(II), nickel(II), cobalt(II), and manganese(II) complexes do not fluoresce at room temperature in solution or in the solid state. At a low temperature (77°K) in a rigid glass the first group of complexes mentioned also shows fluorescence. The cobalt, nickel, and manganese complexes show a barely detectable luminescence under the same conditions. The copper complex shows a definite luminescence at low temperature. It is probable that this latter luminescence is a phosphorescence. The absence of emission at room temperature from the transition metal complexes would be understandable if the emission were phosphorescence; that is, phosphorescence would be quenched at room temperature because of collisional deactivation. The appearance of emission as phosphorescence in a rigid glass would be similarly understandable because of the absence of quenching. Two other transition metal complexes of 8-hydroxyquinoline, the iron(III) and chromium(III) derivatives do not show fluorescence. Impurities or photolytic decomposition products could account for the barely detectable emission observed from the cobalt, nickel, and manganese complexes.

Metallo complexes of quite different ligands show some similarity as well as some differences compared with those of the porphyrin and chlorophyll type. Oxine and variously substituted oxines show only fluorescence of varying quantum yield at 80°K [366]. In closed-shell metal oxinates, the ligand localized fluorescence quantum yield is not significantly affected by an increase in the atomic number of the metal (up to $Z = 49$) (Table 19). However, a marked effect does occur with metals of high atomic number.

Table 19 Emission Characteristics of Oxinates [364]

	Fluorescence		Phosphorescence		
Compound	λ (max)	Quantum Yield	λ (max)	Quantum Yield	Lifetime
Oxine	4290	0.28	Absent[a]		
Oxine anion	4855	0.28	Absent[a]		
Oxine cation	4810	0.043	Absent[a]		
$Zn(Ox)_2$[b]	5075	0.32	Absent[a]		
$Cu(Ox)_2$	4950	0.012	Absent[a]		
$Cd(Ox)_2$	4760	0.21	Absent[a]		
$Pb(Ox)_2$	5025	0.013	6170	0.005	
$Ni(Ox)_2$	Absent[a]		Absent[a]		
$Al(Ox)_3$	4880	0.30	Absent[a]		
$Fe(Ox)_3$	Absent[a]		Absent[a]		
$Co(Ox)_3$	4830	0.01	Absent[a]		
$Bi(MeOx)_3$[b]	4695	0.04	6665	$\leq$0.001	
$Zn(diBrOx)_2$[b]	5100	0.25	Absent[a]		
$Zn(diIOx)_2$[b]	5130	0.025	Absent[a]		
$Zn(ThioOx)_2$[b]	5130	0.0072	Absent[a]		
$Zn(seIOx)_2$	4950	0.003	6710	0.002	95 μsec
$Co(seIOx)_3$	4695	0.0011	Absent[a]		
$Cd(seIOx)_2$	Absent[a]		6100	0.01	135 μsec
$Cu(seIOx)_2$	Absent[a]		Absent[a]		

[a] Not greater than 0.0003.
[b] Ox: oxinate; diBrOx: 5,7 dibromooxinate; diIOx: 5,7 diiodooxinate; ThioOx: thiooxinate; selOx: selenooxinate; MeOx: 4-methyloxinate.

Phosphorescence is observed for the lead(II) and bismuth(III) complexes; the total quantum yield, however, undergoes a substantial reduction. The introduction of heavy atoms on the ligand portion of the zinc(II) oxinate causes a decrease in the intensity of fluorescence that is particularly marked for the iodo derivative (Table 19). Changing the oxinate to progressively heavier atom substitution for oxygen, thioxinate and selenooxinate, also causes a marked decrease in the intensity of fluorescence compared with the oxinates (Table 19). Except for the selenooxinate, however, there appears to be no increase in the intensity of phosphorescence. Thus the probability of internal conversion is significantly increased. Transition metal oxinates show a markedly reduced quantum yield of fluorescence and apparently no phosphorescence (see Table 19). If the metallo complex is a selenooxinate, some noticeable differences appear. Although the total quantum yield is generally smaller than for the oxinates, phosphorescence is observed for some nontransition closed-shell metals and at least for nickel(II). Both the total and phosphorescence quantum yields increase between the zinc(II) and cadmium(II) derivatives (fluorescence appears to be absent for the cadmium derivative and very weak for the zinc chelate).

Another very important class of metal complexes are those derived from rare-earth ions. In such complexes there is generally a rather high probability of *intra*molecular energy transfer between the ligand and the ion [367–370]. When rare-earth complexes are excited, a variety of spectral emissions can occur. The emission can be essentially line emission from the rare-earth ion, molecular band fluorescence and phosphorescence, or combinations of both band and line emission. In the latter case, one or the other of emissions can predominate. The principal emission depends on both the trivalent rare-earth ion and the type of ligand in the complex. The principal ligands that have been studied include benzoylacetone, dibenzoylmethane, tribenzoylmethane, 8-hydroquinoline and methyl derivatives, *o*-hydroxybenzophenone, and acetylacetone. In certain cases the instability of the complexes presents problems since spurious emission may occur from the chelating molecule and its anion. Moreover, the presence of intermediates including mono- or di-chelated species complicate the results [370,371]. Absorption spectra of nearly all the rare-earth ions are known for the gaseous, crystal, and solution states. The line-like fluorescence originates from transitions within the $4f$ levels. The anhydrous chlorides of all the rare-earth ions show a line fluorescence by direct excitation of the ions [372]. If, however, the hydrated crystals are excited, only five of the rare-earth ions emit. When coordinated with water, only gadolimium(III), terbium(III), and europium(III) emit strongly, samarium(III) and dysprosium(III) weakly, and others show only very weak or no fluorescence [373]. If D_2O is used instead of H_2O, however, dramatic enhancement in the fluorescence intensity of certain ions occurs [374–378]. This also is true for other deuterated solvents such as methanol. For example, the emission of europium(III), as $EuCl_3$ is enhanced 19 times in D_2O versus H_2O. If the anion is changed, the degree of enhancement also changes but the intensity is always greater in the deuterated solvent—either water or methanol [376]. It appears that the enhancement is not the same for emission from the different ion states [377]. Further, the lifetime of fluorescence of europium ion (unchelated) varies over a certain temperature range. This temperature-dependent quenching can be accounted for on the basis of either interaction of the ion with hydrogen vibrations of the solvents or thermal population of the next higher ion level (the $^5D_1$) [378].

The possible number of transitions within the rare-earth ion depends on the electron configuration. Thus it is possible to separate the ions into broad classes according to the number of transitions possible. In the first group, I, are the lanthanum(III) and lutetium(III) ions in which the f orbitals are empty and filled, respectively, and no transitions can originate from states resulting from f-orbital configurations. The gadolinium(III) ion with an f^7 configuration is also included in this group since no transitions in the visible or near infrared regions exist. In the second group, II, are the cerium(III) and ytterbium(III) ions, where because of a single f electron or thirteen f electrons, only one transition can occur. Finally, in the third group, III, are all of the other +3 valence ions in which multiple transitions can occur. In the latter group it would seem that many state origins would exist and therefore, many emissions. However, it is known that this is not true and emissions originate from only a few states designated as *resonance* levels. This circumstance provides a fortunate situation for interpreting the final results in the chelated complexes.

The location of the lowest triplet level of many rare-earth ion complexes is known [370,379,380]. The average location depends significantly upon the nature of the organic chelating agent. Table 20 shows the average location of the lowest triplet state for several different rare-earth chelates agents. A given rare-earth complex with various ligands may show only band emission, line emission, or a combination of both. In certain cases, such as lanthanium, lutetium, and gadolinium, only molecular band emission results no matter what chelating agent is involved [370,379,380]. The reasons for such varying results will be discussed shortly.

The chelating anions of particular interest include benzoylacetonate, dibenzoylmethide, tribenzoylmethide, 8-hydroxyquinolate, and some others. The effect of rare-earth ions on enhancing phosphorescence of certain organic chelating agents was first considered some time ago [381]. The lifetimes of phosphorescence of various rare-earth complexes of dibenzoylmethane show no simple relation with atomic number or size of the metal ion. Table 21 shows the lifetimes for some complexes of dibenzoylmethane. The

Table 20 Location of Triplet States for Some Rare-Earth Complexes [379]

Metallo Chelate[a]	Average Location of Triplet-State Origin[b] (cm^{-1})	Å
Benzoylacetonate (La, Sm, Gd, Dy, Tm, Yb, Lu)	21,480	(4655)
Dibenzoylmethide (La, Sm, Eu, Gd, Dy, Tm, Yb, Lu)	20,520	(4870)
Tribenzoylmethide (Gd, Dy)	20,765	(4815)
Anthranilate[c] (La, Sm, Eu, Gd, Tb, Dy, Tm, Yb, Lu)	22,135[c]	(4520)
8-Hydroxyquinolate (La, Gd, Lu)	17,760	(5630)
2-Methyl-8-hydroxyquinolate (La, Gd, Lu)	18,000	(5555)

[a] All metal ions are in the +3 oxidation state.
[b] Maximum deviation throughout, 500 cm^{-1} or less.
[c] This corresponds to the maximum of phosphorescence and not to the origin.

Table 21 Lifetimes of Phosphorescence for Some Dibenzoylmethide Complexes [381]

Metal	Atomic Numbers of Metal	τ_P (sec in alc)	Origin of Phosphorescence (cm^{-1})	(A)
Al	13	0.70	20,900	(4785)
Sc	21	0.39	20,600	(4855)
Y	39	0.24	20,350	(4915)
Lu	71	0.10	20,350	(4915)
La	57	0.06	20,250	(4940)
Gd	64	0.002	20,350	(4915)

effectiveness of the atomic number in promoting spin-orbit coupling depends to a significant extent upon the closeness of approach of the optical electrons to the nucleus. It will be recalled that spin-orbit coupling depends strongly on both the atomic number and on the distance of the electrons from the nucleus. The distance of approach will depend upon the electron configuration to some extent, as is apparently the case for the lutetium and lanthanum complexes where despite the higher atomic number of lutetium, the lanthanum complex has a shorter lifetime (see Table 21). The gadolinum complex exhibits no fluorescence and its phosphorescence has a short lifetime. In this case, the results are consistent with those to be expected because of the paramagnetic nature of the ion; that is, spin-orbit coupling must be effectively complete because of the strong inhomogeneous magnetic field provided by the ion. Further, there is little variation in the location of the triplet level of certain of the rare-earth complexes (Tables 20 and 21). The average position of the triplet level of complexes with dibenzoylmethane is 20,520 cm^{-1} (4880 Å) with a maximum deviation of 250 cm^{-1} (Table 21) [370,379,381].

If salts of dibenzoylmethane are compared within a series containing metals of similar electronic configuration (same periodic group), then the phosphorescence lifetimes decrease with increasing atomic number [381]. This is interesting in comparison to the porphyrin complexes where, for example, the calcium(II), strontium(II), and barium(II) complexes do not follow such a trend. In part, this is due to the considerably different geometric characteristics of the complexes. This comparison, however, brings up a valuable point; that is, generalizing on the consequences of certain considerations such as the effect of atomic number on spin-orbit coupling must be qualified, even within a series containing metal ions of a parallel electronic configuration.

The general spectral characteristics of the chelates may now be considered. The chelates of the first group, I, of rare-earth ions show only molecular band emission. Figure 34 will be helpful in considering the results. In Group I

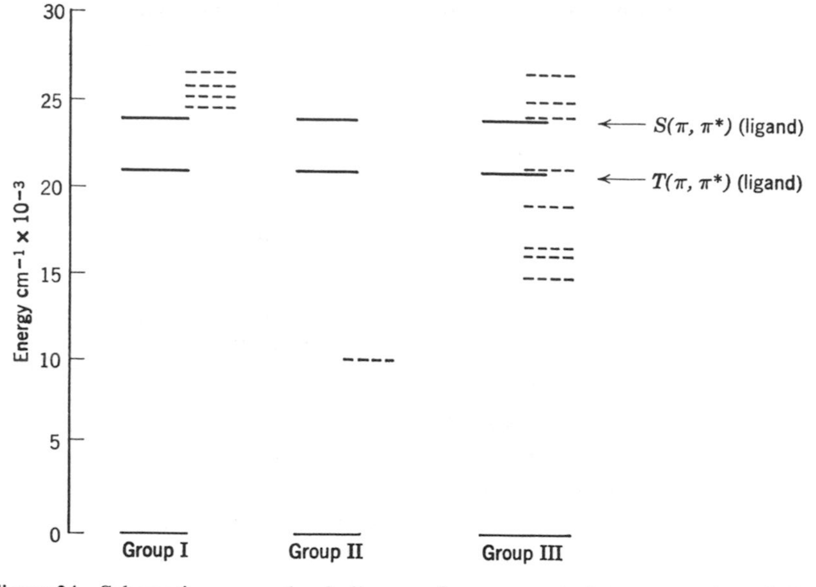

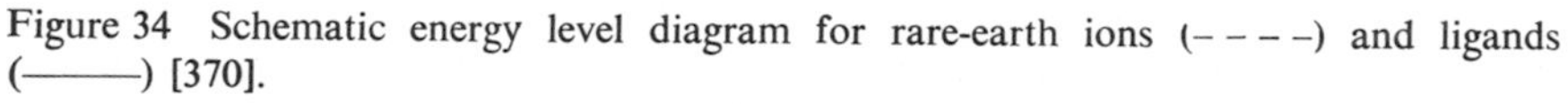
Figure 34 Schematic energy level diagram for rare-earth ions (– – – –) and ligands (———) [370].

the rare-earth ion levels are above the possible lowest singlet and triplet levels of the ligands. No linelike emission results but only molecular band emission is seen. In these cases, the principal effect of the metal substitution is to perturb the ligand. This perturbation is manifested in several ways including effects on the intensity and position of absorption and emission, on the nature of the emission (whether fluorescence or phosphorescence), and on the phosphorescence decay time. In addition, these factors will vary depending on the ligand and magnetic characteristics of the ion. In the complexes with dibenzoylmethane, all three ions of Group I show phosphorescence, but only in the case of gadolinium is this the sole type of emission [367]. The relative intensity of fluorescence compared to that of phosphorescence depends on the solvent. The lifetime of the triplet emission is shortest for gadolinium and longest for lutetium. Complexes with the benzoylacetonate, tribenzoylmethide, anthranilate, and the 8-hydroxyquinolate ions also show both molecular fluorescence and phosphorescence [379].

Of the chelates in the Group II, cerium(III) and ytterbium(III), the ytterbium complexes have been studied in the most detail. Complexes of ytterbium with ligands as those given above result in both line and molecular band emission [379].

The emission of the rare earths that comprise the Group III depends on the ligand [367,372,379]; for example, dysprosium complexes with benzoylacetone and tribenzoylmethane give line emission (fluorescence). However, complexes of the same metal ion with dibenzoylmethane and 8-hydroxyquinoline give molecular band emission only, while complexes with anthranilate ion give a combination of band and lines. Other rare-earth ions with these ligands give line, band, or a combination, depending upon the ligand [379]. All emission data referred to above for Groups II and III are in rigid glasses at low temperature (77°K). Europium salicylaldehyde and *m*-nitrobenzoylacetonate complexes show solely line emission and the intensity of the fluorescence in benzene increases as the temperature decreases to a greater extent than does that of europium nitrate in water over the same range. In addition, the intensity of fluorescence of the solid increases vastly (150-fold) upon temperature reduction to 90°K, whereas that of solid europium chloride hexahydrate changes very little (two-fold). The fluorescence intensity of the nitrobenzoylacetonate is considerably greater than that of the aldehyde complex. The effect of solvent upon the intensity is quite pronounced since the intensity is considerably weaker in alcohol solutions than in benzene; however, the intensity in alcohol can be increased by cooling. In the case of the acetonate complex at low temperatures (90°K), the quantum yield of fluorescence of europium *m*-nitrobenzoylacetonate is high (approximately 0.85). Although the quantum efficiency varies greatly with temperature, the lifetime does not. This indicates that the principal quenching occurs before excitation of the ion. Finally, the efficiency of intramolecular energy transfer from the ligand to the ion is decreased the more ionic is the complex.

The results given earlier for Groups II and III can be explained in terms of transfer of energy from the lowest triplet state of the ligand to rare-earth ion levels and ultimately to the resonance level [370,379]. On the basis of this approach for the Group I complexes, the ion states could not be occupied since they lie higher than the lowest triplet of the ligand. Further, in the case of Yb^{+3} in Group II, the resonance level lies below any of the ligand triplets; therefore, it can be occupied with a resultant line emission (bands appear also). It would appear that because of the large energy gaps occurring in this case, the intercombination between the states is sufficiently slow that the rate constant for emission from the triplet state can be competitive with that one responsible for occupation of the ion state. In the third group, Figure 34 shows that the ion resonance level may be below, equal to, or above the possible triplet states. Thus either lines, molecular bands, or combinations could exist [379]. Again, it would appear that the presence of lines only instead of a combination of lines and molecular bands would depend on the relative magnitude of the appropriate rate constants as noted earlier. Moreover, since the presence of an ion resonance level below the various ligand triplets can result in either lines or combinations of lines and bands for a given rare-earth ion, some specific properties of the ligands must be important. It would appear that variation in the state mixings is a dominant feature. The importance of state mixing is further emphasized by the fact that with two different rare-earth ions whose resonance levels are below those of the triplet of a given ligand, only lines will result in one case whereas a combination of bands and lines will result in the other.

Other possibilities for energy transfer may be as likely [382–384]. Based on calculations of lifetime data, it appears as if the triplet level from which energy transfer occurs (to an ion level) is not the one from which phosphorescence occurs (the lowest triplet) [382]. The transfer may be from a higher vibrational level of the lowest triplet or from a higher triplet [382]. There is some incongruity in these conclusions as originally expressed since energy transfer was assumed to occur from the lowest triplet. Another mechanism could be involved particularly with reference to those cases in which the lowest triplet level is below the ion resonance level (the emitting level). Energy transfer could occur from a higher ligand triplet state to a higher ion level followed by internal conversion to the ion resonance level which then would be quenched by the lower chelate triplet (no resonance fluorescence) [383]. Also energy transfer could occur from the singlet state of the complex to the rare-earth ion and be quenched by the lower chelate triplet state [384]. It has been possible to directly excite an ion resonance level and show that if there is a lower lying triplet state, ion resonance fluorescence is quenched; however, if there is no lower triplet state, the resonance fluorescence is observed [383,384].

The effect of quenchers on certain europium and erbium chelates offers some further information. Most of the quenching in the europium chelates appears to occur only when the energy matching criterion between the triplet of the chelate and that of the quencher is proper, regardless of the relative location of the europium levels [385]. Chelated ion emission decreases with increasing quencher concentration to a value of approximately 10 percent of the initial value. If it is assumed that the unquenched ion emission does not come from another species whose triplet is lower than that of the quencher, then energy transfer occurs to ion levels higher than the lowest triplet of the ligand and the ion levels are not quenched by the ligand triplet. Such results indicate that mechanisms involving energy transfer from a triplet state of the ligand to rare-earth ion levels rather than from the ligand singlet to rare-earth ion levels are more probable [385].

In most of the considerations given heretofore, it is assumed that the phosphorescence that arises along with the sensitized rare-earth ion emission is from the same species that does give rise to the ion emission. It can be shown that for certain gadolinium and samarium complexes, spurious phosphorescences result. These most likely occur from ligand negative ions [386]. Thus it is important that such a possibility be taken into account in analysis of all data that will eventually lead to a mechanistic proposal pertinent to energy transfer.

Four possible paths for intramolecular energy transfer are

$$S_1(\text{chelate}) \rightsquigarrow T_1(\text{chelate}) \rightsquigarrow \text{rare-earth ion level} \longrightarrow \text{emission [371,379]} \quad (107)$$

$$S_1(\text{chelate}) \rightsquigarrow \text{rare-earth ion level } (>T_1) \rightsquigarrow T_1 \rightsquigarrow \text{rare-earth ion level } (<T_1) \longrightarrow \text{emission [385]} \quad (108)$$

$$S_1(\text{chelate}) \rightsquigarrow T(\text{chelate}, >T_1) \rightsquigarrow \text{rare-earth ion level } (>\text{resonance level}) \rightsquigarrow \text{rare earth ion level (resonance)} \longrightarrow \text{emission [383]} \quad (109)$$

$$S_1(\text{chelate}) \rightsquigarrow \text{rare-earth ion level} \longrightarrow \text{emission [384]} \quad (110)$$

The paths involving the transfer from the triplet, primarily (107) and (108), seem to be most favored. However, the others cannot be ruled out, particularly for some complexes in which the ligand levels could require it. Terbium, and europium chelates show three principal sources of loss of electronic energy [387]. These include losses in the singlet or triplet ligand levels, the upper rare-earth ion levels, and the rare-earth ion emitting level. For example, in the case of several different europium chelates the efficiency of emission from the $^5D_0$ ion level is appreciably higher if the $^5D_0$ level is directly excited compared with either excitation into the ligand or higher ion levels [387]. The losses from the upper ion levels appear to occur via direct nonradiative transitions to the ground state.

In certain terbium chelates, ion fluorescence is strong at low temperatures. This is particularly noticeable in those cases in which the ligand triplet states lie energetically close to the $^5D_4$ ion emitting level. However, if there is a large energy separation such as for the acetylacetonate in which the triplet is appreciably higher than the $^5D_4$ level, the fluorescence lifetime is constant throughout the temperature range. These and similar considerations for some europium chelates provide good evidence for the fact that thermal quenching of the emitting ion level occurs via excitation of the lowest triplet of the ligand. This may occur even if the triplet is appreciably higher in energy than the emitting ion level (1500–2500 cm^{-1}) [387].

The foregoing results provide support for one of the mechanisms of quenching of ion emission discussed earlier [383], at least for some cases. The quenching mechanism referred to [383], however, suggested that the lowest triplet should be below the ion resonance level. The discussion above indicates that even this requirement is unnecessary, again, at least in certain cases. Actually, the quenching of the rare-earth ion emission by a ligand triplet also could occur in mechanisms (107) and (108) as well as (109) and (110) although such a specific suggestion was not originally given [371,379,385] relative to the former two mechanisms. Thus it appears as if the simple, first-postulated mechanism [370,379] for energy transfer (107), may be too restrictive. Further, it is likely that variations of state mixings, as noted earlier, are indeed important.

Mention was made earlier of rare-earth complexes with phthalocyanine [363]. The europium, gadolinium, and ytterbium complexes show only ligand band fluorescence at approximately 6800 Å with quantum yields of 15, 4, and 1 percent, respectively. In the case of the europium and gadolinium complexes, the ion resonance level is above the ligand singlet; however, for ytterbium the ion level is below the ligand singlet. In spite of this, it appears as if the ion level must be quenched and by intramolecular energy transfer (probably to the ligand triplet).

Rare-earth ion emission can occur not only from stable chelates but also from solution mixtures of rare-earth ions and aldehydes and ketones [388–390]. In one case [388], two different rare-earth salts dissolved in such pure solvents such as acetophenone, benzophenone, and acetone give absorption spectra essentially the same as those of the pure solvent. More concentrated solutions show absorptions typical of the ions. Excitation by wavelengths principally absorbed by the solvent results in characteristic line fluorescence of the rare-earth ion. It is postulated that a weakly bonded ketone rare-earth ion complex is formed in which, upon excitation of the ketone, energy transfer occurs to the ion. Also, intermolecular energy transfer [389] can result in ion emission. When the $\pi^* \leftarrow n$ transition of benzophenone is excited in a mixture of benzophenone and europium *tris* hexafluoracetylacetonate, characteristic europium lines (red) are observed [389]. This would indicate an intermolecular triplet-triplet energy transfer between the sensitizer and the complex followed by intramolecular energy transfer to the rare-earth ion states (providing no complex with the sensitizer exists). The intensity of the presumed sensitized emission is affected by temperature, concentration, and the presence of oxygen. Also, solutions of 21 aromatic aldehydes and ketones have been shown to be able to transfer energy to terbium and europium ions [390]. The efficiency is significantly greater at room temperature than at 77°K. Also, the ratio of ion emission to the carbonyl emission decreases with decreasing concentration of the rare-earth ion [390]. Addition of hydrocarbons with triplets lower than the sensitizer can quench the ion emission. It appears that the carbonyl-rare earth ion energy transfer process is diffusion controlled and the lowest triplet of the carbonyl is the state from which energy is transferred [309].

In general, the efficiency of emission from chelated rare-earth ions is greater than that from unchelated ones, although it can still be low. This appears to be due to the fact that the radiative rate constant is increased by the ligand in contrast to any appreciable protection against quenching. Nevertheless, it has been found that substitution of bulky groups such as trifluormethyl and pentafluormethyl on the chelate itself, increase the quantum efficiency of the ion emission [391,392]. Further, "synergic agents" [391,392] such as trioctylphosphine oxide can coordinate directly to the rare-earth and provide what appears to be an insulating sheath—that is, insulation from coordination with the solvent, such as water, which acts as a quencher. The quantum efficiency can be increased considerably using such synergic agents and a hydrocarbon solvent [391,392].

Little work has been published on the sandwich or π complexes, the best known of which are those between transition metals and the ligands cyclopentadiene or benzene. Ferrocene apparently shows a weak phosphorescence [393,394,134,395]. This was first observed at 77°K [393] and later confirmed at 20°K in a variety of rigid solvents including methane, nitrogen, and argon [394]. The lifetime of the phosphorescence is approximately 1 sec. In both cases, [393] and [394], emission apparently occurs only upon excitation into the second excited singlet state (maximum at 3240 Å). Energy transfer experiments utilizing ferrocene are in harmony with the latter observation; that is, no ferrocene-photosensitized *cis-trans* isomerization of piperylene occurs when excitation is into the first excited state of ferrocene (maximum 4400 Å) but it does occur upon excitation into the second excited state [395]. The intersystem crossing quantum yield for ferrocene when it is used as a low-energy sensitizer is approximately 0.003 [396].

Ruthenocene and osmocene do not appear to emit [396]. Nickelocene shows no emission at wavelengths shorter than 10,000 Å [393]. Dibenzene chromium likewise shows no emission up to 10,000 Å either in fluid solution at room temperature or in a rigid medium at low temperature, although it possesses at least one very intense absorption band in the region of excitation (nor is there any noticeable photodecomposition) [397]. Titanocene dichloride and dibromide, however, show a phosphorescence emission in the red at approximately 6200 Å [396]. Thus, transition metals, when complexed with cyclopentadiene or benzene (only), appear not to fluoresce and show only a very weak or no phosphorescence. Most of the excitation energy must be internally converted and in the singlet manifold, at least for ferrocene based on the energy transfer experiments.

B. CHARGE TRANSFER COMPLEXES

In this section we shall be concerned with those complexes that involve hydrocarbons as electron donors and a variety of other molecules as acceptors (e.g., anthracene-trinitrobenzene).

The earliest studies of the emission spectra of complexes involved trinitrobenzene and various aromatic hydrocarbons at low temperature (77°K).

Because the emission of the complex resembled that of the free donor hydrocarbon, it was assumed that the emission was actually only that of the phosphorescence of the donor [398,399]. However, if the trinitrobenzene (TNB)-anthracene and TNB-phenanthrene complexes are excited at room temperature in fluid solution, emission is also observed [400]. The emission spectrum is a mirror image of the absorption spectrum. Many other examples of such mirror-image relationships also exist for various other hydrocarbons with a variety of donors [401,402]. Moreover, the lifetime of the emission of many complexes at both room and low temperature is between 10^{-8} and 10^{-9} sec (see Table 22). The information above provides very strong evidence for the conclusion that the emission is not phosphorescence but is, in fact, the charge transfer fluorescence. The final piece of evidence that confirms this is as follows. If the emission were a phosphorescence from the free hydrocarbon, its energy should not change to any significant degree in different complexes involving the same donor hydrocarbon and various acceptors. If the emission is a charge transfer fluorescence, however, the energy of the emission should depend on the location of the charge transfer absorption. Thus as the electron affinity of the acceptor changes, the absorption band should move to longer wavelengths with increased electron affinity of the acceptor as will the emission. Data on six different hydrocarbon donors and six different acceptors are in complete accord with the conclusion that the emission represents the charge transfer fluorescence.

In many instances, the charge transfer fluorescence is overlapped by phosphorescence [402,404], as in the case of aromatic hydrocarbon-TNB complexes. The emission band system is found to consist of two parts, a strong charge transfer fluorescence and a weak emission nearly identical to the phosphorescence of the free uncomplexed hydrocarbon. Since the energy of the charge transfer fluorescence depends in part on the electron affinity of the acceptor, a change in acceptor presumably could move the fluorescence

Table 22 Lifetimes of Emission of Some Donor-Acceptor Complexes at Room and Low Temperature[a] (nsec)

Acceptor / Donor	Trinitrobenzene 20°C	Trinitrobenzene −190°C	Tetrachlorophthalicanhydride 20°C	Tetrachlorophthalicanhydride −190°C
Hexamethylbenzene	2.6		7.7	11.0
Durene		3.6	8.1	6.8
Anthracene	5.0		4.3	
Naphthalene			2.6	2.3
Phenanthrene	3.6		3.0	
Benz(a)anthracene			7.3	

[a] J. Czekalla, A. Schmillin, and K. Mager, *Z. Electrochem.*, **61**, 1053 (1957); **63**, 623 (1959).

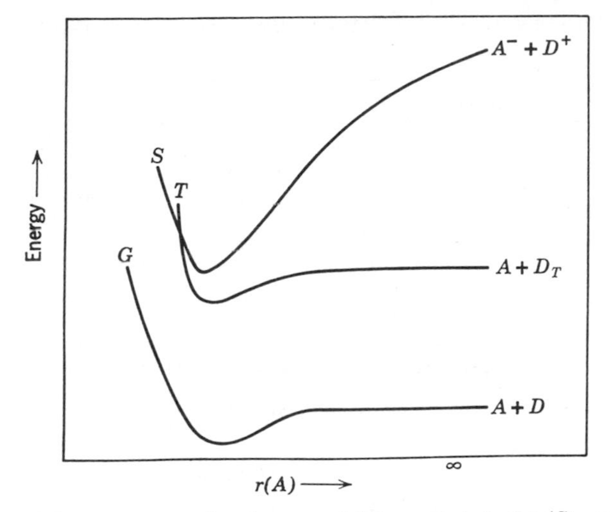

Figure 35 Potential energy curves for the ground (G), excited singlet (S), and triplet (T) states for a donor (D) acceptor (A) complex.

relative to the phosphorescence so that they would be separated. This has been observed in the series of complexes in which hydrocarbon donors are complexed with tetrachlorophthalicanhydride [402]. Careful investigation of the location and fine structure of the phosphorescence spectrum shows that if differs from that of the free uncomplexed hydrocarbon. In the crystalline complex, the phosphorescence spectrum is red shifted and the vibrational frequencies lowered compared to that of the free hydrocarbon [402,404]. The frequency lowering is more pronounced in the crystal than in solution [402]. Further, the lifetime of the phosphorescence of the complex is shorter than that of the free hydrocarbon, especially so in the case of the crystal (see Table 23). These considerations indicate that the observed phosphorescence emission may not be that of the free uncomplexed hydrocarbon as first thought [403] and that complete dissociation does not occur. If dissociation did occur, then the simplified energy scheme given in Figure 35 would be appropriate. Intersystem crossing would occur near the crossing of the excited singlet (S) and triplet (T) states which is dissociative for the triplet state. Thus the complex might be expected largely to dissociate and emission to occur from the uncomplexed hydrocarbon (see region of $r = \infty$ in Figure 35). One possible explanation for the fact that the emission does not occur from the uncomplexed hydrocarbon is that the surplus energy of any dissociative triplet is dissipated and the complex reforms in its triplet state and emission occurs from the reformed complex triplet [403]. Another possibility is that the emission process, phosphorescence, is largely localized on the hydrocarbon component. It would appear that with better knowledge of other possibilities

Table 23 Phosphorescence Lifetimes of Some Hydrocarbons and Their Complexes[a]

	Donor[a]	Complex			
		Tetrachloro-phthalicanhydride		Tetrabromo-phthalicanhydride	
		Solution[a]	Crystal	Solution[a]	Crystal
Durene	5.9 sec	0.25 sec	0.002 sec	0.0095 sec	0.0012 sec
Hexamethylbenzene	5.6	0.0085	0.0011	0.0047	0.001
Naphthalene	2.4	0.16	0.45	0.29	0.029
Phenanthrene	3.6	1.8	0.25	0.36	0.019

[a] In *n*-propyl ether-isopentane (3:1) at −190°C; crystal at same temperature.

for the shape of the appropriate potential energy curves and their crossing points, different mechanisms could evolve.

The current thought is that the so-called "donor" phosphorescence is indeed that of the donor [405–408] but that a donor-acceptor complex of well-defined structure exists in the excited state [407]. The cause of the change in lifetime will be considered shortly.

Polarization of the charge transfer bands of complexes of aromatic hydrocarbons with tetrachlorophthalicanyhydride is along the intermolecular axis and perpendicular to the planes of the rings [407]; that is, the polarization of the 0–0 band of the hydrocarbon emission is parallel to that of the exciting light (into the charge transfer band) and since naphthalene phosphorescence is out-of-plane polarized, the charge transfer band is also polarized perpendicular to the molecular plane(s). However, polarization of bands other than the 0–0 is not the same as for free naphthalene indicating a coupling of the triplet of the complex with other triplets. There is a general decrease in the lifetime of the triplet state of naphthalene in naphthalene-phthalicanhydride complexes when the bromoanhydride is the acceptor compared with the chloroanhydride. Complexes of naphthalene and acenapthene with trinitrobenzene, and acenapthene with halosubstituted phthalic anhydrides, show an increase in the quantum yield ratio Φ_P/Φ_F [405] compared to the uncomplexed hydrocarbon. This is postulated to occur because of an increase in the gross probability for the intersystem crossing between the excited singlet and triplet states of the donor hydrocarbon and of the radiative (phosphorescence) process of the donor, the former of the two being considered the more important [405]. Further, the gross increase in intersystem crossing is postulated to arise principally because of a different, more probable intersystem crossing process involving an intermediate charge transfer state transfer state ($S_C \rightarrow$ charge transfer state $\rightarrow T_D$) where the subscripts D and C signify donor and complex, respectively.

The ratio of Φ_P/Φ_F for naphthalene-tetrahalophthalicanhydride complexes increases as the atomic number of the halogen increases [405]. This indicates the presence of a heavy atom enhancement effect in addition to the enhancement effect resulting from the charge transfer complexation. Complexes of perdeutero aromatic hydrocarbons show longer hydrocarbon phosphorescence lifetimes than those of perprotonated hydrocarbons [409]. From the actual lifetime data it can be concluded that the heavy atom in a charge transfer complex (hydrocarbon-tetrahalophthalicanhydride) enhances the $S_0 \leftarrow T_1$ radiative probability and has little affect on the $S_0 \leftsquigarrow T_1$ radiationless probability of the complex. If this is true, it is different from the effect the heavy atom, deuterium, has on the $S_0 \leftsquigarrow T_1$ process in the uncomplexed hydrocarbon (see Chapter 9-B).

The phosphorescences of naphthalene in complexes with propyl iodide and tetrabromoanhydride show very similar vibrational features, although with broadening compared to that of uncomplexed naphthalene [408]. This can be interpreted as indicating that the primary result of the complexing is to increase the mixing (of the singlet and triplet states) already present in the hydrocarbon without the introduction of additional mixing schemes. However, if the hydrocarbon donor is 1- or 2-chloronaphthalene, the 0–0 band and other out-of-plane symmetric bands are enhanced relative to the in-plane bands [408]. Chloronaphthalene has subspectra of (a) the naphthalene type and, (b) the type that arises from vibronic and spin-orbit interaction. If a charge transfer state is mixed with the donor triplet, the out-of-plane bands could be enhanced. We recall that the charge transfer band is polarized parallel to the 0–0 band of the hydrocarbon (that is, perpendicular to the plane or out-of-plane). This then would account for the change in band intensities noted above. This interpretative approach assumes a sandwich or parallel stacked geometry for the donor and acceptor. Polarization data [408] is also in accord with such a mixing scheme involving a charge transfer state when a halonaphthalene is the donor.

In conclusion, the phosphorescence of the hydrocarbon donor in a donor-acceptor complex appears to be essentially that of the uncomplexed hydrocarbon donor. Even in the excited state, the complex maintains a well-defined structure. The charge transfer absorption does not obtain the major source of its intensity from donor or acceptor bands or contact charge transfer interactions. It appears that when unsubstituted aromatic hydrocarbons are complexed with tetrahalophthalic anyhydrides, the inherent singlet-triplet mixing of the hydrocarbon is increased and no new couplings occur. Both the probability for intersystem crossing and radiative $S_0 \leftarrow T_1$ phosphorescence may be increased. Lifetime data require that only the $S_0 \leftarrow T_1$ radiative probability be increased. Additional couplings involving a charge transfer state appear when halonaphthalenes (donors) are complexed with halophthalic anhydrides.

Chapter 14 Singlet-Triplet Transitions and Enhancement by External Perturbation

No attempt will be made to cover all cases involving singlet-triplet absorption but, rather, representative situations will be presented with a view to elucidate the process and its implications. This will also apply to examples of external perturbations.

A. SINGLET-TO-TRIPLET TRANSITIONS

If the phosphorescence of organic compounds represents the radiation emission accompanying the transition of a molecule from an excited triplet state to the ground-state singlet, it follows that the corresponding absorption process involving a transition from the normal singlet to an excited triplet state should also be observable, although it will usually be very weak. The low intensity of most singlet-to-triplet ($T \leftarrow S_0$) absorption spectra demands long path lengths of concentrated solutions or of the pure liquid for their observation. Even then, detection of the forbidden transition depends on adequate separation from the long wavelength edge of the stronger singlet-to-singlet ($S_1 \leftarrow S_0$) absorption spectrum. Care must be taken to avoid interpreting the absorption of an impurity as the $T_1 \leftarrow S_0$ absorption. One of the criteria for $T_1 \leftarrow S_0$ absorption is the near correspondence of the 0–0 absorption band with that of the 0–0 phosphorescence band (for the compound in dilute solid solution). Normally, the same type of approximate mirror-image relationship between the $T_1 \leftarrow S_0$ absorption bands and the phosphorescence bands exists as for the normal $S_1 \leftarrow S_0$ absorption spectrum and the fluorescence spectrum.

The intensity of the $T_1 \leftarrow S_0$ absorption is governed by, among other factors, the spin-orbit interaction, and in a series of like molecules the intensity of this absorption will increase with the atomic number of substituting atoms. Figures 36 and 37 show some typical singlet-triplet absorption curves for a halohydrocarbon [410] and a heterocyclic [411]. In the case of halohydrocarbons, in general, the effect of the increased atomic number of the halogen is to increase the intensity (integrated area or f number) of the absorption. This is in accord with what would be expected from increased spin-orbit coupling.

In the case of pyrazine, it is possible that the transition could be either of $\pi^* \leftarrow \pi$ $(^3B_{3u} \leftarrow A_{1g})$ or $\pi^* \leftarrow n$ $(^3B_{2u} \leftarrow A_{1g})$ type (see also Chapter 12). Theoretical calculation predicts that absorption would correspond to an intercombinational $\pi^* \leftarrow n$ type [411]. However, this is not unequivocal proof. A substitutional perturbation is predicted to lower the energy of the π, π^* triplet state and raise the energy of the n, π^* triplet state (thus produce a blue shift of the singlet-to-triplet absorption) [412,413]. A dimethyl pyrazine shows a blue shift of approximately 300 cm^{-1} compared with the parent molecule. If the transition were of $\pi^* \leftarrow n$ type, a blue shift of 100 cm^{-1} would be predicted while a red shift of 1000 cm^{-1} would be predicted for one of $\pi^* \leftarrow \pi$ type. The phosphorescence is also blue shifted by a comparable amount. Thus the transition is assigned as one to the n, π^* triplet state [411]. The absorption and emission spectra show a dominant vibration of about 615 cm^{-1}. This is significant in terms of the nature of the transition. The excitation of the n-orbital electron (sp^2 hydrid) would be expected to result in some rehydridization in the excited state. Thus the principal vibration excited would be one of angular distortion since the carbon-nitrogen-carbon angle is associated with the hydridization of σ orbitals on the nitrogen [411]. This should apply to all $\pi^* \leftarrow n^*$ transitions in diazines ($S_1 \leftarrow S_0$ as well as $T_1 \leftarrow S_0$). Also, chloropyrazine shows both an enhancement and blue shift of the $T_1 \leftarrow S_0$ transitions relative to that for pyrazine.

Calculation predicts that for pyridine, the $T_{n,\pi^*} \leftarrow S_0$ transition will be at higher energy than the $T_{\pi,\pi^*} \leftarrow S_0$ [411]. Perturbation studies with O_2 [414] show that the location of the lowest $T \leftarrow S_0$ transition is in harmony with the prediction and that it should be of the $T_{\pi,\pi^*} \leftarrow S_0$ type.

Biacetyl shows $T_1 \leftarrow S_0$ absorption in various solvents [415]. The band has no structure and occurs near 5000 Å. The intensity depends on the solvent and is strongest in heptane and weakest in pyridine among the solvents studied. Generally, solvents containing oxygen and nitrogen lower the intensity, as is also true for the lowest $S \leftarrow S_0$ absorption. The dichloro- and dibromo-substituted biacetyl apparently show no $T_1 \leftarrow S_0$ absorption. This may arise because of a red shift and intensification of the tail of the closely adjacent $S_1 \leftarrow S_0$ transition thus burying any $T_1 \leftarrow S_0$ transition.

It is also possible to observe $T \leftarrow S_0$ absorption in crystals. For example, singlet-to-triplet transitions exist in crystalline p-dichloro, p-dibromo, and

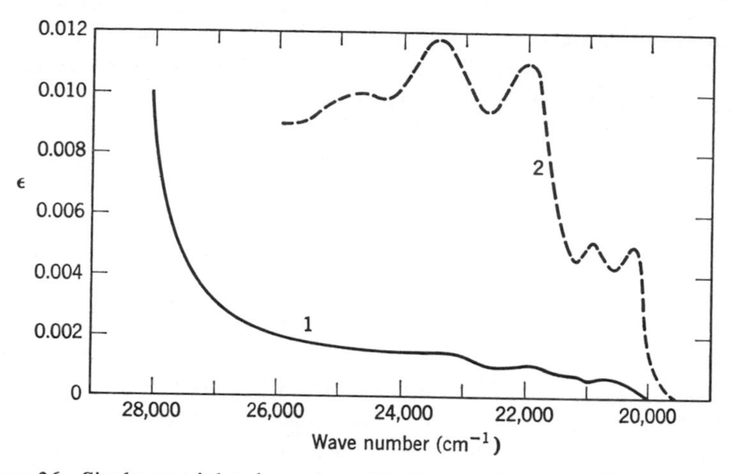

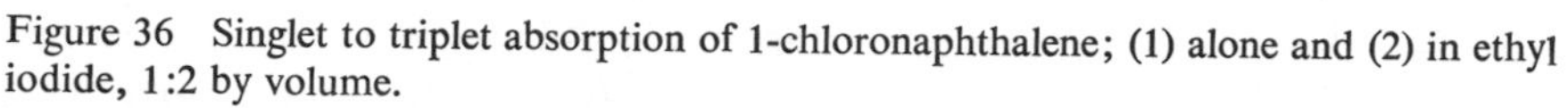
Figure 36 Singlet to triplet absorption of 1-chloronaphthalene; (1) alone and (2) in ethyl iodide, 1:2 by volume.
Reprinted by permission from American Institute of Physics, M. Kasha, *J. Chem. Phys.*, **20**, 71 (1952).

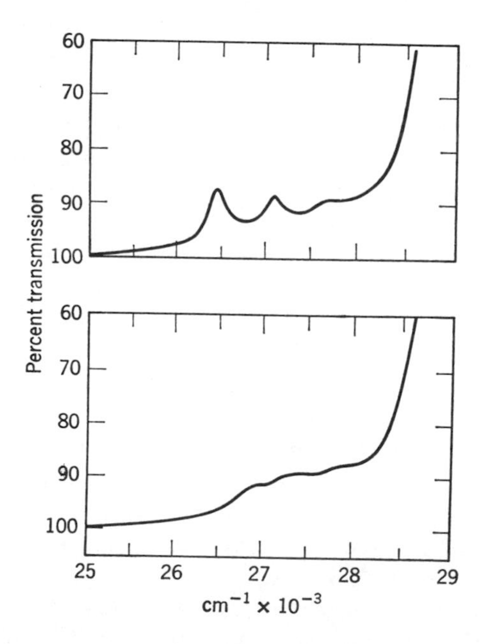

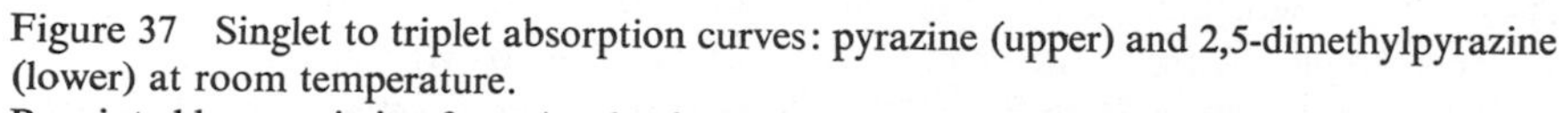
Figure 37 Singlet to triplet absorption curves: pyrazine (upper) and 2,5-dimethylpyrazine (lower) at room temperature.
Reprinted by permission from Academic Press, Inc., L. Goodman and M. Kasha, *J. Molec. Spec.*, **2**, 58 (1958).

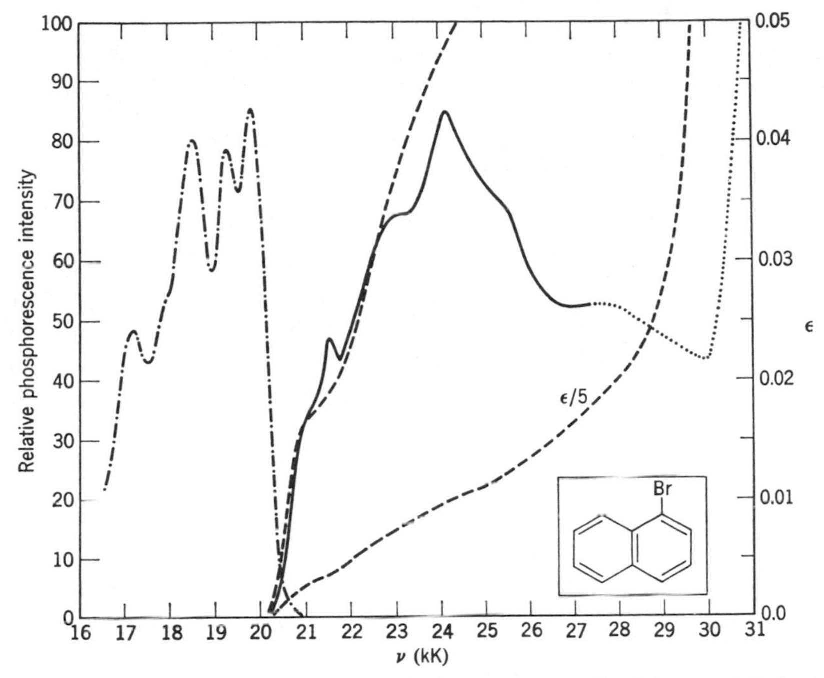

Figure 38 ———, phosphorescence excitation spectrum of α-bromonaphthalene (20 vol percent) excited by a 500-W tungsten light or a 500-W xenon lamp and corrected for the wavelength variation of the photon flux from the lamps. – – – –, the single-triplet absorption spectrum of α-bromonaphthalene (40 vol percent). – . – . –, the phosphorescence spectrum of α-bromonaphthalene (20 vol percent). All spectra were obtained at 77°K by using solutions of α-bromonaphthalene in a 3:1 mixture of ether and isopropyl alcohol. Reprinted by permission from American Institute of Physics, W. Rothman, A. Case, and D. R. Kearns, *J. Chem. Phys.*, **43**, 1067 (1965).

p-diido benzenes [416]. From polarization data of the 0–0 bands it is possible to assign the transition as $B_{3u} \leftarrow A_g$ (polarized perpendicular to the plane) for the dichloro and dibromo cases.

Another interesting method to determine $T \leftarrow S_0$ absorption spectra is that involving the use of phosphorescence excitation spectra—that is, the determination of the changes in intensity of phosphorescence as a function of the exciting wavelength in the region of the $T \leftarrow S_0$ absorption. For example, 1-bromonaphthalene exhibits a moderately resolved excitation spectrum, thus absorption, in the 4900–3700 Å region at 77°K [417,418] (Figure 38). Similar excitation spectra can be obtained for a wide variety of aromatic ketones and aldehydes [249] and acyl bromides [245]. In many cases it is possible to identify both the $T_{\pi,\pi^*} \leftarrow S$ and $T_{n,\pi^*} \leftarrow S$ transitions. The $T \leftarrow S_0$ absorption data along with phosphorescence (and sometimes external perturbation studies) permit assignment of the nature of the lowest triplet state. Further, intersystem crossing quantum yields can be determined utilizing quantitative absorption data for the $T \leftarrow S_0$ absorption [418].

B. ENHANCEMENT OF SINGLET-TRIPLET TRANSITIONS BY EXTERNAL PERTURBATIONS

The enhancement of spin-orbit coupling by heavy atoms, with the consequent increase in the probability of $T \leftarrow S_0$ transitions, can be effected by external perturbation with molecules containing heavy atoms as well as by internal perturbation by heavy atoms within the absorbing or emitting molecule. The first example of an external perturbation was the effect of ethyl iodide as a solvent on the $T_1 \leftarrow S_0$ absorption of 1-chloronaphthalene (Figure 38) [410]. The $T_1 \leftarrow S_0$ absorption of 1-chloronaphthalene alone, which extends into the violet region of the spectrum, is so feeble that the liquid appears colorless. The absorption bands are barely detectable in the long wavelength tail of the first $S_0 \leftarrow S_0$ transition in the ultraviolet. Pure ethyl iodide is also colorless, but a mixture of one part of 1-chloronaphthalene and two parts of ethyl iodide by volume is distinctly yellow. The spectrum of the mixture shows well-defined vibrational structure in the violet spectral region similar in appearance and intensity to the $T \leftarrow S_0$ absorption spectrum of 2-iodonaphthalene. There is no evidence of compound formation of the ordinary kind. The generality of this phenomenon has been well established. The principal question involves the nature of the mechanism responsible for

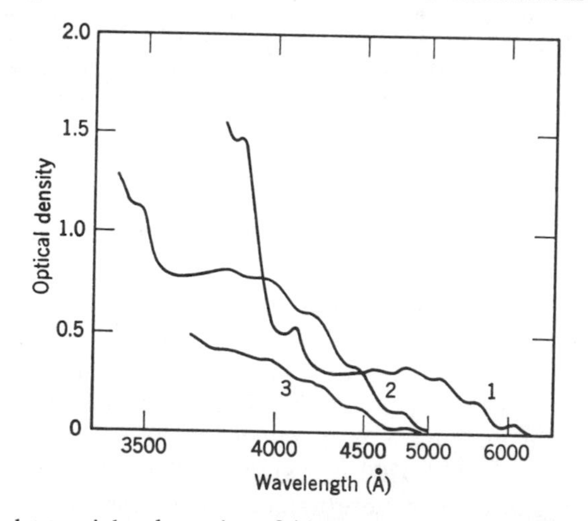

Figure 39 Singlet to triplet absorption of (1) hexatriene and (2) butadiene in chloroform solutions with oxygen at 130 atm and (3) butadiene in ethyl iodide solution.
Reprinted by permission from The Royal Society, D. F. Evans, *Proc. Royal Soc.* (*London*), **255A**, 55 (1963).

the result. Part of the problem revolves around the fact that several different classes of molecules or atoms produce a similar effect.

The first mechanism proposed for the action of ethyl iodide involved an increase in the spin-orbit coupling by a magnetic perturbation via the iodine atom on collision with the 1-chloronaphthalene [410]. However, evidence of the formation of weak charge-transfer complexes between halo-aromatic compounds and propyl iodide [419] raises the possibility that charge transfer states of such complexes may participate in the $T \leftarrow S_0$ enhancement. Also, the degree of enhancement of $T \leftarrow S_0$ absorption with a group of alkyl-halide solvents presumably parallels the electron acceptor ability of the solvents [420].

Considerable enhancement of $T \leftarrow S_0$ absorption bands of many aromatics, heterocyclic, acetylinic, and olefinic compounds (see Figure 39) can be effected by introducing O_2 at high pressure. At the same time, stronger absorption bands appear at shorter wavelengths than the $T \leftarrow S_0$ bands, but at longer wavelengths than the normal bands of the organic molecule alone [92,414,421]. These latter O_2-induced bands have been attributed to charge transfer bands (from the organic molecule as donor and the O_2 as acceptor) [414,421]. Perturbing molecules in the doublet state, as NO, as well as in the triplet state, as O_2, are effective in enhancing $T \leftarrow S_0$ absorption [421]. As the result of other studies [422,423], it has been concluded that in the enhancement of the $T \leftarrow S_0$ absorption by O_2, the magnetic moment of the perturber, O_2, exercises an insignificant effect, or one of minor importance. In one mechanism [422], the effect depends on the formation of a charge transfer complex between the O_2 and the organic molecule. Further, it is the mixing between the singlet and triplet states that produces the increased absorption and this occurs through the mediation of a charge transfer excited state. Interaction between a charge transfer excited state and the triplet state of the organic molecule in the complex causes intensity to be borrowed by the $T \leftarrow S_0$ band from the charge transfer band. In another mechanism [423], a collision complex forms. Furthermore, excited states of the complex exist—that is, doublets for perturbing molecules that are doublets and triplets for perturbing molecules that are triplets. Under electron exchange, the matrix element interacting the excited states does not vanish except by accident and a transition is possible to the lower of the resulting mixed states. By first-order perturbation theory, the transition moment is proportional to the singlet $\leftarrow$ singlet transition energy of the perturbed molecule. Further, the small energy terms due to the interaction of the two molecules are neglected and the matrix element is considered small compared to the singlet-triplet separation in the perturbed molecule. Thus, for example,

$$M_{T \leftarrow S_0} = \left(\frac{H_{12}}{E_S - E_T}\right) M_{S \leftarrow S_0} \tag{111}$$

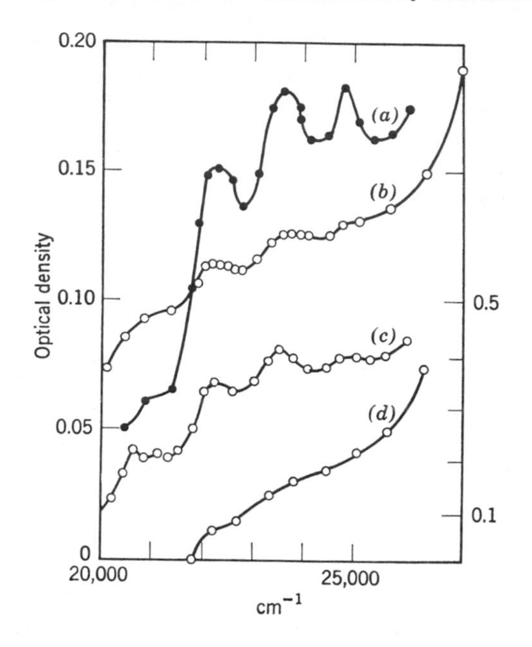

Figure 40 Absorption spectra of 1-chloronaphthalene with different perturbing solvents: (*a*) 1-chloronaphthalene—ethyl iodide mixture; (*b*) 1-chloronaphthalene—xenon (143 atm); (*c*) 1-chloronaphthalene—oxygen (30 atm, optical density scale at right side of diagram); (*d*) pure 1-chloronaphthalene.

Reprinted by permission from Pergamon Press, Inc., A. Grabowski, *Spectro. Chim. Acta* **19**, 307 (1963).

where $M_{T \leftarrow S_0}$ is the transition moment of the $T \leftarrow S_0$ absorption in the perturbed molecule, H_{12} is the matrix element mixing the excited states of the complex under electron exchange, $E_S - E_T$ is the singlet-triplet separation in the perturbed molecule, and $M_{S \leftarrow S_0}$ is the transition moment of the singlet transition of the perturbed molecule. A similar exchange idea involving rare-gas matrices has been suggested [206].

Nevertheless, observations of enhancement of singlet-triplet absorption of 1-chloronaphthalene by xenon at pressures from 100–200 atms [424] suggest that a true "heavy atom effect" may exist [424]. The effects of ethyl iodide, oxygen, and xenon on the enhancement are compared in Figure 40. If the extinction coefficients of the chloronaphthalene are computed on the assumption that the absorption is to be wholly ascribed to a 1:1 complex, the relative effects of xenon:ethyl iodide:oxygen are approximately in the ratio 1:10:100. It is quite possible that the enhancement caused by ethyl iodide and oxygen involves charge transfer complexes with the aromatic molecule, as already described. It would seem doubtful that xenon could form a complex of sufficiently long life for enhancement to occur via a charge transfer mechanism. In addition, the electron structure of xenon would not seem favorable for such a mechanism to be involved. The effect of xenon compared with that of ethyl iodide or oxygen may indicate the relatively smaller magnitude of a collisionally induced heavy atom magnetic perturbation in contrast to the greater perturbation via charge transfer. If this is so, a small proportion of the observed effect of ethyl iodide may be attributed to the heavy atom effect, and a magnetic component in the effect of oxygen is to be anticipated because of the paramagnetic nature of this molecule.

Several other examples of perturbation of the $T_1 \leftarrow S_0$ transition exist. For example, enhancement of absorption occurs in the case of biacetyl when ethyl iodide is the solvent [415]. No vibrational structure is seen. Also, ethyl iodide causes a notable enhancement in the case of *trans*-stilbene [425]. In this case, some vibrational structure does appear which is also noted when O_2 is the perturbing agent.

It is quite probable that two mechanisms can operate in external enhancement of $T_1 \leftarrow S_0$ absorption—one involving charge transfer states and the other a true magnetic effect. It is likely that both operate when alkyl halides are the perturbers with the magnetic effect a minor contributor. Also, both probably operate when O_2 is the external perturbing agent. Here again, it is likely that the magnetic effect is minor. On the other hand, when xenon is the perturbing agent, the magnetic effect may well be of relatively greater significance.

So far we have concentrated on $T \leftarrow S_0$ absorption. It is known, however, that it is possible to externally perturb the $S_0 \leftarrow T_1$ radiative process (phorescence). A major area of concern is the perturbation effects of solvents containing heavy atoms. The lifetimes of certain phthalimides are generally lower in propyl bromide than in any other of about 20 solvents [426]. Several other molecules appear to have both their lifetime and intensity of phosphorescence modified by a solvent system containing a heavy atom [427]; for example, a series of dinitronaphthalenes show an increase in intensity and a decrease in lifetime of phosphorescence when the solvent is a mixture of an alcohol and 20 percent ethyl iodide compared with the values in pure alcohols at 77°K. A similar phenomenon occurs for N,N-dimethyl-4-nitroaniline, coumarin, and acid fluorescein [427]. It is probable that this effect is real although the lifetimes of at least certain phthalimides undergo a large variation depending on the nature of the solvent [426]. Also, it is interesting that the lifetimes of all the phthalimides mentioned above in the ethyl-iodide solvent system are about the same (0.005–0.008 sec). This is somewhat puzzling.

A series of 1-monohalonaphthalenes in various propyl-halide solvents (cracked glasses) exhibit notable differences [419] in lifetimes, depending

Table 24 First Observable Lifetimes of Some Compounds in Different Solvents

Compound	EPA (sec)	Propyl Chloride (sec)	Propyl Bromide (sec)	Propyl Iodide (sec)
Np	2.6	0.52	0.14	0.076
1-FNp	1.4	0.17	0.10	0.029
1-ClNp	0.23	0.075	0.059	0.023
1-BrNp	0.014	0.0073	0.0069	0.0063
1-INp	0.0023	0.0014	0.0012	0.009

upon the solvent (see Table 24). The lifetimes are reported as first observed because of the nonexponential nature of the decay in the cracked glass systems. This may arise because of equilibrium of species involved in the charge transfer complex [419]. However, there are other considerations such as the non-uniformity of the glasses (see later discussion). The decrease in lifetime produced by substitution of chlorine on naphthalene is less than that produced by the external perturbation of naphthalene by iodine (as propyl iodide). In addition, the heavier the halogen of the propyl halide, the shorter is the lifetime of the halonaphthalene. These data provide good evidence for a true heavy atom perturbation. This would be an additional perturbation to that proposed in terms of charge transfer mixing of the singlet and triplet states of the aromatic hydrocarbon by the alkyl iodide-hydrocarbon complex [422] (see earlier discussion). The effect of various alkyl halides on the singlet-to-triplet absorption intensity is also in harmony with the low-temperature lifetime data. This is further support for the proposition that a heavy atom effect is operative in addition to the proposed charge-transfer mixing [419].

Considerable care must be exercised in relating changes of spectral properties with the nature of the heavy atom in an external perturbing solvent. The nonexponential nature of the decay curves [419,428] indicates the non-uniformity of the various regions of the glasses. The different halides may modify the density and geometry characteristics of the glassy matrix. If the mechanism of external spin-orbit coupling is of the electron-exchange types, these have exponential dependency on internuclear distances. Thus, for example, factors other than atomic number must be considered. Furthermore the relative effect of an external heavy atom perturbation on the rate constants of triplet emission, k_P and internal conversion, k_P^{IC}, is $k_P \gg k_P^{IC}$ [428]. The effect on k_P^{IC} compared with the intersystem crossing rate constant, k_{IS}, is not clear because of the lack of quantitative data [428].

Another type of external perturbing agent is a solid matrix of a rare gas such as argon, krypton, or xenon [183] (also see Chapter 11-C). In the case of benzene, the lifetime of phosphorescence in methane, argon, krypton, and xenon (at 4.2°K) is 16, 16, 1, and 0.07 sec, respectively. The phosphorescence in krypton and xenon shows a strong band near 29,500 cm^{-1} which is probably the 0–0 band. This band is either weak or absent in the other matrices (lower atomic or molecular weight solvents) and other intensity differences exist between the matrices. Further, the quantum yield ratio of fluorescence to phosphorescence ($\Phi_F\Phi/_P$) is effectively zero in argon, krypton, and xenon but greater than one in methane. The result for perdeuterated benzene in the same matrices is quite parallel to that of perprotonated benzene. Any external perturbation produced by methane and argon appears to be relatively small while for krypton and xenon it is larger than the existing intramolecular one. Thus the spectral results in argon or methane would be those resulting from the *lack* of any external perturbation. It has been suggested that the intramolecular phosphorescence mechanism is highly vibrationally induced [107,187]. The behavior of the 0–0 band and the effect of deuteration (see Chapter 11) in the two sets of matrices support this view [107].

The mechanism for the perturbation by rare-gas matrices involves, in part, an exchange interaction [107,206]; that is, there is a second-order process in which there is exchange interaction between the hydrocarbon and the rare-gas solvent. The exchange mechanism permits the $T \leftarrow S_0$ transition of the hydrocarbon to borrow intensity from transitions of the solvent. The normally spin-forbidden transitions will be "allowed" because of the strong spin-orbit coupling in the high atomic number rare-gas atom.

Chapter 15 Triplet-Triplet Transitions and External Quenching of the Triplet

A. TRIPLET-TRIPLET TRANSITIONS

Triplet-triplet absorption can occur via excitation from the lowest triplet to the higher ones. This is observable in many cases because the lifetime of the lowest triplet is sufficiently long to allow for the existence of a reasonable concentration of molecules in the triplet state.

The observation of triplet-triplet absorption is commonly accomplished by flashing a sample in fluid solution followed by scanning and recording [429]. The flash-excitation method and its application to solutions permit observation of triplet-triplet spectral decay processes and chemical reactions of excited molecules. The technique can be used to study states having lifetimes longer than approximately 3×10^{-6} sec [429]. Many molecules have been studied by this technique including aromatic hydrocarbons, and their derivatives [430,431], porphyrins [432], chlorophylls [429,433,434], aldehydes and ketones [430,434], and dyes [433,435,436]. In addition to the triplet-triplet absorption, the decay kinetics of the triplet state can be ascertained.

Some typical triplet-triplet absorption spectra are shown in Figures 41–44. In addition, Table 14 shows the location of excited triplet levels for some molecules determined by triplet-triplet absorption. The wavelengths of the absorption origins can, of course, be determined from the data in Table 25. The maximum percent conversion to the triplet varies greatly, being as low as a few percent and as high as 95 percent. For the linear polyacenes except benzene, there are two strong transitions that progressively move to longer

Table 25 Location of Some Excited Triplet States[a] (cm^{-1}) [430]

Compound	T_1	T_a (weak)	T_b (strong)	T_c (weak)	T_d (strong)
Benzene	29,400				
Naphthalene[b]	21,300		45,400		59,800
Anthracene[b]	14,700	36,070	38,280		50,400
Naphthacene	10,250	31,250	31,990	42,150	45,280
Pentacene	8000	27,000	28,400	34,000	40,790
Phenanthrene[b]	21,600	40,830	42,390		
Benz(*a*)anthracene	16,500	35,020	37,120		48,350
Chrysene[b]	19,800		37,340		44,740
Benzo(*c*)phenanthrene	20,600		39,940		45,600
Triphenylene[b]	23,800		47,160		52,700
Pyrene[b]	16,800	36,050	40,840		44,560
Dibenz(*a,h*)anthracene	18,300		37,080	41,880	49,000
Perylene	12,600		33,090		48,300
Coronene	19,100	38,160	40,790		
1-Methylnaphthalene	21,200		44,900		
1-Hydroxynaphthalene	20,500	39,730	46,180		
2-Bromonaphthalene	21,100		45,050		
Benzaldehyde	25,200	48,500	56,450		
Acetophenone	26,600		61,530		
Benzil	21,600	42,160	47,900		
Benzoic acid	27,200		58,450		
Biphenyl	22,800		49,930		
Biacetyl	19,700		51,250		

[a] The weak and strong notation beside the triplet states T_a, T_b, etc., indicate the relative intensity of the transition $T_n \leftarrow T_1$ where in the original text, the states were enumerated T_2, T_3, etc. However, this author has changed this designation since, by implication, these would be the energy order of the states. Triplets have been found between the triplets given (see other footnotes and text).

[b] New $T \leftarrow T$ absorption bands have been found for several hydrocarbons as follows: anthracene: 477, 517, 789, 805, 843, and 887 mμ; pyrene: 775 and 870 mμ; chrysene: 910 mμ; picene: 805 and 910 mμ; phenanthrene: 653, 725, and 820 mμ, reference 201. Also, for naphthalene at approximately 390 and 402 mμ; phenanthrene at approximately 460 and 490 mμ; triphenylene at 410 mμ (band maxima), R. Keller and S. Hadley, *J. Chem. Phys.*, **42,** 2382 (1965). The 490 mμ band in phenanthrene probably corresponds to the 510 mμ band from reference 430.

wavelengths as the number of condensed rings increases [430]. In addition, in several of the hydrocarbons other weaker transitions exist. The first strong transition (in the 17,000–27,000 cm^{-1} region) shows some vibrational structure (see Figure 41). The second strong transition is in the 33,000–38,000 cm^{-1} region. As implied above, no triplet-triplet absorption has been found for benzene. Tentative assignments of these excited triplet transitions of relatively strong intensity are ${}^3B^-_{1g}$ and ${}^3A^-_{1g}$, respectively (x and y axes in plane, x axis parallel to long axis of molecules). These assignments were

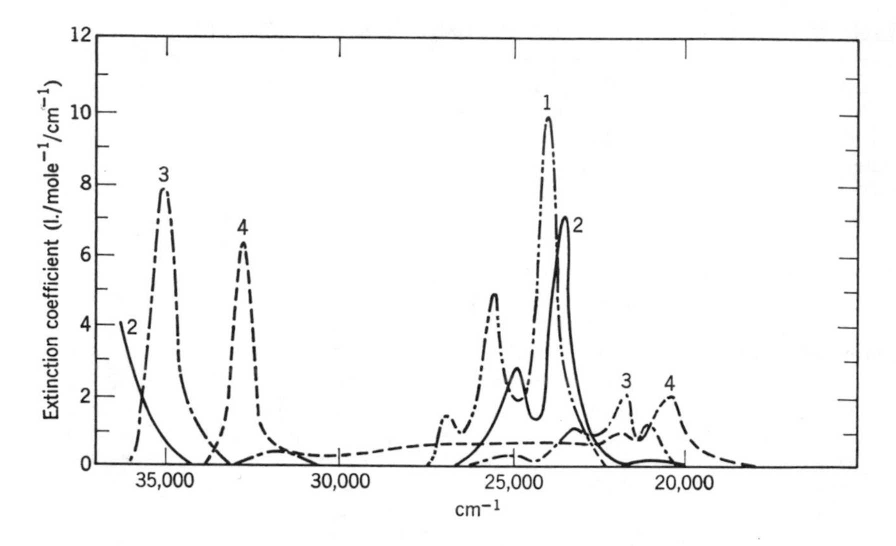

Figure 41 Triplet-triplet absorption of (1) naphthalene; (2) anthracene; (3) naphthacene; (4) pentacene.
Reprinted by permission from The Royal Society, G. Porter and N. Windsor, *Proc. Royal Soc.* (*London*), **245A,** 238 (1958).

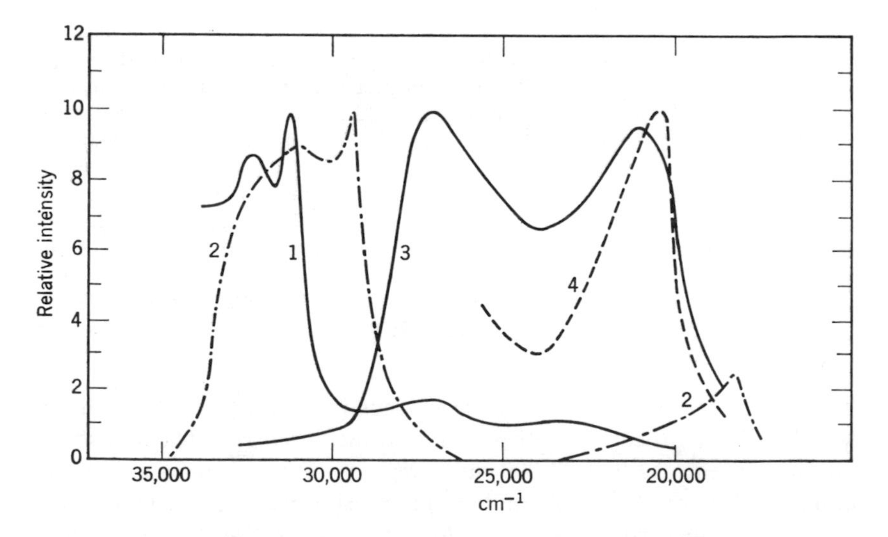

Figure 42 Triplet-triplet spectra of (1) benzaldehyde, (2) benzophenone, (3) benzoin; (4) benzil.
Reprinted by permission from The Royal Society, G. Porter and N. Windsor, *Proc. Royal Soc.* (*London*), **245A,** 238 (1958).

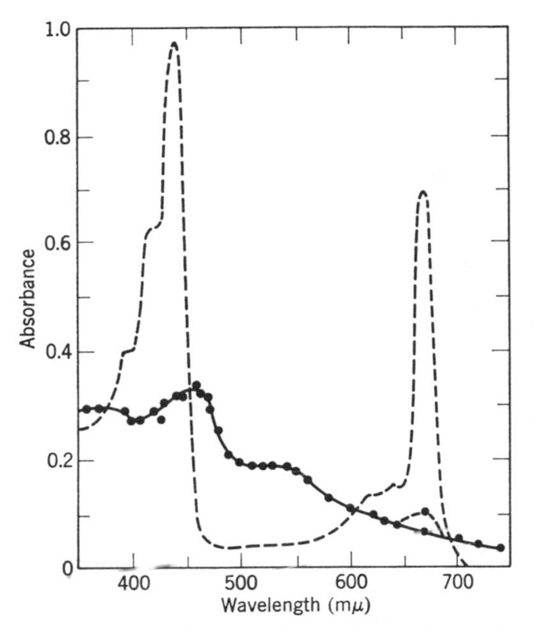

Figure 43 Absorption spectra of chlorophyll *a* singlet-singlet absorption (- - - -), and triplet-triplet absorption (●—●—●).
Reprinted by permission from Journal of the American Chemical Society, H. Linschitz and K. Sarkanen, *J. Amer. Chem. Soc.*, **80**, 4826 (1958).

primarily based on energy and intensity predictions [214]. More recent calculations [437–439] have accounted for discrepancies initially found between theoretical and experimental results; that is, two relatively intense transitions are expected in the visible and near ultraviolet for the polyacenes (see Table 25). Early theoretical predictions [214] indicated the lower energy one to be stronger than the high energy one which is not in agreement with experiment. When more extended configuration interaction is added that includes singly and doubly excited configurations [437–439], the relative intensity order is modified and is in agreement with experiment [430].

Calculations using extended configurational interaction of the type described above have been quite successful in predicting the energy and polarization of some transitions of alternate hydrocarbons [439]. Theoretical predictions [439] and experimental measurements on $T \leftarrow T$ transitions of the hydrocarbons themselves [178] and charge transfer complexes [440] show that the lowest energy allowed transitions (in the visible) are polarized along the long axis of the molecules. This is also in agreement with earlier predictions [214]. Further, forbidden triplets have been predicted [214,437,439] including some lying between the lowest triplet and singlet states. For example,

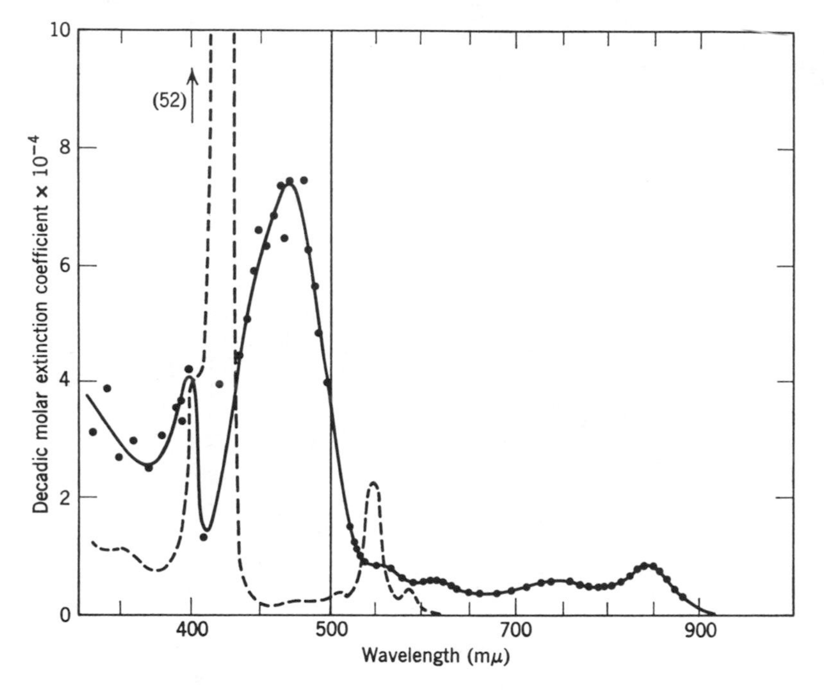

Figure 44 Absorption spectra of zinc tetraphenyl porphyrin, singlet-singlet (- - - -) and triplet-triplet (●—●—●).
Reprinted by permission from Journal of the American Chemical Society, L. Pekkarinen and H. Linschitz, *J. Amer. Chem. Soc.*, **82**, 2407 (1960).

the second triplet of anthracene is predicted to be approximately 10,000 cm^{-1} [214,439] above the first triplet and it has been found at 11,200 cm^{-1} above the first triplet (26,050 cm^{-1} above the ground state). This puts it close to the lowest excited singlet. These higher excited triplets are particularly important with reference to the mechanism for intersystem crossing as discussed in Chapter 11-C and D.

The reasons for the lack of observation of triplet-triplet transitions in benzene in fluid solutions are not completely clear. Evidence indicates, however, that the lifetime of the lowest triplet state in fluid solution (hexane) is at least as short as that of the lowest excited singlet ($\sim 2 \times 10^{-8}$ sec); therefore, the triplet state decays so rapidly that the concentration of triplet molecules would be vanishingly small. The second triplet of benzene has been located by O_2 and NO perturbation studies [441]. The first one, $^3B_{1u}$, is at 29,674 cm and the second one, $^3E_{1u}$, is at 36,560 cm^{-1}. Furthermore, the lowest triplet of benzene is slightly distorted from hexagonal symmetry; this distortion is greater than for the lowest excited singlet state, $^1B_{2u}$ [442].

Other aromatic hydrocarbons also show triplet-triplet absorption [430]. All of the tetracyclic aromatic hydrocarbons have two strong absorption regions similar to the linear hydrocarbons. Benz(*a*)anthracene seems to show the most marked vibrational structure whereas chrysene and benzo(*c*)-phenanthrene show little. Also, the latter two differ from the others in that the lower energy strong transition is more intense than the higher energy one. The oscillator strengths of the transitions decrease as the number of "bends" in the structure increase. This is somewhat similar to the case of the normal singlet-singlet transitions. Larger aromatic hydrocarbons such as dibenz-anthracene and coronene also show triplet-triplet transitions. Biphenyl and fluorene show one strong absorption in the 4000 Å region with fluorene having one additional weaker transition at longer wavelength.

Many derivatives of these hydrocarbons show triplet-triplet absorption [430]. Alkyl, halo, hydroxy, and amino derivatives of benzene show no absorption. The only benzene derivatives that show triplet-triplet absorption are those containing a carbonyl group such as benzaldehyde, acetophenone, benzoic acid, and benzamide. In the last two cases, however, the transitions are considerably weaker than in the other derivatives that show absorption. None of the spectra shows much vibrational structure, but benzaldehyde and benzophenone exhibit the most. Two aliphatic ketones, acetone and biacetyl, also show triplet-triplet absorption but with no structure.

Many derivatives of naphthalene and anthracene show triplet-triplet absorption [430]. The lower energy, strong transition (~4000–3300 Å) of many derivatives of naphthalene (such as bromo, methyl, and hydroxy) looks very similar to naphthalene. Other derivatives (such as sulfonic acid and chloro-) are quite similar except for some variation in wavelength or clarity of structure. In all cases the intensities of the transitions are very comparable. Particularly in 1-naphthol, a moderately strong longer wavelength transition occurs (~5000 Å). In a monochloro- and adi chloro-anthracene, the band intensities are quite comparable and they occur at nearly the same position as in anthracene (~4350 Å).

Both chlorophylls *a* and *b* show triplet-triplet absorption (see Figure 43 for chlorophyll *a*). Chlorophyll *a* has less detailed structure than does chloro-phyll *b*. The former has two maxima (~4620 and 5300 Å) and the latter has five maxima (~3800, 4450, 4850, 5500, and 6050 Å). The strongest maxima occur at 4620 and 4850 Å, respectively [429]. The absorption continues to about 8000 Å and then decreases to a flat low value to about 10,000 Å [432]. Tetraphenyl porphyrin shows four maxima (~3900, 4300, 6900, and 7800 Å). The zinc derivative also shows four maxima that are all red shifted relative to the parent porphyrin (see Figure 44) [432]. In a general way, the absorption spectra of zinc protoporphyrin and magnesium phthalo-cyanine bear a resemblance to those of the chlorophylls [433]; that is, a strong blue band (similar to the strong blue or Soret band of the singlet-singlet transition) exists and absorption continues to the red with some band structure. In general, however, the porphyrins show lower triplet-triplet absorption in the yellow-green region than do the chlorophylls. Also, the infrared band in the porphyrins is more distinctive than in the chlorophylls. No triplet-triplet absorption occurs for the divalent transition-metal complexes of porphyrins (Cu, Fe, Co, Ni) or chlorophylls (Cu) [432,433]. This is most likely attributable to the short lifetime of the triplet state [349, 350] with an accompanying small triplet state concentration and therefore, vanishingly small absorption.

Several dyes exhibit triplet—triplet absorption spectra [434,435]. Retinene, a *p*-toluidene complex of retinene, rhodopsin [434], and acridine orange [435] show absorption. Retinene has one very strong band ($\varepsilon \sim 80{,}000$) with a maximum at approximately 4500 Å—the singlet-singlet maximum is about one half as intense and is at shorter wavelengths (~3750 Å). The *p*-toluidene complex has a strong band ($\varepsilon \sim 45{,}000$) to the red of retinene (at ~5000 Å) and a minor maximum at shorter wavelength (~3500 Å) [434]. In retinene, the amount of conversion to the triplet varies with the solvent, being as low as 1.7 percent and as high as 11 percent (in this case the conversion is equivalent to quantum yield defined as percent triplet formed from excited singlet) [434]. Acridine orange shows a strong triplet-triplet absorption band and some further weaker absorption, the shape and wavelength of which depend on the solvent [405]. In acidic solvents, acridine orange has a strong band near 4900 Å with further rather undefined absorption extending toward 3500 and 10,000 Å. In basic solvents, a slightly weaker maximum occurs near 4300 Å with absorption extending to shorter and longer wavelengths; however, rather clear but significantly weaker maxima occur near 5800 and 3600 Å.

Finally, it appears that molecules such as β-carotene, *p*-benzoquinone, and *p*-nitroaniline do not show triplet-triplet absorption. Anthraquinone does, however, show triplet-triplet absorption [443].

B. EXTERNAL QUENCHING OF THE TRIPLET

In this part we shall be particularly concerned with the effects of external heavy metal ions on the triplet state of a solute as determined by flash-excitation procedures. This method permits the study of quenching of triplet states in fluid solution in contrast to the highly viscous or rigid media necessary for the observation of phosphorescence. Triplet-triplet absorption is absent in the internally complexed transition metal chelates of porphyrins and chlorophylls (see previous section). This is associated with the expected short lifetime of the triplet because of high spin-orbit coupling.

The lifetimes of the triplet states of anthracene and chlorophylls vary significantly in the presence of quenchers. They are notably shortened in the presence of *external* nickel(II), cobalt(II), and copper(II) ions but others such as manganese(II) and neodymium(II) show only, minor effect [429]. From lifetime data it is possible to evaluate bimolecular quenching constants involving a molecule in the triplet excited state (and the quenching metal ions) [432]. In the case of anthracene, the quenching constants for copper(II), chromium(III), nickel(II), and cobalt(II) are in the neighborhood of 10^8 l mole^{-1} sec^{-1}. For the porphyrins, the values are smaller except for copper(II). The quenching constant for nickel(II) on the triplet state of the zinc complex of tetraphenyl porphyrin, in pyridine solution, is the same as that for the parent porphyrin (no metal complexed).

Rare-earth ions are weaker quenching agents than manganese(II) not only for anthracene but also for the porphyrins and chlorophylls. The manganese(II) ion quenches less than other transition-metal ions and particularly less than the copper(II) ion (approximately 100–200 times less depending on the solvent). The addition of water to tetrahydrofuran causes a particularly marked decrease in the quenching constant for nickel(II), chromium(III), and cobalt(II) which levels off with increasing concentration of water. Addition of water to pyridine solutions has less effect. In aqueous solution, potassium iodide is comparable to the more effective transition metals as a quencher of the triplet state but zinc(II) ion has no effect. The near absence of quenching effect by gadolinium(III) and manganese(II), both of which have high spin magnetic moments, shows that the quenching by nickel(II) chromium(III), and copper(II) is not caused by a simple magnetic perturbation.

Further evidence for the above conclusion comes from observations on the quenching of the fluorescence of β-naphthalene sulfonic acid by various transition and rare-earth ions [432] The quenching appears to involve a charge transfer intermediate formed from the triplet molecule and the quenching ion. The potential energy curve of the charge transfer intermediate is assumed to cross that of the ground state metal ion-organic molecule complex; therefore, the excitation energy is converted to vibrational energy in the ground state with consequent dissociation. An alternative mechanism involves the decomposition of the charge transfer complex to a pair of one-electron oxidation-reduction products which then return to the initial state by relatively slow back reactions.

Further evidence for the existence of an intermediate charge transfer complex in quenching by metal ions comes from studies on the anthracene triplet [436]. In the quenching of the anthracene triplet by copper(II) chloride in pyridine, the value of the bimolecular quenching constant k in the equation

$$\frac{-dC^*}{dt} = kC^*Q \qquad (112)$$

(where C^* is the concentration of triplet anthracene and Q is the concentration of the quencher) is intermediate between those found in the presence of ethylenediamine and *o*-phenanthroline. These changes in rate result from the changes in catalytic [436] or quenching effects resulting from replacement of pyridine in the $CuCl_2 \cdot 2$ pyridine complex by the other ligands (the ethylenediamine copper complex showing the lowest k value). Charge transfer reactions involving reduction of inorganic ions by another inorganic ion solvated in different ways show marked dependence on the nature of the solvate (for example, whether it is an aquo or pyridyl complex). The effects noted are parallel to those found for triplet quenching. This parallelism provides additional evidence for a charge transfer intermediate in the triplet quenching process.

A different quenching mechanism involves energy transfer on collision of the type [444,445]

$$M^*(T) + Q(\text{multiplet}) \rightarrow M(S) + Q(\text{multiplet}) \qquad (113)$$

where $M^*(T)$ and $M(S)$ are the donor or quenchable molecules in their excited triplet and ground-state singlet, respectively, and Q is the quencher molecule. When the triplet energy level of the quencher (acceptor) is significantly lower than that of the donor, the energy transfer is diffusion controlled (and apparently no long-range transfer occurs). When the triplet-state energy levels are similar, the energy transfer process is less likely. If the triplet level of the quencher molecule is higher than that of the donor, no quenching occurs. These results are reflected in the quenching rate constants. This mechanism should result in excitation of the quencher triplet and thus triplet-triplet absorption should be observable in the quencher (or acceptor) molecule. This has been observed in many cases in which the quenchers are a hydrocarbon or halo derivative and the donors are hydrocarbons or their halo derivatives and ketones [445]. It is interesting that quenching of hydrocarbon triplets by hydrocarbon quenchers apparently does not require that an intermediate complex be formed [445].

For comparable quenchers (e.g., paramagnetic ions) acting on aromatic hydrocarbons there does appear to be a difference in concept regarding mechanisms of quenching. One mechanism was discussed earlier in this section. Another mechanism involves the general process as given in (113) where the multiplicity change may occur without either energy transfer or other change in the quencher molecule [444]. This process is independent of the magnetic suspectibility of the quencher (providing the multiplet state of the quencher is not of singlet character). Different modes of interaction could exist. One is that in which a relatively stable collision complex (potentially possible with quenchers as oxygen, nitric oxide, and aromatic triplets) is formed followed by dissociation into products with altered spin and for which there is a statistical probability for the overall process. The other is one in which the energy of the complex with respect to reactants is small compared with kT energy. The difference in quenching rates would then depend on the lifetime of any collision complex and the strength of the spin-spin interaction. Well-shielded electrons of quencher ions would be expected to interact less than unshielded electrons of quenching ions with electrons of the excited molecule. Therefore a combination of the factors of lifetime of the complex and spin-spin interaction could account for the observed variable quenching rates of metal ions of the first-transition series and ions of the lanthanides. It is true that in the two major mechanisms, one involving charge transfer and the one being presently discussed, a complex of some sort is postulated. The problem of characterizing it may be largely only a sematic one. The details of the language used to describe the ultimate fate of the complex again differ but it does not appear to be serious.

The one consideration that may have to be resolved is the fact that the quenching rates for diamagnetic quenchers can be comparable to those quenchers containing unpaired electrons [446]. This could presumably be explained by energy transfer involving

$$M^*(T) + Q(S) \rightarrow M(S) + Q(T) \tag{114}$$

which is a parallel mechanism to that discussed above (appropriate to equation 113) [445]. The variation in quenching rates therefore arises from possible differences in the lifetime of any complex and the spin-spin interactions. Thus it would appear likely that at least two principal mechanisms operate depending upon the nature of the quencher. The only real problem is whether the particular mechanism as represented by (114) involves a complex [432] or does not [444].

Chapter 16 Molecular Excitons and Enhancement of Phosphorescence by Aggregation

We shall only be concerned with exciton theory relative to its application to molecular aggregates and deductions pertinent to enhancement of phosphorescence. For details, the excellent works of Kasha [447,450], McRae [447,451,452], and Simpson et al. [453,454] should be consulted. We shall consider an aggregate of molecules in which an excited electron is strongly bound to a specific molecular center. The intermolecular overlap is considered to be small so that molecular units maintain the identity in the aggregate and a perturbation approach is therefore valid. As a beginning point, the excited states of the molecular aggregate may be described in terms of the excited states of the component molecules. One description of the excited molecular aggregate involves a superposition of the collective excited states, which describes the excitation in terms of a traveling wave packet of excitation known as an *exciton*.

We shall consider the case of linear polymers (or composite molecules) and strong coupling. The polymer is presumed to consist of N molecular units. Before excited-state interaction, degeneracy exists that is N fold in character. After the perturbation (electrostatic) resulting from the intermolecular dipole-dipole interaction of the transition dipoles, an exciton band of N individual exciton states is produced. Another way to describe this is that the upper state of each monomer 0–0 transition is expanded by N-fold aggregation into an N-fold band of levels or states. The band width is directly proportional to the intensity of the corresponding 0–0 component of the monomer absorption [447]. Figure 45 shows exciton bands and states appropriate to various geometrical arrangements of the molecules within aggregates. Schemes II and III represent the arrangements that are most likely to be approached in real aggregates. We are interested in the energy of exciton

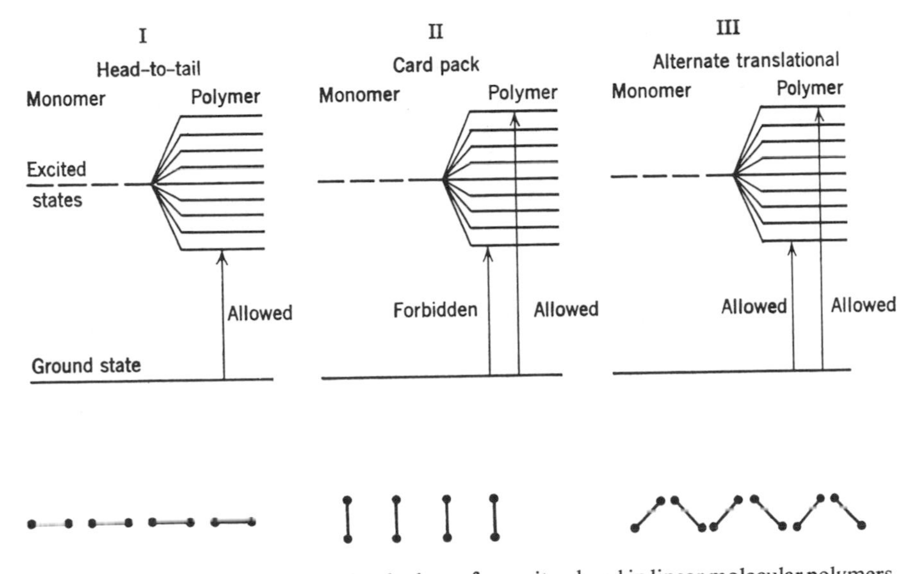

Figure 45 Diagrammatic energy level scheme for exciton band in linear molecular polymers with various geometric arrangements of transition dipoles [448].
Reprinted by permission from Academic Press, Inc., M. Kasha, *Radiation Res.*, **20**, 55 (1963).

states, the band width, and the dipole selection rules, particularly for the singlet ↔ singlet transitions. Quantitative equations for determining the band width and transition moment integrals have been developed [447]. Also, appeal to an electrostatic vector model permits a qualitative development of the energy of the exciton states and, of selection rules [448].

In Figure 45 the vertical arrows represent those transitions that are allowed (ground state to exciton state). The band width for the singlet states of aggregates of dye molecules can be quite large ($\sim$1000 cm^{-1}) depending on the geometrical arrangement of the molecules within the aggregate system [447]. On the other hand, the triplet band width is virtually zero because of the forbidden nature of the spin intercombination selection rule (see following discussion for the dimer case). However, the order of magnitude of the splitting in the singlet band is comparable to that present between the lowest singlet and triplet states of several dye monomers. Because of this, the intersystem rate constant would be expected to increase. In addition, the rate constant for fluorescence is important. In the cases I and III, the rate constant for fluorescence is effectively unchanged from that of the monomer since the lowest transition is allowed. In case II, however, the lowest transition is forbidden and therefore, the rate constant for fluorescence is significantly decreased. Thus the enhancement of phosphorescence for the aggregate relative to the monomer is expected in all three cases. The maximum effect will occur in case II because of the forbidden character of the $S_0 \leftrightarrow S_1$ transition and fluorescence will be quenched. In cases I and III, the enhancement is primarily the result of the exciton splitting (see above and also the discussion of the dimer cases immediately following).

A parallel discussion exists for the case where the aggregate is composed of two molecules (dimer) [448,455] and one case will be considered in some more detail than the preceding. We shall draw analogies to the N molecular unit case.

The situation for the double-molecule case in which the transition dipoles are parallel is shown in Figure 46. There is no splitting of the ground-state singlet but there is for the excited states (Figure 46). The splitting is, in general, for parallel, oblique, and in-line dipoles,

$$\Delta E = \frac{2\,|M|^2}{r_{ab}^3}(\cos\alpha + 3\cos^2\theta) \tag{115}$$

where $\Delta E = E(S'') - E(S')$, M is the transition moment for the transition in the monomer, r_{ab} is the distance between the centers of the two molecules a and b, α is the angle between the polarization axes of the molecular units for the transition considered, and θ is the angle between the polarization axes and the line connecting the molecular centers. For example, for oblique transition dipoles r_{ab}, θ, and α are defined as follows:

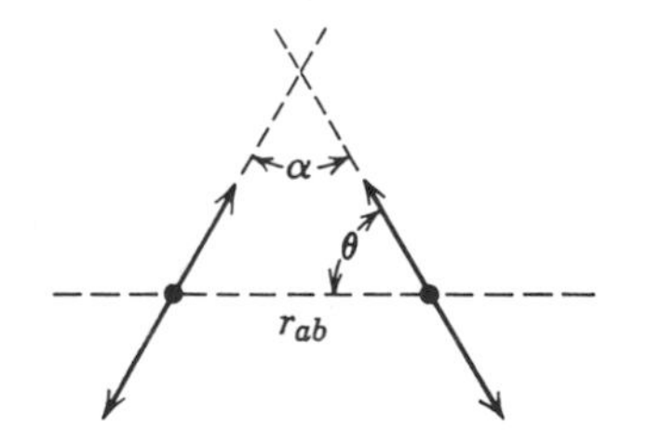

The transition moments can be in or out-of-phase. For the parallel dipole case (Figure 46), the energy level corresponding to the out-of-phase relation is lower than the one in which they are in-phase. The transition moment is the vector sum of the individual transition moments. From Figure 46 it can be seen that excitation to S' is forbidden (dipole cancellation) while that to S'' is allowed. Furthermore, this means that the lowest allowed singlet transition in the dimer will be blue shifted from the monomer or constituent molecule transition. This case of parallel dipoles (Figure 46) corresponds to the case II, card-pack arrangement, for the N molecular polymer (Figure 45).

For other cases, in-line or head-to-tail and oblique, there are some differences in results parallel to those of Case I and III for the N molecular

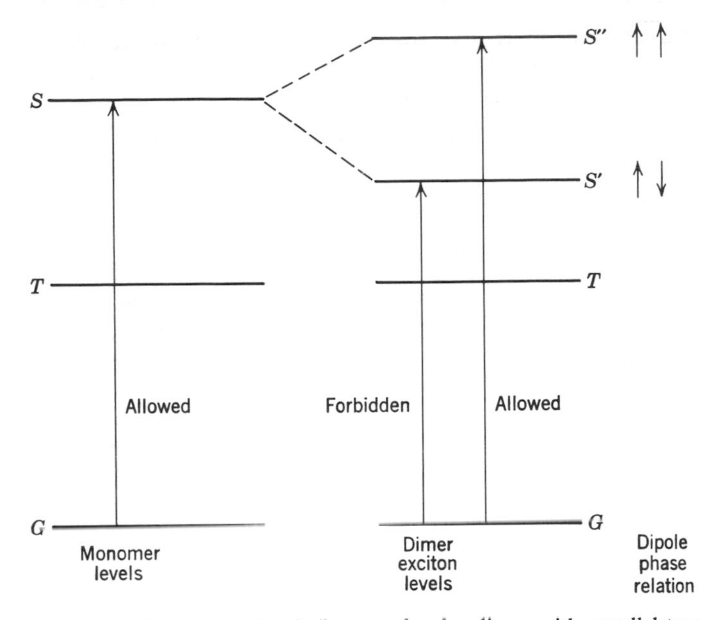

Figure 46 Schematic exciton energy levels for a molecular dimer with parallel transition dipoles [455].

polymer. For the in-plane or head-to-tail case, the in-phase arrangement of the transition moment dipoles leads to an electrostatic attraction and thus is associated with the lower energy level S' (for the parallel case, S' was associated with the out-of-phase arrangement). Furthermore, this in-phase arrangement gives a nonzero resultant transition moment for S' and the out-of-phase arrangement gives a zero resultant moment for S''. Thus the transition to S' is allowed and the one to S'' is forbidden. This corresponds to case I of the N molecule polymer (Figure 45). This also results in a red shift of the dimer transition compared with the monomer.

In the oblique arrangement, the in-phase case is associated with S' and the out-of-phase with S''. Because of the oblique arrangement, however, there is a resultant moment for transitions to both S' and S''. This corresponds to case III of the N molecule polymer (Figure 45).

The splitting of the original monomer states is dependent upon the transition dipole moment (115), and thus the f number or allowedness of the transition. Because of this, there is essentially no splitting of the triplet since the $T_1 \leftarrow S_0$ transition in the monomer is highly forbidden (therefore, small f and M). For the parallel arrangement (Figure 46), excitation is principally to the upper exciton state, S'', followed by rapid internal conversion to S'. Since the transition from $G \leftarrow S'$ (and $S' \leftarrow G$) is forbidden, fluorescence would be quenched. Further, intersystem crossing to the triplet will then be even more competitive than for the monomer and there will be a relatively high intensity of phosphorescence. Thus in the dimer it would be expected that the fluorescence would be quenched and phosphorescence enhanced relative to the monomer. This situation has been observed in several cases [440,452,453].

An exciton model also can be employed to explain an increase in the phosphorescence-to-fluorescence ratio for covalently bonded composite molecules such as diphenylmethane and triphenylmethane [455]. Other possible interactions to increase the triplet-state occupation such as increased spin-orbit coupling are possible; however, a relatively constant lifetime of phosphorescence tends to indicate that an increase in spin-orbit coupling is not a strongly contributing factor.

Chapter 17 pK_a of Excited States

An interesting consideration is the effect of electronic excitation on certain physical chemical properties of molecules. Of particular interest is the change in the pK_a of certain compounds in the excited states compared with the ground state. Of primary interest in this regard are the results for phenols, naphthols, amines, acids, and bases [456–466]. Table 26 summarizes data for a considerable number of molecules.

The pK_a of acridine in the ground state is 5.5 whereas in the first excited state it increases to 10.6 [459]. Thus acridine is a very much stronger base in the excited state than in the ground state. However, pyridine has about the same pK_a in both the ground and first excited states [459]. The molecule 2-naphthylamine shows a change from pK_a = 4.1 in the ground state to −2 in the excited state, and thus becomes a considerably stronger acid. Apparently indole becomes a stronger acid in the excited state. In Chapter 12 it was noted that in base solutions the fluorescence of indole is quenched. There is evidence to indicate that this occurs because of proton transfer while the molecule is in the excited singlet state.

The ammonium salts of aromatic amines are very strong acids in the first excited state with pK values between −2 and −6. Based on the reaction

$$(\mathrm{ArNH_3^+})^*\cdot\mathrm{HOR} \underset{k_r}{\overset{k_f}{\rightleftharpoons}} \mathrm{ArNH_2^*}\cdot\mathrm{H_2OR^+} \qquad (116)$$

where the asterik indicates the electronic excited state and k_f and k_r are the rate constants for the forward and reverse reactions, respectively, and $k_f \gg k_r$. Acidic alcohol solutions of the ammonium compound show fluorescence of the amine at room temperature. If the temperature is lowered, however, the

Table 26 pK_a **Values in Various Electronic States**[a]

Molecule	$pK_a^{S_0}$	$pK_a^{S_1}$		$pK_a^{T_1}$	
Phenol	10	3.6[b]	4.0[b]	[c]	8.5[c]
p-fluoro	9.9	3.5	4.4		8.7
m-fluoro	9.2		3.8		8.5
p-chloro	9.4	3.5	3.2		8.0
m-chloro	9.1	4.0	3.0		7.6
o-chloro	8.5	3.3			
p-bromo	9.3	2.9	3.1		7.7
m-bromo	9.0		2.8		8.5
p-methyl	10.3	4.1	4.3		8.6
m-methyl	10.1	4.2	4.0		8.7
o-methyl	10.3	5.3			
p-ethyl	10.0	4.3	4.3		
m-ethyl	9.9	4.5	4.1		
o-ethyl	10.2	4.5			
p-methoxy	10.2	4.7	5.6		8.6
m-methoxy	9.7	2.7			
o-methoxy	10	5.2			
p-ethoxy	10.1		5.3		8.4
m-ethoxy	9.5		4.4		8.5
p-hydroxymethyl	9.8	3.0			
m-hydroxymethyl	9.8	3.0			
o-hydroxymethyl	9.9	2.9			
Benzoic acid	4.2	9.5[d]			
2-Chlorobenzoic acid	2.9	8.1[d]			
4-Sulfobenzoic acid	3.7	9.1[d]			
2-Naphthol	9.5	3.1		8.1	7.7
2-Naphthylamine	4.1		−2[e]	3.3[f]	3.1[f]
N,N-Dimethyl-1-naphthylamine	4.9			2.7[f]	2.9[f]
1-Naphthoic acid	3.7	10.0[d]		3.8[f]	4.6[f]
2-Naphthoic acid	3.7	11.5[d]		4.0[f]	4.2[f]
Quinoline	5.1			6.0[f]	5.8[f]
Acridine	5.5	10.6[g]		5.6	
o-Phenanthroline	4.85				6.7[h]
o-Phenanthroline-H^+	−1·4				3.8[h]

[a] S_0 is the ground state, S_1 the first excited singlet state, and T_1 is the lowest triplet state.

[b] First column from [465] and second column from [464] except when noted. Generally, [464] uses Förster cycle and 0–0 bands, see text.

[c] First column from flash, triplet-triplet data and second from phosphorescence data. All data above 2-naphthol from [464].

[d] From [466].

[e] From [463].

[f] From [462]

[g] From [459].

[h] From [294].

fluorescence of the ammonium compound appears and the intensity of that of the amine decreases. The temperatures at which this occurs depend on the nature of the alcohol. This effect presumably arises because the configurational changes of the alcohol solvent associated with the proton transfer become rate determining at different temperatures depending on the alcohol.

Hydroxy substituted benzenes and naphthalenes also show large variation in pK_a values between the ground state and first excited state (Table 26) [464]. The 2-naphthol shows a Δ pK_a of 6.3 between the two states in both aqueous and glycerine solutions. Phenol shows a somewhat greater Δ pK_a (6.8) in glycerol solution compared with aqueous solution.

Three techniques are commonly employed for determining pK_a values: (a) variation of the quantum yield or fluorescence intensity of the undissociated molecule as a function of pH, (b) application of a Förster cycle [456,457] to the energy of the absorption maxima of the undissociated and dissociated species, and (c) application of the Förster cycle using 0–0 band energies for the undissociated and dissociated species. The 0–0 band energy is estimated by averaging the energy of absorption and fluorescence band maxima [463].

Before considering one of these, approach (c), some additional comments are important. Each of these approaches gives different excited state pK_a values [464]; for example, for *p*-bromophenal the pK_a values in the excited singlet are 3.0, 4.7, and 3.1, and for *p*-methylphenol the pK_a values are 4.6, 5.7, and 4.3 by methods (a), (b), and (c), respectively [464]. Parallel variations occur for other molecules such as 2-naphthol and *m*-methoxyphenol. The method that should best take into account the variation in the solvent cage between the ground and excited states, the nonequilibrium Franck-Condon state, is method (c). Consequently some care should be exercised in using excited state pK_a data by determining the method by which it was obtained.

The procedure (c) requires that the pK_a of the molecule in the ground state be known. The determination of excitated state pK_a values of any acid-conjugate base system is based on the energy diagram shown in Figure 47 and thermodynamic quantities from the equilibria,

$$\begin{aligned} ROH + H_2O &= RO^- + H_3O^+ \\ ROH^* + H_2O &= (RO^-)^* + H_3O \end{aligned} \tag{117}$$

The appropriate equations are

$$\Delta E + E_d^* = \Delta E' + E_d \tag{118}$$

where E_d and E_d^* are the heats of dissociation in the ground state and excited state and ΔE and $\Delta E'$ are the energy differences between the ground and first excited singlet states for the undissociated acid and anion, or conjugate base, respectively (Figure 47)

$$E_d - E_d^* = (\Delta G - T\,\Delta S) - (\Delta G^* - T\,\Delta S^*) \tag{119}$$

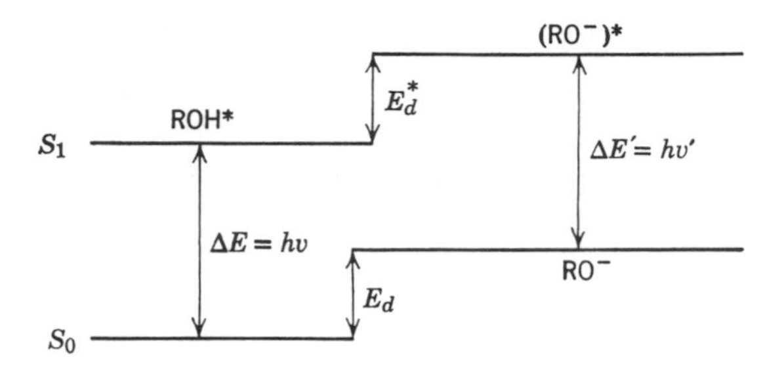

Figure 47 Schematic thermochemical diagram for determination of excited-state pK_a values [123].

If $\Delta S = \Delta S^*$, then

$$\Delta G - \Delta G^* = -RT \quad \text{in} \quad \frac{K_a}{K_a^*} = \Delta E - \Delta E' \tag{120}$$

where the K values are the dissociation constants in the appropriate state. Then

$$\text{p}K_a - \text{p}K_a^* = \frac{\Delta E - \Delta E'}{2.303RT} \tag{121}$$

where pK_a^* refers to the excited state. In the case of pK_a values for the excited singlet state, the ΔE and $\Delta E'$ can be estimated from the longest wavelength absorption maximum and the fluorescence maximum [463] by

$$\Delta E = h\upsilon = \frac{h\upsilon_A + h\upsilon_F}{2} \tag{122}$$

$$\Delta E' = h\upsilon' = \frac{h\upsilon_{A'} + h\upsilon_{F'}}{2} \tag{123}$$

where the subscript A refers to absorption and F to fluorescence. Equation 121 can also be written

$$\text{p}K_a - \text{p}K_a^* = 2.10 \times 10^{-3}\,\Delta\bar{V} \tag{124}$$

where $\Delta\bar{V}$ is the frequency difference in cm^{-1} of the 0–0 bands. Thus from a known value for the pK_a of the ground state (p$K_a^{S_0}$), the pK_a^* of the excited singlet (p$K_a^{S_1}$) can be determined (see Table 26). To reiterate, this approach is general for any acid and conjugate base system including conventional acids, phenols, naphthols, and others (see Table 26).

The pK_a of the excited singlet state of substituted phenols, 2-naphthol and 2-naphthylamine, is considerably lower than in the ground singlet state and thus these molecules become much stronger acids in the S_1 state. For the organic acids, however, the pK_a is considerably higher in the excited singlet state than in the ground state and they thus become much weaker acids in the S_1 state.

The effect of the position of substitution on the pK_a of phenol in the ground singlet state and excited singlet state can be quite different. For example, the maximum difference in the ground state pK_a values for the three methoxy phenols is 0.5 while in the first excited singlet it is 2.5. For the three ethyl phenols, however, the maximum difference in pK_a values is about the same in both the ground and first excited singlet states. Thus the effect of substituents on charge densities, and the consequences derived from these in the excited state does not seem to parallel that for the ground state. In addition, as previously indicated, in some cases a molecule normally considered an acid in the ground state becomes a base (or very weak acid) in the first excited singlet state. For example, the pK_a value of both 1- and 2-naphthoic acid changes from approximately 4 in the ground state to 10–12 in the excited singlet state. Some calculations on naphthol and phenol indicate that changes in pK for the excited state relative to the ground state are to be expected. Certainly more effort should be made to understand the differences in the effects of substituents on the ground and excited states.

If deuterium oxide is substituted for the water, phenols and carboxylic acids show an increase of 0.4–0.8 pK units in their pK_a values in both the ground and excited singlet states (see Table 27) [466]. For substituted phenols the pK_a increase is slightly greater (~0.1 pK unit) in the excited singlet state compared with the ground state. In addition, this change from D to H can have a noticeable affect on both the shape and quantum yield of emission [467]. When both the protonated and unprotonated excited species are fluorescent, the shape of the emission spectrum is different in H_2O and D_2O. This can arise because of the fact that for the reaction

$$RH^* + H_2O \underset{k_2}{\overset{k_1}{\rightleftharpoons}} (R^-)^* + H_3O^+ \tag{124}$$

the ratio k_1/k_2 is near unity for H^+ while it is considerably less than unity for D^+ [467]. Thus, more anion fluorescence will be present for the D_2O solvent case and the shape of the emission will change relative to that in H_2O. In the case where $(R^-)^*$ is nonfluorescent, the quantum yield of fluorescence of RH is higher in D_2O than in H_2O. This could arise because of a slower rate of proton transfer during the lifetime of the excited state in D_2O (to form a nonfluorescent species). If there is no ionizable proton, the difference in quantum yields between D_2O and H_2O is noticeably less. Several examples of the effect D_2O versus H_2O has on quantum yields are given in Table 27.

In the particular case of the determination of the pK_a of an excited triplet state, the ΔE and $\Delta E'$ values are usually determined differently than for the case of an excited singlet state. Here the ΔE and $\Delta E'$ represent the difference

Table 27 Effect of D_2O Relative to H_2O on Fluorescent Quantum Yields [467]

Compound	Quantum Yield Ratio $\frac{D_2O}{H_2O}$
Anisole	1.05
Phenol	1.27
L-Tyrosine	1.25
L-Tryptophan	1.65
1-Methylindole	1.09
Indole	1.29
1-Naphthalene sulfonate	1.00
5-Amino-1-naphthalene sulfonate	3.04

in energy between the lowest triplet and ground state of the acid and conjugate base, respectively [462]. These can be determined by determining the location of the 0–0 band of the phosphorescence of the two forms (or short wavelength limit). From the known pK_a of the ground state the pK_a of the triplet can be determined from an equation [462] parallel to that given earlier

$$pK_a^{S_0} - pK_a^{T_1} = \frac{\Delta E_{HA} - \Delta E_A}{2.303RT} \tag{125}$$

where superscript S_0 and T_1 refer to the ground and lowest triplet states, and the subscripts HA and A refer to the acid and conjugate base, respectively. Table 26 gives some comparative pK_a values for both the excited singlet and triplet states. From the data in Table 26 it is worth noting that the pK_a in the triplet state is similar to that of the ground state while the pK_a for the excited singlet is considerably different from that of the ground state. The *o*-phenanthrolene molecule shows the greatest difference in the pK_a of the ground and triplet states (~2) [294] while *m*-methoxyphenol shows the greatest difference between the excited singlet and ground states (~7) [464]. Generally the pK_a of the triplet is not very sensitive to the position of a given substitution on a phenol (Table 26).

It should also be noted that pK_a values of the triplet state can be determined by direct observation of the triplet-triplet absorption spectra of the acid and conjugate base [462].

Chapter 18 Intermolecular Energy Transfer and Photochemistry

The variety of molecules and systems studied involving energy transfer and photochemistry is extremely large. Only certain examples will be discussed in order to illustrate certain points of interest.

A. AROMATIC HYDROCARBONS

Among the hydrocarbons anthracene is probably one of the most thoroughly studied molecules. If anthracene solutions are irradiated in the absence of oxygen, the product is dianthracene. If oxygen is present, however, the product is the bridged peroxide. Further, the irradiation of anthracene in CCl_4 gives detectable quantities of the 9-chloro and 9,10-dichloro derivatives [468]. Flash photolysis studies show that anthracene in the triplet state is deactivated by reaction with oxygen at nearly every encounter. This reaction with oxygen does not occur with unexcited anthracene. These and other data show that photodimerization results from anthracene being excited to singlet state while the oxidation (to the peroxide) occurs when anthracene is in the triplet state. Other linear polyacenes such as naphthacene, also exhibit photoperoxide formation parallel to that for anthracene. The reactions indicate that some charge localization occurs in the 9 and 10 position in the triplet state which could be viewed in this case as a classical biradical corresponding to a spectroscopic triplet. It should be noted, however, that this classical picture is not a complete or accurate description and should not be taken as a general relation.

Anthracene photooxidation occurs fastest in carbon disulfide. The exact reason for this is not known but it appears that some sort of energy transfer occurs that facilitates the formation of anthracene triplets [468]. Further, the presence and nature of substitution and the nature of the solvent play important roles in the photochemistry. For example, in 9,10-diphenylanthracene the quantum yield of photooxidation depends both on the oxygen and hydrocarbon concentration [469], whereas for anthracene itself the yield is quite independent of the oxygen concentration. This plus data on the fluorescence yield and lack of dimerization in systems without oxygen indicate that the phenyl substituents prevent solvent interactions because of steric reasons. This is further supported by the fact that heavy atom containing solvents enhance the rate of photooxidation. Therefore, a likely mechanism involves an increased singlet-triplet conversion by oxygen perturbation that always results in photooxidation [468]. Finally, the amount of quenching by oxygen and the amount of peroxide formed are quite parallel in benzene and some of its derivatives [470]. The latter, however, is not true for anthracene itself. At low concentrations, 9-methyl anthracene shows a considerably higher quantum yield of fluorescence in the absence of oxygen than with the oxygen present although the quantity of peroxide formed (in the presence of the oxygen) is extremely small [471]. Thus oxygen is quenching fluorescence although the expected chemical reaction does not occur.

B. CARBONYLS

Another important type of molecule participating in photochemistry is that containing the carbonyl group. Here we shall be concerned primarily with *inter*molecular triplet-triplet energy transfer. This does not mean that intermolecular transfer is excluded when hydrocarbons are involved. For example, in Chapter 15 it was noted that the quenching of a hydrocarbon triplet could be accomplished by other hydrocarbons or their derivatives and that this involved triplet-triplet energy transfer. In fact, in this same chapter reference was also made to ketones. However, it seems more appropriate to stress the participation of ketones in energy transfer and photochemical phenomena here.

1. Energy Transfer

The earliest observation on triplet-triplet energy transfer principally involved aldehyde or ketone donors and hydrocarbons or their derivatives as acceptors [472]. If, say, benzaldehyde and naphthalene are in a rigid glass solution at 77°K, irradiation at a wavelength where the absorption of naphthalene is nonexistent still results in phosphorescence of naphthalene. The irradiation is in a region where the aldehyde does have absorption. Naphthalene alone irradiated at the same wavelengths does not emit phosphorescence. The mechanism involves excitation of the donor (benzaldehyde) triplet with subsequent sensitization of the acceptor (naphthalene) triplet followed by phosphorescence emission of the acceptor. It would be expected that the mean lifetime of phosphorescence of the donor would be shortened by virtue of the radiationless transfer to an acceptor and this is true for many combinations of sensitizers and acceptors. In addition, in some cases the quantum yield of emission of the acceptor is greater by sensitization than by direct excitation. The relative probability of transfer from donors to acceptor varies; for example, for the benzophenone-naphthalene system the relative probability is 0.75, whereas for carbazole-naphthalene it is 2.3 [472]. The energy transfer process cannot occur by the trivial mechanism of emission reabsorption since the acceptor has no absorption in the region of the donor emission (see below for further discussion).

Another method for detecting triplet energy transfer involves the observation of triplet-triplet absorption of the acceptor molecule [473]. For example, a mixture of benzophenone-naphthalene or biacetyl-benz(*a*)-anthracene, when flashed in the region of the donors only (the ketones), results in the observation of triplet-triplet absorption for the acceptors. In addition, no fluorescence of the acceptors is noted thus proving that the formation of the triplet state of the acceptor does not occur via its singlet state. The energy transfer process can be represented as

$$D^*(\text{triplet}) + A(\text{singlet}) \rightarrow D(\text{singlet}) + A^*(\text{triplet}) \qquad (126)$$

where D and A represent donor and acceptor, respectively. Experimental evidence indicates the mechanism for energy transfer is best explained by an exchange interaction involving overlap of electron charge clouds.

Considerable interest regarding triplet energy transfer centers around the system benzene-biacetyl. In this case, benzene is the donor and biacetyl is the acceptor. A number of molecules that have triplet states higher than that of biacetyl sensitize the phosphorescence of biacetyl in solution [474]. It is also possible to sensitize biacetyl *fluorescence* with benzene in solution [475]. Further, benzene sensitizes biacetyl phosphorescence in the gas phase giving a moderate increase in the quantum yield ratio of phosphorescence to fluorescence [476,477] for biacetyl. In fluid solution, however, it is found that benzene does not sensitize the phosphorescence of biacetyl [478]. The similarity of the phosphorescence quantum yields of biacetyl in solution, vapor, and rigid media suggests that yield of triplet should be comparable in the same phases. Also, since the phosphorescence of biacetyl can be sensitized in a rigid medium, it would seem likely that the energy transfer process would occur in fluid solution at almost each collision. Based on the preceding results it appears that the most likely explanation for the lack of sensitization of phosphorescence of biacetyl in solution is that the lifetime of the benzene triplet is very short. The lifetime is estimated to be comparable to that of the singlet ($\sim 2 \times 10^{-8}$ sec). Other information based primarily on the concentration dependence of the intensity ratio of biacetyl phosphorescence to benzene fluorescence (instead of biacetyl fluorescence, which would be the usual case [475,476]), indicates that some triplet transfer from benzene to biacetyl does occur [479]. With one assumption the lower limit of lifetime of the triplet benzene is estimated to be $\sim 10^{-6}$ sec. The difference in the estimated lifetimes of the benzene triplet likely resides in the inherent difference in the sensitivity of the methods employed.

Finally, triplet energy transfer can occur in plastics at room temperature as well as at low temperature (77°K) [480]. The results on the lifetime of decay of naphthalene at various concentrations and in the presence of benzophenone apparently rule out the formation of donor-acceptor complexes. However, the results do not support or refute the fact that the mechanisms of transfer are by exchange interactions.

2. Photoreactions

Since the early observations of photochemical activity of benzophenone-alcohol systems exposed to sunlight [481], a vast literature has accumulated on the photochemistry of benzophenone and other ketones and aldehydes such as benzaldehyde. In this section we shall review some of the salient points in the modern photochemical discussion of some of these reactions.

One of the interesting problems is the photochemistry of benzophenone in various alcohols in the *absence* of oxygen. Our first concern is with the primary steps

$$\phi_2CO + h\nu = \phi_2CO^* \tag{127}$$

$$\phi_2CO^* + RH = \phi_2COH\cdot + R\cdot \tag{128}$$

The points to be considered are the nature of the excited state of benzophenone and the proof of the existence of the ketyl radical. Flash-photolysis experiments on benzophenone confirm that the transient produced is the ketyl radical [482]. The initial photochemical reactions of benzophenone could be the following [482]:

$$\phi_2CO + h\nu \rightarrow \phi_2CO^* \text{ (singlet)} \tag{129}$$

$$\phi_2CO^*(S) \rightarrow \phi_2CO^* \text{ (T, triplet)} \tag{130}$$

$$\phi_2CO^*(T) \rightarrow \phi_2CO + h\nu \tag{131}$$

$$\phi_2CO^*(S) + RH \rightarrow \phi_2COH\cdot + R\cdot \tag{132}$$

$$\phi_2CO^*(T) + RH \rightarrow \phi_2COH\cdot + R\cdot \tag{133}$$

The important question to resolve is whether reaction (132) or (133) is the correct one. It is known that naphthalene can quench the triplet state of benzophenone [445]. When benzophenone is flashed in the presence of naphthalene ($\sim 10^{-3}$ *M*), no ketyl radical is detected. Also benzophenone, when irradiated in the presence of naphthalene in isopropyl alcohol, does not produce any precipitate of benzopinacol. It thus appears quite certain that reaction (132) is unimportant and that the important spectroscopic state is the triplet in the primary photochemical step (128 or 133).

The next problem concerns the following steps that lead to the product benzopinacol. It appears that ketyl radicals dimerize to give the benzopinacol. The detailed behavior of the reaction appears to depend on the relative quantity of oxygen present; that is, if air is excluded during the reaction (except for that initially present in the reaction vessel), the reaction proceeds as follows in isopropyl alcohol [483]:

$$\phi_2CO^* + MeCHOH \rightarrow \phi_2COH\cdot + Me_2COH\cdot \tag{134}$$

$$Me_2COH\cdot + \phi_2CO \rightarrow \phi_2COH\cdot + Me_2CO \tag{135}$$

$$2\phi_2COH\cdot \rightarrow \phi_2\underset{\displaystyle OH}{\overset{}{\underset{|}{C}}}-\underset{\displaystyle OH}{\overset{}{\underset{|}{C}}}-\phi_2 \tag{136}$$

In systems containing some initial residual oxygen and irradiated with low-intensity light, the quantum yield is about one-half. The quantum yield for benzopinacol is 0.93 at 3660 Å (at high intensity) in oxygen-free solutions. The quantum yield for acetone is 0.92 at 3660 Å and 0.99 at 3130 Å and is independent of the quantity of oxygen in solution. If oxygen is "totally absent," then the detailed reaction mechanism appears to require an intermediate of the type $\phi_2COHOC\phi_2\cdot$ which forms a dimer that then splits to produce benzopinacol and two benzophenone molecules [484]. If oxygen is present during the reaction, the yield of benzopinacol is reduced to zero even though the yield of acetone is not reduced. In addition, hydrogen peroxide is produced and benzophenone is not consumed in the reaction. Mechanisms for the reaction have been suggested [483].

Biacetyl also undergoes photochemical reaction in hexane containing oxygen [484]. The reaction appears to produce acetic acid. The mechanism presumably involves the triplet state of biacetyl reacting with the solvent by hydrogen extraction to produce a radical that then reacts with oxygen to form a peroxide. Following this, hydrogen extraction from the solvent occurs giving a product that splits to give two molecules of acetic acid. The triplet state of biacetyl is quenched by a photochemical reaction involving hydrogen abstraction from such molecules as primary and secondary amines, phenols, and aldehydes as well as alcohols [485].

A considerable number of other reactions utilizing benzophenone as a sensitizer have been investigated. Ethyl pyruvate can undergo photochemical decomposition to acetaldehyde, carbon monoxide, and a small amount of carbon dioxide. In the presence of benzophenone as a sensitizer, the same products result but the quantum yield is higher [486]. The decomposition of diazomethane is photosensitized by benzophenone to give triplet methylene radical. The addition of triplet methylene to cyclohexene gives products that differ primarily in the relative quantities formed compared to those formed by the addition of singlet methylene (formed by direct irradiation of diazomethane). Further, many carbonyl molecules, including benzophenone, acetone, biacetyl, and naphthaldehyde, have been examined as triplet photosensitizers for *cis-trans* isomerization [487]. Energy transfer occurs at nearly every collision between a triplet and acceptor molecule if the transfer is exothermic (triplet of acceptor below that of donor). For example, 9-anthraldehyde (low-lying triplet) will *not* isomerize *cis*-piperylene while biacetyl will.

Of a large number of substituted benzophenones irradiated in an alcohol, some can be photoreduced to pinacols and some cannot [264]. For example, *m*- and *p*-methyl-, *o*-, *m*-, *p*-chloro-, *o*, *p*-dimethoxy- and *p*-methoxybenzophenones give pinacols; while nitro-, *o*-hydroxy-*p*-methoxy-, *o*-amino-, *p*, *p*′-bis(dimethyl amino)-, *o*,*p*-dihydroxybenzophenones give no (or very low quantum yields) pinacols. In certain cases, such as the *o*-amino, hydroxy or methyl derivatives, it appears that on irradiation, photoenolization occurs. This may be responsible, at least in part, for the lack of reactivity since the enols are apparently not photoreduceable to the pinacol. If the above groups are replaced by methoxy or carboxy groups, the *inter*molecular hydrogen abstraction capability is restored and photoreduction can occur. Although this may account for the results in some of the cases, it is probably not a satisfactory explanation for the behavior of such molecules as *p*-amino- and *p*,*p*′-bis(dimethylamino)benzophenone. In these cases, it is postulated that the lowest triplet state is not of n, π^* character but of π, π^* type. Presumably the difference in charge distribution of the π, π^* would be such that the diradical type distribution expected for the n, π^* case would not exist and the chemical reactivity of the oxygen site (to extract hydrogen) would be absent. Other evidence based on racemization experiments involving aldehydes and ketones and optically active alcohols indicates that certain ones of the ketonic sensitizers do not give racemization [271]. In one case, 2-acetonaphthone, it is thought that the phosphorescence has its origin from a π, π^* triplet [266]. It is generally thought that the unreactive group of ketones and aldehydes have π, π^* triplet states lowest [271]. Certain anthraquinones undergo photochemical reaction (1-chloro, and 1,5-dichloro) while others do not (1,8-, 1,5-, 1,4-dihydroxy). It is probable that either or both photoenolization and the nature of the triplets are responsible for the results.

C. GENERAL PHOTOCHEMISTRY AND PHOTOCHROMISM

One unique approach to photochemistry involves irradiation into the multiplicity forbidden $T_1 \leftarrow S_0$ absorption band; for example, if the system

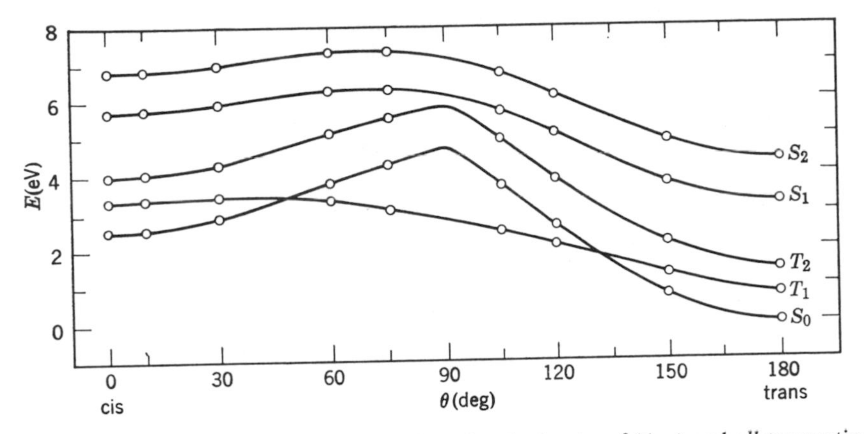

Figure 48 Energies of the ground state S_0 and excited states of 11-*cis* and *all-trans* retinals as a function of angle of twist about 11–12 bond.
Reprinted by permission from MacMillan (Journals) Ltd., K. Inuzuka and R. S. Becker, *Nature*, **219**, 383 (1968).

benzene-nitric oxide is excited with light in the 2900–3400 Å region, both *o*-nitrophenol and 2,4-dinitrophenol are formed [488]. The quantum yield of each of these products is about 0.012 when the mixed system is irradiated with light of 3130 Å. Also, if dry benzene with oxygen at 100 atmospheres is irradiated in the same spectral region as above, phenol and probably *o*-benzoquinone are produced. The quantum yield of phenol is approximately 0.03. In each case it appears that the benzene is excited to the triplet and reacts with the O_2 or nitric oxide in the ground state.

There are quite a few examples of photochemically induced *cis-trans* isomerizations such as that of *p*-dimethylaminocinnamic acid nitrile [*p*-$(CH_3)_2N—C_6H_4—CH{\equiv}CH—CN$)]. The *cis* and the *trans* forms are interconvertible and the sum of the quantum efficiencies is unity [489]. It is postulated that in both cases the intermediate state is a triplet of very short lifetime.

There is now a more quantitative insight into the mechanism of *cis-trans* isomerization of retinals employing a polyene aldehyde, $(C{=}C)_5CH = 0$, as a model [490]. Figure 48 shows the potential energy curves of the ground and excited states of 11-*cis* and all-*trans* retinal. These are obtained by first calculating state energies within the self-consistent field molecular orbital-configurational interaction (SCF-MO-CI) framework and employing a bond-order correction [490]. Then, the state energies are calculated as a function of the angle of twist around the 11–12 bond. From Figure 48, it can be seen that the lowest triplet potential energy curve is unique in that there is essentially no barrier to *cis* → *trans* isomerization. Furthermore, the triplet potential energy curve crosses the ground-state curve at approximately

45 and 135 degrees. In view of this crossing it might be expected that excitation of the 11-*cis* isomer would result in essentially no emission but that photoisomerization would occur to the *all-trans* isomer. The latter would be expected to occur principally via the triplet state. On the other hand excitation of the *all-trans* isomer would more likely result in some emission (fluorescence) and little or no isomerization. The latter would be expected as a result of the difference in the barrier for *trans* → *cis* isomerization compared to that for *cis* → *trans*. These predictions are in harmony with the experimental results [340,491,492].

Photochemical studies of the phenyloxiranes (I)

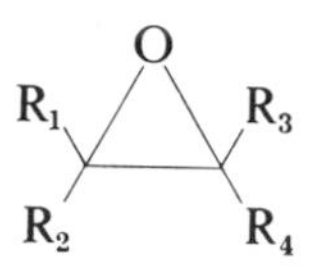

where R_1, R_2, R_3, R_4 = phenyl or hydrogen at 77°K show that a colored intermediate is the primary product [493] for the di-, tri-, and tetraphenyl derivatives. Furthermore, with additional data from the dinaphthyloxirane ($R_2 = R_3$ = naphthyl $R_1 = R_4$ = hydrogen) the intermediate can be shown to undergo photochemical reaction to final photoproducts (as well as recyclization to the original oxirane) [493].

The nature of the final photoproducts clearly indicates that for all polyphenyl substituted oxiranes, splitting of bonds in pairs occurs where all pair combinations are possible [493]; that is, a C—C bond and one or the other of the C—O bonds split as well as no C—C bond but both C—O bonds split. It is interesting that in the latter case, oxygen atoms should be ejected. Both diphenyl [493,494] and monophenylcarbenes [493] can be photochemically produced and each show emission. The emission from the diphenyl carbene appears to be a fluorescence [495] and the ground state for both carbenes is triplet [494].

Considerably more is now known about the photochemistry of molecules showing photochromism [496–511]. In a general sense, any compound undergoing thermally reversible photochemistry might be called *photochromic*, even if none or both of the forms are colored, since their absorption spectra will differ. Nonetheless the term is normally employed only when one species is colored. Although the phenomenon has been quite intensively studied, information on the mechanism of the process is scarce and a comprehensive review emphasizes that "no general theory is available to explain all phototropic (photochromic) reactions" [497]. It has been proposed that only a few basic reaction mechanisms are operating in the known cases of organic photochromism and, moreover, that one of these mechanisms can explain the great majority of the cases [509]. This proposal is presented in what follows.

The requirements for photochromism of a pure compound A in solid state or an inert solvent are (a) on absorption of light it changes into a photocolored form B; (b) B undergoes a reasonably fast thermal back reaction to A without appreciable side reactions. Thus among the basic possibilities for the process are (a) *photoisomerization*—either a multistep photoisomerization $A \rightarrow B + C \rightarrow D$ followed by the reverse process $D \rightarrow B + C \rightarrow A$, or a one step $A \rightarrow B$ followed by $B \rightarrow A$; (b) *photodissociation*—$A \rightarrow B + C$ followed by recombination $B + C \rightarrow A$; (c) *photoionization*—$A \rightarrow A^+ + e$ followed by electron capture $A^+ + e \rightarrow A$; (d) *population of a metastable electronically excited state*, such as a triplet—$A \rightarrow A^*$ followed by return to the electronic ground state $A^* \rightarrow A$.

First, brief mention shall be made of the latter three cases since they apply to only a few known classes of photochromic compounds.

Photodissociation is a fairly common process but the fragments are usually highly reactive and fail to recombine to give the starting compound in good yield. This kind of recombination is facilitated if the process is carried out in a solid or if it is an ionic reaction. A known case of the first type in solids is the photochromism of 2,3,4,4-tetrachloro-1(4H)-naphthalenone [510]. The latter type of reaction, ionic, is closely connected with the well-known [511] phenomenon of dramatic changes in basicity or acidity on electronic excitation. This has been studied so far in most detail on nitrogen or oxygen acids and bases such as phenols or amines and has not been connected with photochromism both because the proton transfer reactions in these compounds are extremely fast and because the changes in the wavelengths of absorption are small. However, it is quite straightforward to extend this concept to the pseudobasicity of carbonium ions. Here the differences in wavelength of absorption of the covalent and ionic forms are large and a simple Förster-cycle [511] type reasoning predicts a large decrease in pseudobasicity and a strongly increased dissociation in the excited state. At present the only class of compounds known to exhibit this type of photochromic behavior are the triphenylmethane and trinaphthylmethane dye derivatives [497].

Photoionization followed by electron capture seems likely to occur only in solids. This process has been suggested to explain the photochromism of sydnones [512].

Population of a metastable electronically excited state is not commonly considered a case of photochromism. A well-known procedure to produce this kind of photochromism is to use high-intensity flash techniques to invert the electronic state occupation in favor of the lowest triplet followed by triplet-triplet absorption.

Photoisomerization involving the multistep mechanism is not very common.

It is related to the dissociative one (b) but the ion is only an intermediate leading to an isomer of A. A case of photochromism by this mechanism is probably found in ω-nitrotoluene [513].

Although other cases are undoubtedly possible, we shall focus our interest on the one-step photoisomerization A $\rightarrow$ B followed by isomerization B $\rightarrow$ A. This would *a priori* seem to offer fewer complications and a greater possible variety of recation conditions.

The number of known types of photochemical reactions, A $\xrightarrow{h\nu}$ B, whose thermal back reactions proceed spontaneously, B $\rightarrow$ A, at a reasonably high rate at or below room temperature is not very large. One such possibility is *cis-trans* isomerization which is indeed believed to be responsible for the photochromism of formazanes [514], azo compounds [515], and thioindigo dyes [497]. However, spontaneous *cis-trans* isomerization of a C=C double bond (the back thermal reaction) is usually quite slow.

It appears that *valence isomerization* in the six-membered ring simply and formally depicted by

$$\text{(six-membered ring: open form)} \rightleftharpoons \text{(six-membered ring: closed form)} \tag{137}$$

is the common factor permitting explanation of the greatest majority of the known cases of organic photochromism [509] as shown below. Three general cases of this common feature are possible if no fragmentation is to occur.

$$\underset{\text{I}}{\left[T_1{-}B_1{-}M,\ T_1{=}T_2,\ T_2{-}T_3{=}B_2\right] \rightleftharpoons \left[T_1{=}B_1\ M,\ T_2{=}T_3{-}B_2\right]};\quad \underset{\text{II}}{\left[B_1{=}T_1{-}B_3,\ B_2{=}T_2{-}B_4\right] \rightleftharpoons \left[B_1{-}T_1{=}B_3,\ B_2{-}T_2{=}B_4\right]};\quad \underset{\text{III}}{\left[T_2{=}T_1{-}B_1,\ T_3{=}T_4{-}B_2\right] \rightleftharpoons \left[T_2{-}T_1{=}B_1,\ T_2{=}T_3,\ T_4{=}B_2\right]}$$

Here, T denotes a tervalent atom or group such as =CR—, =N—, =N$^+$(R)—, =N$^+$(O)—, etc., B is a bivalent one such as —CR$_2$—, —CO—, —O—, —NR—, etc., and M a monovalent one (in all known cases this is hydrogen although others may be envisaged). According to the nomenclature proposed by Woodward and Hoffmann [516], the first two cases, I and II, represent sigmatropic reactions while Case III is an electrocyclic reaction. The rules proposed by these authors give more information about the stereochemistry of the process with which we are not concerned here (e.g., the product of the back reaction may be stereoisomeric with the starting compound).

The relative stability of the two forms depends on circumstances. If an aromatic ring is fused to one side of the six-membered ring, the stability difference will generally markedly increase; that is, the isomeric form having a double bond across the fused location will be favored over the other.

Case I is very often equivalent to photoenolization. Strong evidence exists for photoenolization being responsible for the photochromism of *o*-nitrobenzyl compounds [497,517,518] and salicylaldehyde anils [498,499,502,505]. The same kind of reversible photochemistry occurs in *o*-benzylbenzophenone and derivatives [519], in *o*-hydroxybenzoyl compounds [520], and seems to complicate the *trans-cis* photoconversion of *o*-hydroxy derivatives of azo compounds [521]. It has also been invoked to explain the absence of photopinacolization of *o*-amino and *o*-hydroxybenzophenone, and also of 1-hydroxyanthraquinone [264].

For case II, no examples of this case seem to have been described thus far although again many possibilities can be visualized, such as

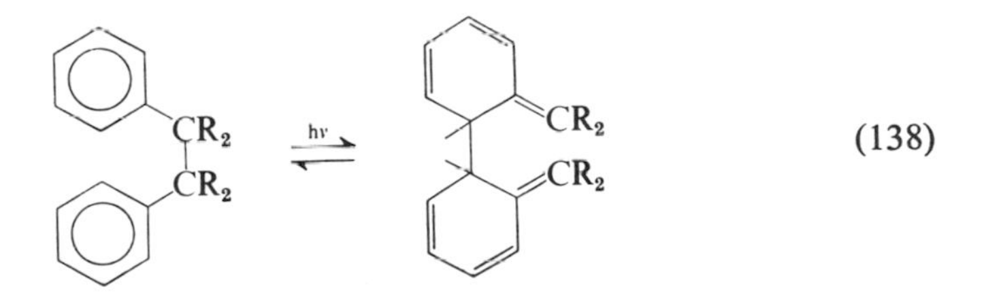

(138)

Case III includes the well-known reversible photochemical and thermal transformations of hexatrienes and cyclohexadienes, of cyclohexadienones and unsaturated ketenes [522], usually not thought of as photochromism because absorption changes occur in the ultraviolet region, or because the thermal back-reaction often requires heating or both. Both of these factors are greatly modified if one or more benzene rings is condensed to the six-membered ring as discussed above. The reversible hexamethylstilbene-hexamethyldihydrophenanthrene photocyclization has been reported [523] and the same mechanism has been suggested independently to account for the photochromism of bianthrone although competitive reactions complicate the results [524,525]. The long-known but previously unexplained photochromism of fulgides and fulgic acids seems also to occur via the same mechanism, III, [506],

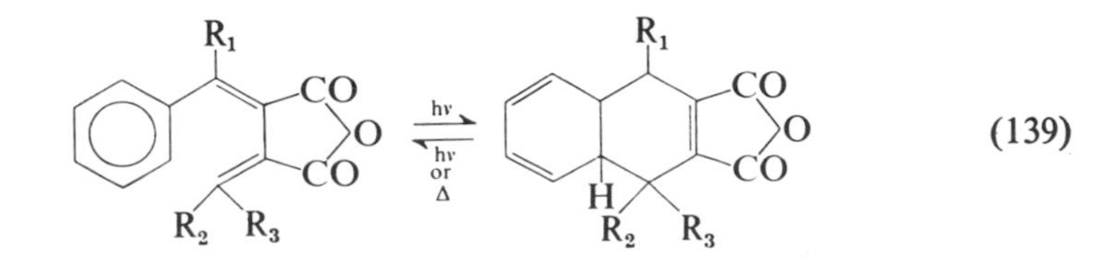

(139)

where, for example, $R_1 = R_2 = H$, $R_3 = Ph$. Among benzo derivatives, the 2-H-chromenes [501,503,504] and the well-known spiropyrans [496,500,503,508] serve as prime examples of this case, III, for pyrans see [504],

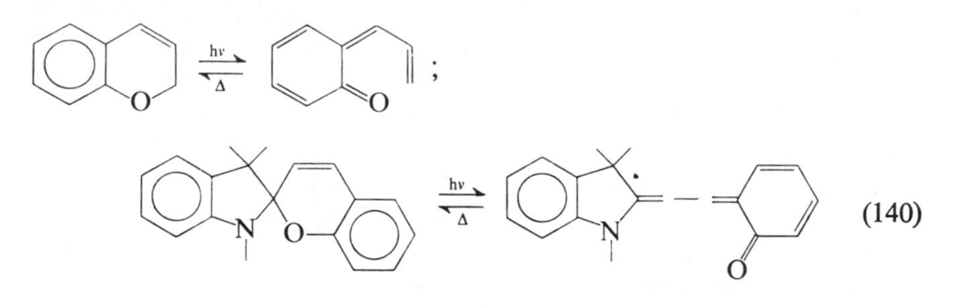

(140)

References

[1] J. R. Platt, *J. Chem. Phys.*, **17,** 484 (1949).
[1a] R. G. Parr, *The Quantum Theory of Molecular Electronic Structure.* Benjamin, New York, 1963.
[2] L. S. Brooker, *J. Amer. Chem. Soc.*, **73,** 5332 (1951); **73,** 5350 (1951).
[3] N. S. Bayliss, *J. Chem. Phys.*, **18,** 292 (1950).
[4] N. S. Bayliss and L. Hulme, *Aust. J. Chem.*, **6,** 257 (1953).
[5] N. S. Bayliss and E. G. McRae, *J. Phys. Chem.*, **58,** 1002 (1954); **58,** 1006 (1954).
[6] E. G. McRae, *J. Phys. Chem.*, **61,** 562 (1957).
[7] E. G. McRae, *Spectrochim. Acta.*, **12,** 192 (1958).
[8] Y. Ooskika, *J. Phys. Soc. Japan*, **9,** 594 (1954).
[9] H. C. Longuet-Higgins and J. A. Pople, *J. Chem. Phys.*, **27,** 192 (1957).
[10] O. E. Weigang, Jr , *J. Chem. Phys.*, **33,** 892 (1960).
[11] O. E. Weigang, Jr., and D. P. Wild, *J. Chem. Phys.*, **37,** 1180 (1962).
[12] W. W. Robertson, O. E. Weigang, Jr., and F. A. Matsen, *J. Molec. Spec.*, **1,** 1 (1957).
[13] R. S. Becker, *J. Molec. Spec.*, **3,** 1 (1959).
[14] N. Mataga, Y. Kaifu, and M. Koizumi, *Bull. Chem. Soc. Japan*, **29,** 465 (1956).
[15] N. Mataga, Y. Kaifu, and M. Koizumi, *Bull. Chem. Soc. Japan*, **29,** 115 (1956).
[16] E. Lippert, *Z. Physik*, *Chem.* (*N.F.*), **125,** 1956; *Z. Naturforsch*, **10a**, 541 (1955).
[17] R. B. Woodward, *J. Amer. Chem. Soc.*, **63,** 1123 (1942).
[18] L. F. Fieser and M. Fieser, *Steroids*, Reinhold, New York, 1959.
[19] G. J. Brealey and M. Kasha, *J. Amer. Chem. Soc.*, **77,** 4462 (1955).
[20] G. C. Pimentel, *J. Amer. Chem. Soc.*, **79,** 3323 (1957).
[21] R. S. Becker, *J. Molec. Spec.*, **3,** 1 (1959).
[22] W. W. Robertson, S. E. Babb, and F. S. Matsen, *J. Chem. Phys.*, **26,** 367 (1957).
[23] W. W. Robertson and A. D. King, *J. Chem. Phys.*, **34,** 1511 (1961); ibid., **34,** 2190 (1961); W. W. Robertson O. E. Weigang, Jr., and F. A. Matsen, *J. Molec. Spec.*, **1,** 1 (1957).
[24] W. W. Robertson, *J. Chem. Phys.*, **33,** 362 (1960).
[25] H. G. Drickamer and J. C. Zahner, *Adv. Chem. Phys.*, **4,** 161 (1962), and references therein.
[26] D. R. Stephens and H. G. Drickamer, *J. Chem. Phys.*, **30,** 1364 (1959).
[27] D. R. Stephens and H. G. Drickamer, *J. Chem. Phys.*, **30,** 1518 (1959).
[28] J. R. Gott and W. G. Maisch, *J. Chem. Phys.*, **39,** 2229 (1963).
[29] H. H. Jaffe and M. Orchin, *Theory and Applications of Ultraviolet Spectroscopy*, Wiley, New York, 1962.
[30] J. Platt, *J. Chem. Phys.*, **8,** 1168 (1950); J. Platt, *J. Opt. Soc. Amer.*, **43,** 252 (1953).
[31] J. Collomon and K. Innes, *J. Molec. Spec.*, **10,** 166 (1963).
[32] R. Shimada and L. Goodman, *J. Chem. Phys.*, **43,** 2027 (1965).
[33] T. M. Ballester, J. Riera, and L. Spialter, *J. Amer. Chem. Soc.*, **86,** 4276 (1964), and Reference 43.
[34] K. Inuzuka, *Bull. Chem. Soc. Japan*, **33,** 500 (1960).

[35] M. Ito, H. Tsukioka, and S. Imanishi, *J. Amer. Chem. Soc.*, **82,** 1559 (1960).
[36] H Klevens and J. Platt, *J. Chem. Phys.*, **17,** 470 (1949).
[37] E. Clar, *Aromatische Kohlenwasserstoffe*, 2nd ed., Springer-Verlag, Berlin (1952), and English Translation (1964), Academic Press, N.Y.
[38] W. Moffitt, *J. Chem. Phys.*, **22,** 1820 (1954).
[39] W. Moffitt, *J. Chem. Phys.*, **22,** 320 (1954).
[40] D. P. Craig and L. E. Lyons, *J. Chem. Phys.*, **20,** 1499 (1952), gives a summary of experimental results.
[41] D. S. McClure, *J. Chem. Phys.*, **22,** 1256, 1668 (1954).
[42] R. N. Jones, *Chem. Rev.*, **41,** 353 (1947).
[43] J. R. Platt, *J. Chem. Phys.*, **19,** 1418 (1951); see also References 1 and 36.
[44] D. P. Craig and P. C. Hobbins, *J. Chem. Soc.*, 2303 (1955).
[45] J. Sidman, *J. Chem. Phys.*, **25,** 115 (1956).
[46] A. Bree and L. Lyons, *J. Chem. Soc.*, 5206 (1960).
[47] R. S. Becker, I. S. Singh, and E. A. Jackson, *J. Chem. Phys.*, **38,** 2144 (1963).
[48] W. E. Wentworth and R. S. Becker, *J. Amer. Chem. Soc.*, **84,** 4263 (1962).
[49] R. S. Becker and W. E. Wentworth, *J. Amer. Chem. Soc.*, **85,** 2210 (1963).
[50] R. S. Becker and E. Chen, *J. Chem. Phys.*, **45,** 2403 (1966).
[51] J. Michl and R. S. Becker, *J. Chem. Phys.*, **46,** 3889 (1967).
[52] G. Hoijtink and P. Zandstra, *Molec. Phys.*, **3,** 533 (1960).
[53] K. Bowden, E. Broude, and E. Jones, *J. Chem. Soc.*, 948 (1946).
[54] S. Gronowitz, *Arkiv Kemi*, **13,** 239 (1958); L. Chierice and G. Pappalardo, *Gazz. Chim. Ital.*, **88,** 453 (1958).
[55] G. Leandri, A. Mangini, F. Montanari, and R. Passerini, *Gazz. Chim. Ital.*, **85,** 769 (1955).
[56] F. Boig, G. Casta, and I. Osvar, *J. Org. Chem.*, **18,** 775 (1953).
[57] S. Mason, *J. Chem. Soc.*, 493 (1962); 1204 (1959); 1247 (1959).
[58] Reference 29, pp. 126–127 for more detail.
[59] H. P. Stephenson, *J. Chem. Phys.*, **22,** 1077 (1954).
[60] L. Goodman and R. Harrell, *J. Chem. Phys.*, **30,** 1371 (1959).
[61] L. Goodman, *J. Molec. Spec.*, **6,** 109 (1961).
[62] R. Hirt, F. King, and J. C. Cavagnol, *J. Chem. Phys.*, **25,** 574 (1956).
[63] G. Brealey and M. Kasha, *J. Amer. Chem. Soc.*, **77,** 4462 (1955).
[64] V. Krishna and L. Goodman, *J. Chem. Phys.*, **33,** 382 (1960); *J. Amer. Chem. Soc.*, **83,** 204 (1961).
[65] S. J. Ladner and R. S. Becker, *J. Chem. Phys.*, **67,** 2481 (1963).
[66] W. T. Simpson, *J. Chem. Phys.*, **17,** 1218 (1959).
[67] H. C. Longuet-Higgins, C. W. Reactor, and J. R. Platt, *J. Chem. Phys.*, **18,** 1174 (1950).
[68] J. R. Platt, *Radiation Biology*, A. Hollaender, Ed., **3.** McGraw-Hill, New York, 1956, Chapter 2.
[69] G. R. Seeley, *J. Chem. Phys.*, **27,** 125 (1957).
[70] M. Gouterman, *J. Chem. Phys.*, **30,** 1139 (1959); *J. Molec. Spec.*, **6,** 138 (1961).
[71] H. Kobayski, *J. Chem. Phys.*, **30,** 1362 (1959).
[72] J. B. Allison and R. S. Becker, *J. Phys. Chem.*, **67,** 2675 (1963), and references therein.
[73] G. Dorough et al., *J. Amer. Chem. Soc.*, **73,** 4315 (1951); **74,** 3974 (1952); **74,** 3977 (1952).
[74] E. I. Rabinowitch, *Photosynthesis and Related Processes*, **I.** Wiley-Interscience, New York (1945); **II,** 1 (1951); **II,** 2 (1956).

[75] C. S. French, *Handbuch der Pflanzenphysiologie*, W. Ruhland, ed. Springer, Berlin, 1960, pp. 252–297.
[76] S. Freed and K. M. Sancier, *Science*, **114**, 275 (1951); **116**, 175 (1952); **117**, 655 (1953); **125**, 1248 (1957); and *J. Amer. Chem. Soc.*, **76**, 198 (1954).
[77] R. Livingston, W. Watson, and J. McArdle, *J. Amer. Chem. Soc.*, **71**, 1542 (1949).
[78] J. Fernandez and R. S. Becker, *J. Chem. Phys.*, **31**, 467 (1959).
[79] J. H. Pinchard, B. Wille, and H. Zechmeister, *J. Amer. Chem. Soc.*, **70**, 1938 (1948), and references therein.
[80] R. P. Linstead, *J. Chem. Soc.*, 2873 (1953).
[81] G. Lewis and M. Kasha, *J. Amer. Chem. Soc.*, **66**, 2100 (1944).
[82] S. Strickler and R. Berg, *J. Chem. Phys.*, **37**, 814 (1962).
[83] W. Melhuish, *J. Phys. Chem.*, **65**, 229 (1961).
[84] E. Bowen, *Trans. Faraday Soc.*, **50**, 97 (1954).
[85] G. Weber and F. Teale, *Trans. Faraday Soc.*, **53**, 646 (1957).
[86] E. Gilmore, G Gibson, and D. McClure, *J. Chem. Phys.*, **20**, 829 (1952); **23**, 399 (1955).
[87] W. Melhuish, *New Zealand J. Sci. and Tech.*, **37B**, 142 (1955).
[88] G. Weber and F. Teale, *Trans. Faraday Soc.*, **54**, 640 (1958).
[88a] J. K. Roy and R. S. Becker, unpublished.
[89] W. Ware and B. Baldwin, *J. Chem. Phys.*, **40**, 1703 (1964).
[90] M. Beer and H. Longuet-Higgins, *J. Chem. Phys.*, **23**, 1390 (1955).
[91] G. Viswanath and M. Kasha, *J. Chem. Phys.*, **24**, 574 (1956).
[92] D. Evans, *J. Chem. Soc.*, 1735, (1960).
[92a] D. S. McClure, *J. Chem. Phys.*, **17**, 665 (1949).
[93] E. Clementi and M. Kasha, *J. Molec. Spec.*, **2**, 297 (1958).
[94] M. Kasha, *Discussions Faraday Soc.*, **9**, 14 (1950).
[95] G. N. Lewis and M. Calvin, *J. Amer. Chem. Soc.*, **67**, 1232 (1945).
[96] G. N. Lewis, M. Calvin, and M. Kasha, *J. Chem. Phys.*, **17**, 804 (1949).
[97] C. Hutchinson and B. Mangum, *J. Chem. Phys.*, **29**, 952 (1958).
[98] M. de Groot and J. van der Waals, *Molec. Phys.*, **3**, 190 (1960).
[99] L. Piette, J. Sharp, T. Kuwana, and J. Pitts, *J. Chem. Phys.*, **36**, 3094 (1962).
[100] D. Shigorin, N. Volkova, A. Pisknov, and A. Gurevich, *Optics and Spec.*, **12**, 369 (1962).
[101] J. Vincent and A. Maki, *J. Chem. Phys.*, **39**, 3088 (1963).
[102] J. Brinen, J. Koren, and W. Hodgson, *J. Chem. Phys.*, **44**, 3095 (1966).
[103] A. Albrecht, *J. Chem. Phys.*, **38**, 354 (1963).
[104] A. Albrecht, *J. Chem. Phys.*, **33**, 937 (1960).
[105] E. Lim and J. Yu, *J. Chem. Phys.*, **47**, 3270 (1967).
[106] M. Gouterman, *J. Chem. Phys.*, **36**, 2846 (1962).
[107] G. Robinson and R. Frosch, *J. Chem. Phys.*, **38**, 1187 (1963); ibid, **37**, 1962 (1962).
[108] G. Hunt, E. McCoy, and I. Ross, *Aust. J. Chem.*, **15**, 591 (1962).
[109] J. Byrne, E. McCoy, and I. Ross, *Aust. J Chem.*, **18**, 1589 (1965).
[110] R. Watts and S. Strickler, *J. Chem. Phys.*, **44**, 2423 (1966).
[111] S. Strickler and R. Watts, *J. Chem. Phys.*, **44**, 426 (1966).
[112] P. Pringsheim, *Ann. Acad. Sci. Tech. Varsovie*, **5**, 29 (1938).
[113] B. Stevens and E. Hutton, *Molec. Phys.*, **3**, 71 (1960).
[114] R. Williams and G. Goldsmith, *J. Chem. Phys.*, **39**, 2008 (1963).
[115] G. Kistiakowsky and C. Parmenter, *J. Chem. Phys.*, **42**, 2942 (1965).
[116] W. Siebrand and D. Williams, *J. Chem. Phys.*, **46**, 403 (1967).
[117] W. Siebrand, *J. Chem. Phys.*, **46**, 440 (1967).

[118] W. Siebrand, *J. Chem. Phys.*, **44,** 4055 (1966).
[119] R. Kellogg and N. Wyeth, *J. Chem. Phys.*, **45,** 3156 (1966).
[120] H. Sun, J. Jortner, and S. Rice, *J. Chem. Phys.*, **44,** 2538 (1966), and references therein.
[121] S. J. Ladner and R. S. Becker, *J. Chem. Phys.*, **43,** 3344 (1965); S. J. Ladner and R. S. Becker, *J. Amer. Chem. Soc.*, **86,** 4205 (1964).
[122] S. Lower and M. ElSayed, *Chem. Rev.*, **66,** 199 (1966).
[123] P. McCartin, *Trans. Faraday Soc.*, **60,** 1 (1964).
[124] G. Porter and L. Steif, *Nature*, **195,** 991 (1962).
[125] G. Porter and L. Steif, *Bull. Soc. Chim. Belges*, **71,** 641 (1962).
[126] R. Livingston and R. Ware, *J. Chem. Phys.*, **39,** 2593 (1963).
[127] N. Mataga, U. Kaife, and M. Kouzumi, *Bull. Chem. Soc. Japan*, **29,** 115 (1956).
[128] E. McRae, *J. Phys. Chem.*, **61,** 562 (1957).
[129] N. Mataga, Y. Kaifu, and M. Kouzumi, *Bull. Chem. Soc. Japan*, **29,** 465 (1956).
[130] E. Lippert, *Z. Physik. Chem.* (*N.F.*), **6,** 125 (1956); *Z. Naturforsch.*, **10a,** 541 (1955).
[131] J. Fernandez and R. S. Becker, *J. Chem. Phys.*, **31,** 467 (1959).
[132] R. Livingston, W. Watson, and J. McArdle, *J. Amer. Chem. Soc.*, **71,** 1542 (1959).
[133] R. Livingston, *Quart. Rev.*, **14,** 174 (1960).
[134] R. S. Becker, Univ. of Houston, unpublished.
[135] V. Evstingneev, A. Gavrilova, and A. Krasnovski, *Dokl. Akad. Nauk. SSSR*, **70,** 261 (1950).
[136] J. Allison and R. S. Becker, *J. Phys. Chem.*, **67,** 2675 (1963).
[137] S. Broyde and S. Brody, *J. Chem. Phys.*, **46,** 3334 (1967).
[138] S. Ladner and R. S. Becker, *J. Phys. Chem.*, **67,** 2481 (1963).
[139] M. ElSayed, *J. Chem. Phys.*, **38,** 2834 (1963); **36,** 573 (1962).
[140] E. Lim and J. Yu, *J. Chem. Phys.*, **45,** 4742 (1966).
[141] E. Shpolskii and R. Personov, *Optics and Spec.*, **8,** 172 (1960), and references therein; T. Bolotnikova, *Optics and Spec.*, **7,** 138 (1959).
[142] M. Beer, *J. Chem. Phys.*, **25,** 745 (1956).
[143] M. Furst, H. Kallman, and F. Brown, *J. Chem. Phys.*, **26,** 1321 (1957).
[144] A. Nikitin and M. Galanin, *Optics and Spec.*, **6,** 226 (1959).
[145] E. Bowen and A. Williams, *Trans. Faraday Soc.*, **35,** 765 (1939).
[146] J. Birks and A. Cameron, *Proc. Roy. Soc.*, **249,** 207 (1959).
[147] J. Laposa, E. Lim, and R. Kellogg, *J. Chem. Phys.*, **42,** 3025 (1965).
[148] W. Ware and B. Baldwin, *J. Chem. Phys.*, **43,** 1194 (1965).
[149] R. S. Becker, Univ. of Houston, unpublished.
[150] C. Karr, *Appl. Spec.*, **13,** 15 (1959).
[151] H. Ley and H. Specker, *Z. Wiss. Phot.*, **38,** 13, 96 (1939).
[152] T. Bolotnikova, *Izv. Akad. Nauk. SSSR*, Ser Fiz, **23,** 29 (1959).
[153] E. Shpoliskii, *Soviet Phys. Uspekhi*, **3,** 372 (1960), and references therein.
[154] M. Nakamizo and Y. Kanda, *Spectrochim. Acta*, **19,** 1235 (1963).
[155] T. Förster and K. Kasper, *Z. Elektrochem.*, **59,** 976 (1955).
[156] K. Kasper, *Z. Physik. Chem.*, **12,** 55 (1957).
[157] A. Dammers-de-Klerk, *Molec. Phys.*, **1,** 141 (1958).
[158] J. Birks and I. Munro, *Luminescence of Organic and Inorganic Materials*, H. Kallmann and G. Sprunch, Eds., Wiley, New York, 1962, pp. 230–234.
[159] B. Stevens, *Spectrochim. Acta.*, **18,** 439 (1962).
[160] K. Kawooka and D. Kearns, *J. Chem. Phys.*, **45,** 147 (1966).
[161] T. Förster, *Pure Appl. Chem.*, **4,** 121 (1962).
[162] J. Ferguson, *J. Chem. Phys.*, **28,** 765 (1958).

[163] R. Hochstrasser, *J. Chem. Phys.*, **136,** 1099 (1962).
[164] J. Murrell and J. Tanaka, *Molec. Phys.*, **7,** 363 (1964).
[165] T. Azumi and S. McGlynn, *J. Chem. Phys.*, **41,** 3131 (1964), and references therein.
[166] C. Parker and C. Hatchard, *Trans. Faraday Soc.*, **59,** 284 (1963), and references therein.
[167] C. Parker, *Advances in Photochemistry*, **2,** 305, Wiley (Interscience), New York (1964).
[168] T. Azumi and S. McGlynn, *J. Chem. Phys.*, **39,** 1186 (1966).
[169] H. Sternlicht, G. Nieman, and G. Robinson, *J. Chem. Phys.*, **39,** 1610 (1963).
[170] V. Krishna, *J. Chem. Phys.*, **46,** 1735 (1967).
[171] E. Lim, C. Lazzara, M. Yank, and G. Sivenson, *J. Chem. Phys.*, **43,** 970 (1965), and references therein.
[172] W. McClain and A. Albrecht, *J. Chem. Phys.*, **44,** 1594 (1966).
[173] H. Offen and R. Elioson, *J. Chem. Phys.*, **43,** 4096 (1965).
[174] H. Offen, *J. Chem. Phys.*, **44,** 699 (1966).
[175] D. Kearns, *J. Chem. Phys.*, **36,** 1608 (1962).
[176] R. Williams, *J. Chem. Phys.*, **30,** 233 (1959).
[177] V. Krishna and L. Goodman, *J. Chem. Phys.*, **37,** 912 (1962).
[178] M. ElSayed and T. Pavopoulos, *J. Chem. Phys.*, **39,** 834 (1963).
[179] T. Azumi and S. McGlynn, *J. Chem. Phys.*, **37,** 2413 (1962), and references therein.
[180] M. ElSayed, *Nature*, **197,** 481 (1963).
[181a] D. McClure, *J. Chem. Phys.*, **20,** 682 (1952).
[181b] D. McClure and S. Koida, *J. Chem. Phys.*, **20,** 765 (1952).
[182] W. Dawson, private communication (1967); W. R. Dawson and M. W. Windsor, *J. Phys. Chem.*, **72,** 3251 (1968); W. R. Dawson and J. Kroop, *J. Phys. Chem.*, to be published (1969).
[182a] M. W. Windsor and W. R. Dawson, *Molecular Crystals*, **4,** 253 (1968).
[183] M. Wright, R. Frosch, and G. Robinson, *J. Chem. Phys.*, **33,** 934 (1960).
[184] G. Robinson, *J. Molec. Spec.*, **6,** 58 (1961).
[185] C. A. Hutchinson and B. W. Mangum, *J. Chem. Phys.*, **32,** 26 (1960).
[186] M. deGroot and J. van der Waals, *Molec. Phys.*, **4,** 189 (1961).
[187] D. Craig, *J. Chem. Phys.*, **18,** 236 (1950).
[188] S. Colson and E. Bernstein, *J. Chem. Phys.*, **43,** 2661 (1965), and E. Bernstein and S. Colson, *J. Chem. Phys.*, **45,** 3873 (1966).
[189] G. Nieman and G. Robinson, *J. Chem. Phys.*, **39,** 1298 (1963).
[190] R. Kellogg, *J. Chem. Phys.*, **44,** 411 (1966).
[191] E. Lim and J. Laposa, *J. Chem. Phys.*, **41,** 3257 (1964).
[192] R. Jones and E. Spinner, *NRC Bulletin*, No. 8, National Res. Council of Canada (1960).
[193] D. Craig and P. Hobbins, *J. Chem. Soc.*, 2309 (1955).
[194] D. Williams and W. Schneider, *J. Chem. Phys.*, **45,** 4756 (1966).
[195] R. Bennett and McCartin, *J. Chem. Phys.*, **44,** 1969 (1966).
[196] C. Parker and C. Hatchard, *Trans. Faraday Soc.*, **59,** 284 (1963).
[197] C. Parker and T. Joyce, *Chem. Commun.*, 234 (1966).
[198] T. Medinger and F. Wilkinson, *Trans. Faraday Soc.*, **62,** 1785 (1966).
[199] B. Stevens, M. Thomaz, and J. Jones, *J. Chem. Phys.*, **46,** 405 (1967).
[200] N. Ham and K. Rudenberg, *J. Chem. Phys.*, **25,** 1 (1956).
[201] M. Windsor and J. Novak, private communication (1967).
[202] B. Stevens, private communication (1967).
[203] E. Lim, *J. Chem. Phys.*, **36,** 3497 (1962).
[204] S. Siegel and H. Jedeekis, *J. Chem. Phys.*, **42,** 3060 (1965).

[205] V. Ermolaev, *Soviet Phys. Usp.*, **6**, 333 (1963).
[206] G. Robinson, *J. Molec. Spec.*, **6**, 58 (1961).
[207] M. O'Dwyer, M. ElBayoumi, and S. Strickler, *J. Chem. Phys.*, **36**, 1395 (1962).
[208] E. Lim, J. Laposa, and J. Yu, *J. Molec. Spec.*, **19**, 412 (1966).
[209] E. Lim, private communication (1967).
[210] S. Strickler, private communication (1967).
[211] R. Kellogg and R. Bennett, *J. Chem. Phys.*, **41**, 3042 (1964).
[212] A. Lamola and G. Hammond, *J. Chem. Phys.*, **43**, 2129 (1965).
[213] J. Ferguson, *J. Molec. Spec.*, **3**, 177 (1959).
[214] R. Pariser, *J. Chem. Phys.*, **24**, 250 (1956).
[215] R. Nurmukhametov and G. Goleov, *Optics and Spec.*, **13**, 384 (1962).
[216] E. Shpolskii, L. Klimova, and R. Personov, *Optics and Spec.*, **13**, 188 (1962).
[217] A. Khesine, *Optics and Spec.*, **10**, 319 (1961).
[218] H. Offen and B. Baldwin, *J. Chem. Phys.*, **44**, 3642 (1966).
[219] P. Jones and S. Seigel, *J. Chem. Phys.*, **45**, 4752 (1966).
[220] M. Nicol, *J. Chem. Phys.*, **45**, 4753 (1966).
[221] B. Baldwin and H. Offen, *J. Chem. Phys.*, **46**, 4509 (1967).
[222] E. Bowen and J. Sahu, *J. Chem. Phys.*, **63**, 4 (1959).
[223] A. Cherkosov, *Optics and Spec.*, **7**, 211 (1959).
[224] A. Cherkosov and R. Bember, *Optics and Spec.*, **6**, 319 (1959).
[225] V. Ermolaev and K. Switashev, *Optics and Spec.*, **7**, 399 (1959).
[226] N. Furst, H. Kallman, and F. Brown, *J. Chem. Phys.*, **26**, 1321 (1957).
[227] A. Cherkasov, *Optics and Spec.*, **6**, 315 (1959).
[228] T. Medinger and F. Wilkinson, *Trans. Faraday Soc.*, **61**, 620 (1965).
[229] J. Sidman, *J. Chem. Phys.*, **25**, 229 (1956).
[230] M. ElSayed and T. Pavlopoulos, *J. Chem. Phys.*, **39**, 1899 (1963).
[231] T. Pavlopoulos and M. ElSayed, *J. Chem. Phys.*, **41**, 1082 (1964).
[232] M. ElSayed, *J. Chem. Phys.*, **43**, 2864 (1965).
[233] J. Roy and L. Goodman, *J. Molec. Spec.*, **19**, 389 (1966).
[234] N. Chowdhury and M. ElSayed, unpublished.
[235] R. Williams, *J. Roy. Inst. Chem.*, **83**, 611 (1959).
[236] D. Hercules and L. Rogers, *Spectrochim. Acta*, 393 (1959).
[237] R. Nurnukhametov, D. Shigorin, Y. Kozlov, and V. Puchov, *Optics and Spec.*, **1**, 327 (1961).
[237a] G. Gabor, Y. Frei, D. Gegiou, N. Kaganowitch, and E. Fischer, *Israel J. Chem.*, **5**, 193 (1967).
[237b] G. Gabor, Y. Frei, and E. Fischer, *J. Phys. Chem.*, **72**, 3266 (1968).
[238] Y. Kanda, R. Shimada, and Y. Takenoshita, *Spectrochim. Acta.*, **19**, 1219 (1963).
[239] V. Ermolaev and K. Svetashev, *Soviet Phys. Usp.*, **3**, 423 (1960).
[240] J. Corkell and I. Graham-Bryce, *J. Chem. Soc.*, 3893 (1961).
[241] Y. Hirshberg, *Anal. Chem.*, **28**, 1954 (1956).
[242] B. Van Duuren, *Anal. Chem.*, **32**, 1436 (1960).
[243] H. Wirth, *Luminescence of Organic and Inorganic Materials*, ed. H. Kallman and G. Spruch. Wiley, New York, 1962, pp. 226.
[243a] J. Sidman, *J. Chem. Phys.*, **29**, 644 (1958).
[244] L. Vanquickenborne and S. McGlynn, *J. Chem. Phys.*, **45**, 4755 (1966).
[245] R. Borkman and D. Kearns, *J. Chem. Phys.*, **46**, 2333 (1966).
[246] J. Brinen, J. Koren, and Hodgson, *J. Chem. Phys.*, **44**, 3095 (1966).
[247] S. McGlynn, F. Smith, and G. Alento, *Photochem. and Photobiol.*, **3**, 269 (1964).
[248] M. Kasha, *Radiation Res. Suppl.*, **2**, 243 (1960).

[249] D. Kearns and W. Case, *J. Amer. Chem. Soc.*, **88,** 5087 (1966).
[250] J. Lhaste, A. Haug, and M. Ptak, *J. Chem. Phys.*, **44,** 648 (1966).
[251] J. Brinen and N. Orloff, *J. Chem. Phys.*, **45,** 4747 (1966).
[252] J. Hollas and L. Goodman, unpublished; however, *see J. Chem. Phys.*, **43,** 2027 (1965).
[253] M. ElSayed, *J. Chem. Phys.*, **41,** 2462 (1964).
[254] H. Dearman and A. Chen, *J. Chem. Phys.*, **44,** 416 (1966).
[255] R. Borkman and D. Kearns, *J. Chem. Phys.*, **44,** 945 (1966).
[256] J. Dubois and F. Wilkinson, *J. Chem. Phys.*, **39,** 899 (1963).
[257] G. Almy and P. Gillette, *J. Chem. Phys.*, **11,** 188 (1963).
[258] W. Kaskan and A. Duncan, *J. Chem. Phys.*, **18,** 427 (1950).
[259] H. Backstrom and K. Sandros, *Acta Chem. Scand.*, **12,** 823 (1958).
[260] D. S. McClure, *J. Chem. Phys.*, **17,** 905 (1949).
[261] H. Okabe and W. Noyes, *J. Amer. Chem. Soc.*, **79,** 801 (1952).
[262] H. Richtol and F. Klappmeier, *J. Chem. Phys.*, **44,** 1519 (1966).
[263] V. Ermolaev and A. Terenin, *Soviet Phys. Usp.*, **3,** 423 (1960), translation.
[264] J. Pitts, H. Johnson, and T. Kuwana, *J. Phys. Chem.*, **66,** 245 (1962).
[265] L. Piette, J. Sharp, T. Kuwana, and J. Pitts, *J. Chem. Phys.*, **36,** 3094 (1962).
[266] V. Ermolaev and A. Terenin, *J. Chim. Phys.*, **55,** 698 (1958).
[267] M. ElSayed, *J. Chem. Phys.*, **38,** 2834 (1963).
[268] D. K. Bredereck, T. Forster, and H. Oesterlin, *Luminescence of Organic and Inorganic Materials*, H. Kallman and G. Spruch, Eds., Wiley, New York, 1962, p. 161.
[269] K. Yoshhara and K. Kearns, *J. Chem. Phys.*, **45,** 1991 (1966).
[270] B. Piantnitskii, *Dokl. Akad. Nauk. SSSR*, **1,** 451 (1956), translation.
[271] G. Hammond and P. Leermakers, *J. Amer. Chem. Soc.*, **84,** 207 (1962).
[272] B. Cohen, H. Baba, and L. Goodman, *J. Chem. Phys.*, **43,** 2902 (1965).
[273] L. Logan and J. Ross, *J. Chem. Phys.*, **43,** 2903 (1965).
[274] B. Cohen and L. Goodman, *J. Chem. Phys.*, **46,** 713 (1967).
[275] D. Evans, *J. Chem. Soc.*, 3885 (1957).
[276] M. Chowdhury and L. Goodman, *J. Chem. Phys.*, **36,** 548 (1962).
[277] E. Lippert, *Angew. Chem.*, **23,** 695 (1961).
[278] V. Krishna and L. Goodman, *J. Chem. Phys.*, **36,** 2217 (1962).
[279] M. ElSayed and R. Brewer, *J. Phys. Chem.*, **39,** 1623 (1963).
[280] J. Sidman, *J. Molec. Spec.*, **2,** 333 (1958).
[281] E. Condon and G. Shortley, *The Theory of Atomic Spectra*. Univ. Press, Cambridge, England, 1959, p. 171.
[282] V. Ermolaev and I. Kotlyar, *Optics and Spec.*, **9,** 183 (1960).
[283] E. Lim and J. Yu, *J. Chem. Phys.*, **45,** 4742 (1966).
[284] E. Lim and J. Yu, *J. Chem. Phys.*, **49,** 3878 (1968).
[285] L. Goodman, *J. Molec. Spec.*, **6,** 109 (1961).
[286] L. Goodman and M. Kasha, *J. Molec. Spec.*, **2,** 58 (1958).
[287] L. Goodman and H. Shull, *J. Chem. Phys.*, **22,** 1138 (1954); **27,** 1388 (1957).
[288] V. Krishna and L. Goodman, *J. Amer. Chem. Soc.*, **83,** 2024 (1961).
[289] M. ElSayed and G. Robinson, *J. Chem. Phys.*, **34,** 1840 (1961).
[290] M. ElSayed and G. Robinson, *Molec. Phys.*, **4,** 273 (1961).
[291] M. ElSayed and G. Robinson, *J. Chem. Phys.*, **35,** 1897 (1961).
[292] E. Lippert, W. Luder, F. Mall, W. Nagele, H. Boss, H. Prigge, and L. Seipold-Blankenstein, *Angew. Chem.*, **73,** 695 (1961).
[293] R. Hochstrasser and C. Marzzacco, *J. Chem. Phys.*, **45,** 4681 (1966).

[294] J. Brinen, D. Rosebrook, and R. Hirt, *J. Phys. Chem.*, **67,** 2651 (1963).
[295] J. Brinen, J. Koren, and W. Hodgson, *J. Chem. Phys.*, **44,** 3095 (1966).
[296] T. Adler, *Anal. Chem.*, **34,** 685 (1962).
[297] V. Zanker and A. Reichel, *Z. Elektrochem.*, **63,** 1133 (1959).
[298] A. Kilimov and V. Voloshina, *Optics and Spec.*, **12,** 362 (1962).
[299] V. Zanker and W. Toerber, *Z. Angew. Phys.*, **14,** 43 (1962).
[300] C. Wheelock, *J. Amer. Chem. Soc.*, **81,** 1348 (1959).
[301] H. Beyer and T. Pyl, *Angew. Chem.*, **68,** 374 (1956).
[302] F. Teale and G. Weber, *Biochem. J.*, **65,** 476 (1956).
[303] A. White, *Biochem. J.*, **71,** 217 (1959).
[304] R. Cowgill, *Arch. Biochem. Biophys.*, **100,** 36 (1963).
[305] E. Fjuimore, *Biochem. Biophys. Acta.*, **40,** 251 (1960).
[306] L. Augenstein and J. Nag-Chauduri, *Nature*, **203,** 145 (1964).
[307] E. Kuntz, F. Bishai, and L. Augenstein, *Nature*, **212,** 98 (1966).
[308] S. Udenfriend, *Fluorescence Assay in Biology and Medicine*. Academic, New York, 1962.
[309] F. Teale, *Biochem. J.*, **76,** 381 (1960).
[310] Y. Vladimirov, *Dokl. Akad. Nauk, SSSR*, **136,** 960 (1961).
[311] L. Stryer, *Bioenergetics Radiation Research*, Supp., **2,** ed. L. Augenstein (1960), p. 432.
[312] F. Teale, *Biochem. J.*, **76,** 381 (1960).
[313] F. Teale, *Biochem. J.*, **80,** 14P (1961).
[314] G. Weber, *Biochem. J.*, **79,** 29P (1961).
[315] L. Longworth, *Biochem. J.*, **81,** 23P (1961).
[316] J. Nag-Chaudhuri and L. Augenstein, *Biopolymers*, Symp., 1, 411 (1964).
[317] E. Yeagers, F. Bishai, and L. Augenstein, *Biochem. and Biophys. Res. Commun.*, **23,** 570 (1960).
[318] H. Rau and L. Augenstein, *J. Chem. Phys.*, **46,** 1773 (1967).
[319] G. Weber and F. Teale, *Discussions Faraday Soc.*, **27,** 134 (1959), and references therein.
[320] V. Shore and A. Pardee, *Arch. Biochem. and Biophys.*, **60,** 100 (1956).
[321] V. Velick, *J. Biol. Chem.*, **233,** 1455 (1958).
[322] G. Weber, *Adv. Protein Chem.*, **8,** 415 (1953).
[323] C. Chadwick, P. Johnson, and E. Richards, *Nature*, **186,** 239 (1960), and references therein.
[324] R. Steiner and A. McAlister, *J. Polymer Sci.*, **24,** 105 (1957).
[325] H. Uehleke, *Naturwiss.*, **45,** 87 (1958).
[326] R. Cohn, R. Shulman, and J. Longworth, *J. Chem. Phys.*, **45,** 2955 (1966).
[327] A. Lamola, M. Gueron, T. Yamane, J. Eisinger, and R. Shulman, *J. Chem. Phys.*, **47,** 2210 (1967).
[328] J. Longworth, R. Rahn, and R. Shulman, *J. Chem. Phys.*, **45,** 2930 (1966).
[329] R. Shulman and R. Rahn, *J. Chem. Phys.*, **45,** 2940 (1966).
[330] R. Rahn, T. Yamane, J. Eisinger, J. Longworth, and R. Shulman, *J. Chem. Phys.*, **45,** 2947 (1966).
[331] B. Cohen and L. Goodman, *J. Amer. Chem. Soc.*, **87,** 5487 (1965).
[332] H. Borrensen, *Acta. Chem. Scand.*, **17,** 921 (1963).
[333] S. Udenfriend and P. Zaltmen, *Anal. Biochem.*, **3,** 49 (1962).
[334] J. Longworth, *Biochem. J.*, **84,** 104P (1962).
[335] V. Kleinwachter, J. Trabnil, and L. Augenstein, *Photochem. and Photobiol.*, **5,** 579 (1966).

[336] L. Argoskin, N. Korolev, I. Kulaev, N. Mesel, and N. Pomoshcinkova, *Dokl. Akad. Nauk. SSSR*, **131,** 1440 (1960).
[337] J. Drobnik and L. Augenstein, *Photochem. and Photobiol.*, **5,** 13 (1966).
[338] J. Drobnik and L. Augenstein, *Photochem. and Photobiol.*, **5,** 83 (1966).
[339] J. Koziol, *Photochem. and Photobiol.*, **5,** 41 (1966).
[340] D. Balke and R. S. Becker, *J. Amer. Chem. Soc.*, **89,** 5061 (1967).
[341] A. Guzzo and G. Pool, *Science*, **159,** 312 (1968).
[342] C. Dhere and O. Biermacher, *Comp. Rend.*, **202,** 442 (1936), and references therein.
[343] A. Stern and Pruckner, *Z. Physik. Chem.*, **185A,** 140 (1939); **180A,** 321 (1937), and references therein.
[344] F. Hourowitz, *Ber. deut. Chem. Ges.*, **68,** 1795 (1935).
[345] V. Albers, H. V. Knorr, and D. Fry, *J. Chem. Phys.*, **10,** 700 (1942); H. Knorr and V. Albers, *Phys. Rev.*, **57,** 347 (1940).
[346] R. Becker and M. Kasha, *J. Amer. Chem. Soc.*, **77,** 3669 (1953).
[347] J. Allison and R. S. Becker, *J. Chem. Phys.*, **32,** 1410 (1960).
[348] R. S. Becker and J. Allison, *J. Phys. Chem.*, **67,** 2662 (1963).
[349] R. S. Becker and J. Allison, *J. Phys. Chem.*, **67,** 2669 (1963).
[350] J. Allison and R. S. Becker, *J. Phys. Chem.*, **67,** 2675 (1963).
[351] C. Dhere and A. Raffy, *Comp. Rend.*, **200,** 1367 (1935), and references therein.
[352] F. Zscheile and P. Harris, *J. Phys. Chem.*, **47,** 623 (1943).
[353] E. Rabinowitch. *Photosynthesis and Related Processes*, **1,** Wiley-Interscience, New York, 1945; **2,** *Part 1*, 1951: **2,** *Part 2*, 1956.
[354] F. Johnson, ed., *The Luminescence of Biological Systems*. Amer. Assoc. Adv. Sci., Washington, D.C. (1955).
[355] H. Gaffron, Ed., *Research in Photosynthesis*, Wiley-Interscience, New York, 1957.
[356] C. French, J. Smith, H. Virgin, and R. Airth, *Plant Physiol.*, **31,** 369 (1956).
[357] V. Gachkovskii, *Chem. Abs.*, **45,** 41b (1951).
[358] M. Gouterman, *J. Molec. Spec.*, **6,** 138 (1961).
[359] I. Singh and R. Becker, *J. Amer. Chem. Soc.*, **82,** 2083 (1960).
[360] P. Latimer, T. Bannister, and E. Rabinowitch, *Science*, **124,** 3222 (1956).
[361] S. Brody and E. Rabinowitch, *Science*, **125,** 555 (1957).
[362] W. Kosonocky, S. Harrison, and R. Standu, *J. Chem. Phys.*, **43,** 831 (1965).
[363] M. Gurevich and K. Solov'yev, *Dokl. Akad. Nauk., SSSR*, **5,** 291 (1961).
[364] W. Ohnesorge and L. Rogers, *Spectrochim. Acta.*, **14,** 27, 41 (1959).
[365] O. Popovich and L. Rogers, *Spectrochim. Acta.*, **16,** 49 (1960).
[366] D. Bhatnagar and L. Forster, *Spectrochim. Acta.*, **21,** 1803 (1965).
[367] S. Weissman, *J. Chem. Phys.*, **10,** 214 (1942); P. Yuster and S. Weissman, *J. Chem. Phys.*, **17,** 1182 (1949).
[368] A. N. Sevechenko and A. Trofimov, *J. Ept'l Theort. Phys.*, (*USSR*) **21,** 220 (1951); A. Sevechenko and A. Morachevsky, *Isvest. Akad. Nauk.*, *SSSR* (Ser. Fiz.), **15,** 628 (1951).
[369] G. Crosby and M. Kasha, *Spectrochim. Acta*, **10,** 377 (1958).
[370] G. Crosby, R. Whan, and J. Freeman, *J. Phys. Chem.*, **66,** 2493 (1962), and references therein.
[371] G. Crosby, *Molec. Crystals*, **1,** 37 (1966).
[372] F. Varsany and G. Dieke, *J. Chem. Phys.*, **31,** 1066 (1959).
[373] G. Dieke and L. Hall, *J. Chem. Phys.*, **27,** 465 (1957).
[374] J. Kropp and M. Windsor, *J. Chem. Phys.*, **39,** 2769 (1963).
[375] J. Kroop and W. Dawson, *J. Opt. Soc. Amer.*, **55,** 822 (1965).
[376] J. Kroop and M. Windsor, *J. Chem. Phys.*, **42,** 1599 (1965).

[377] R. Borkowski, H. Forest, and D. Grafstein, *J. Chem. Phys.*, **42,** 2974 (1965).
[378] J. Kroop and W. Dawson, *J. Chem. Phys.*, **45,** 2419 (1966).
[379] G. Crosby, R. Whan, and R. Alire, *J. Chem. Phys.*, **34,** 743 (1961).
[380] R. Whan and G. Crosby, *J. Molec. Spec.*, **8,** 315 (1962).
[381] P. Yuster and S. Weissman, *J. Chem. Phys.*, **17,** 1182 (1949).
[382] J. Freeman and G. Crosby, *J. Phys. Chem.*, **67,** 2717 (1963).
[383] M. Kleinerman, personal communication, 1964.
[384] M. Kleinerman, *Bull. Amer. Phys. Soc.*, **9,** 265 (1964).
[385] M. Baumik and M. ElSayed, *J. Chem. Phys.*, **42,** 787 (1965).
[386] M. Baumik, L. Ferder, and M. ElSayed, *J. Chem. Phys.*, **42,** 1483 (1965).
[387] W. Dawson, J. Kropp, and M. Windsor, *J. Chem. Phys.*, **45,** 2410 (1966), and references therein.
[388] E. Matovich and C. Suzuki, *J. Chem. Phys.*, **39,** 1442 (1963).
[389] M. ElSayed and M. Bhaumik, *J. Chem. Phys.*, **39,** 2391 (1963).
[390] A. Heller and E. Wasserman, *J. Chem. Phys.*, **42,** 949 (1965).
[391] F. Halverson, J. Brinen, and J. Leto, *J. Chem. Phys.*, **41,** 157 (1964).
[392] F. Halverson, J. Brinen, and J. Leto, *J. Chem. Phys.*, **41,** 2752 (1964).
[393] D. Scott and R. S. Becker, *J. Chem. Phys.*, **35,** 516 (1961).
[394] J. Smith and B. Meyer, *J. Chem. Phys.*, **48,** 5436 (1968).
[395] J. Guillory, C. Cook, and D. Scott, *J. Amer. Chem. Soc.*, **89,** 6777 (1967).
[396] D. Feuerbacher and R. S. Becker, Univ. of Houston (1967), unpublished.
[397] R. S. Becker and J. Bushman, Univ. of Huston, unpublished.
[398] C. Reid, *J. Chem. Phys.*, **20,** 1212, 1214 (1952).
[399] N. Moodie and C. Reid, *J. Chem. Phys.*, **22,** 252 (1954).
[400] A. Bier and J. Ketelaar, *Rec. Trav. Chim.*, **73,** 264 (1954).
[401] J. Czekalla, G. Briegleb, N. Hure, and H. Vahlensieck, *Z. Elektrochem.*, **63,** 715 (1959), and references therein.
[402] G. Briegleb, *Elektronen-Donator-Acceptor-Komplexe*. Springer-Verlag, Berlin, 1961, pp. 82–94, and references therein.
[403] S. McGlynn and J. Boggus, *J. Amer. Chem. Soc.*, **80,** 5096 (1958).
[404] S. McGlynn, J. Boggus, and E. Elder, *J. Chem. Phys.*, **32,** 357 (1960).
[405] N. Christodoyleas and S. McGlynn, *J. Chem. Phys.*, **40,** 166 (1964).
[406] K. Eisenthal and M. ElSayed, *J. Chem. Phys.*, **42,** 794 (1965).
[407] M. Chowdhury and L. Goodman, *J. Amer. Chem. Soc.*, **86,** 2777 (1964).
[408] K. Eisenthal, *J. Chem. Phys.*, **45,** 1850 (1966).
[409] J. Czekalla and K. Mager, *Z. Elektrochem.*, **66,** (1962).
[410] M. Kasha, *J. Chem. Phys.*, **20,** 71 (1952).
[411] L. Goodman and M. Kasha, *J. Molec. Spec.*, **2,** 58 (1958).
[412] L. Goodman and H. Shull, *J. Chem. Phys.*, **27,** 1388 (1957).
[413] L. Goodman and H. Shull, *J. Chem. Phys.*, **22,** 1138 (1954).
[414] D. Evans, *J. Chem. Soc.*, 3885, 1957.
[415] L. Forster, *J. Chem. Phys.*, **26,** 1761 (1957).
[416] G. Castro and R. Hochstrasser, *J. Chem. Phys.*, **44,** 412 (1966).
[417] W. Rothman, A. Chase, and D. Kearns, *J. Chem. Phys.*, **43,** 1067 (1965).
[418] R. Borkman and D. Kearns, *Chem. Commun.*, **14,** 446 (1966).
[419] S. McGlynn, M. Reynolds, G. Daigre, and N. Christodouleas, *J. Phys. Chem.*, **66,** 2499 (1962).
[420] S. McGlynn, R. Sunseir, and N. Christodouleas, *J. Chem. Phys.*, **37,** 1818 (1962).
[421] D. Evans, *Nature*, **173,** 536 (1956); *J. Chem. Soc.*, 1351 (1957); *Proc. Roy. Soc.*, **255A,** 55 (1960).

[422] H. Tsumbora and R. Mulliken, *J. Amer. Chem. Soc.*, **82,** 5966 (1960).
[423] G. Hoytink, *Molec. Phys.*, **3,** 67 (1960).
[424] A. Grabowska, *Spectrochim. Acta*, **19,** 307 (1963).
[425] R. Dyck and D. McClure, *J. Chem. Phys.*, **36,** 2326 (1962).
[426] E. Viktorova, I. Zhmyreya, V. Kolobkov, and A. Sagenenko, *Optics and Spec.*, **9,** 181 (1960).
[427] I. Graham-Boyce and J. Corkell, *Nature*, **186,** 965 (1960).
[428] S. Seigel and H. Judeekis, *J. Chem. Phys.*, **42,** 3060 (1965).
[429] H. Linschitz and K. Sarkanen, *J. Amer. Chem. Soc.*, **80,** 4826 (1958).
[430] G. Porter and M. Windsor, *Proc. Roy. Soc.*, **245A,** 238 (1958), and references therein.
[431] D. Craig and I. Ross, *J. Chem. Soc.*, 1589 (1954).
[432] H. Linschitz and L. Pakkarinen, *J. Amer. Chem. Soc.*, **82,** 2411 (1960).
[433] R. Livingston and E. Fujimori, *J. Amer. Chem. Soc.*, **80,** 5610 (1958).
[434] S. Claesson, L. Lindquist, and B. Holmstrom, *Nature*, **183,** 661 (1959); E. Abrahamson, R. Adams, and V. Wulff, *J. Phys. Chem.*, **63,** 441 (1959); W. Dawson and E. Abrahamson, ibid, **66,** 2542 (1962).
[435] G. Blauer and H. Linschitz, *J. Phys. Chem.*, **66,** 453 (1962).
[436] C. Steel and H. Linschitz, *J. Phys. Chem.*, **66,** 2577 (1962).
[437] G. Hoytink, *J. Appl. Chem.*, **11,** 393 (1965).
[438] R. de Groot and G. Hoytink, *J. Chem. Phys.*, **46,** 4523 (1967).
[439] M. Orloff, *J. Chem. Phys.*, **47,** 239 (1967).
[440] K. Eisenthal, *J. Chem. Phys.*, **46,** 3268 (1967).
[441] S. Colson and E. Bernstein, *J. Chem. Phys.*, **43,** 2661 (1965), ibid., **45,** 3873 (1966).
[442] G. Nieman and D. Tinti, *J. Chem. Phys.*, **45,** 1432 (1967).
[443] W. Neeley and H. Dearman, *J. Chem. Phys.*, **44,** 1302 (1966).
[444] G. Porter and M. Wright, *Discussions Faraday Soc.*, **27,** 18 (1959).
[445] G. Porter and F. Wilkinson, *Proc. Roy. Soc.*, **264A,** 1 (1961).
[446] E. Fujimori and R. Livingston, *Nature*, **180,** 1036 (1957).
[447] E. McRae and M. Kasha, *J. Chem. Phys.*, **28,** 721 (1958), and references therein.
[448] M. Kasha, *Radiation Research*, **20,** 55 (1963), and references therein.
[449] M. Kasha, *Rev. Mod. Phys.*, **31,** 162 (1959).
[450] M. Kasha, *Physical Processes in Radiation Biology*. Academic Press, New York, 1964, p. 17.
[451] E. McRae, *Aust. J. Chem.*, **14,** 329 (1961); **14,** 344 (1961); **14,** 354 (1961), and references therein.
[452] E. McRae and M. Kasha, *Physical Processes in Radiation Biology*. Academic, New York, 1964, p. 23.
[453] W. Simpson, G. Levenson, and W. Curtuis, *J. Amer. Chem. Soc.*, **79,** 4314 (1957).
[454] W. Simpson and D. Peterson, *J. Chem. Phys.*, **26,** 588 (1957).
[455] M. Kasha, H. Rawls, and M. El-Bayoumi, *Bulletin 30*, Inst. Molec. Biophys., Florida State Univ. (1966).
[456] T. Förster, *Z. Electrochem.*, **54,** 42 (1950).
[457] A. Weller, *Z. Electrochem.*, **56,** 662 (1952).
[458] A. Weller, *Z. Electrochem.*, **60,** 1144 (1956).
[459] A. Weller, *Z. Electrochem.*, **61,** 1956 (1957).
[460] A. Weller, *Z. Physik. Chem.*, **3,** 238 (1955).
[461] A. Weller, *Discussions Faraday Soc.*, **27,** 28 (1959).
[462] G. Jackson and G. Porter, *Proc. Roy. Soc.*, **260A,** 13 (1961).
[463] W. Bartok, P. Lucchesi, and N. Snider, *J. Amer. Chem. Soc.*, **84,** 1842 (1962).

[464] E. Wehry and L. Rogers, *Spectrochim. Acta.*, **21,** 1976 (1965).
[465] W. Bartok, R. Hartmann, and P. Lucchesi, *Photochem. and Photobiol.*, **4,** 499 (1965).
[466] E. Wehry and L. Rogers, *J. Amer. Chem. Soc.*, **88,** 351 (1966); E. Wehry and L. Rogers, *J.Amer. Chem. Soc.*, **87** 4234 (1965).
[467] L. Stryer, *J. Amer. Chem. Soc.*, **88,** 5708 (1966).
[468] R. Hochstrasser and G. Porter, *Quart. Rev.*, **14,** 146 (1960), and references therein.
[469] R. Livingston, *J. Chim. Phys.*, **55,** 887 (1958).
[470] E. Bowen and R. Williams, *Trans. Faraday Soc.*, **35,** 765 (1930).
[471] Cherkassov and Vember, *Optics and Spec.*, **6,** 503 (1959).
[472] A. Terenin and V. Ermolaev, *Trans. Faraday Soc.*, **52,** 1042 (1956).
[473] G. Porter and F. Wilkinson, *Luminescence of Organic and Inorganic Materials*, ed. H. Kallmann and G. Spruch. Wiley, New York, 1962, p. 132.
[474] H. Backstrom and K. Sandros, *Acta Chem. Scand.*, **14,** 48 (1960).
[475] J. Dubois and B. Stevens, *Luminescence of Organic and Inorganic Materials*, ed. H. Kallmann and G. Spruch. Wiley, New York, 1962, p. 115.
[476] H. Ishikawa and W. Noyes, *J. Amer. Chem. Soc.*, **84,** 1502 (1962).
[477] H. Ishikawa and W. Noyes, *J. Chem. Phys.*, **37,** 583 (1962).
[478] J. Dubois and F. Wilkinson, *J. Chem. Phys.*, **38,** 2541 (1963).
[479] S. Lipsky, *J. Chem. Phys.*, **38,** 2786 (1963).
[480] K. Eisenthal and R. Murashige, *J. Chem. Phys.*, **39,** 2108 (1963).
[481] G. Ciaminician and P. Sibber, *Ber. deut. Chem. Ges.*, **33,** 2911 (1900).
[482] G. Porter and F. Wilkinson, *Trans. Faraday Soc.*, **57,** 1686 (1961).
[483] J. Pitts, R. Letsinger, R. Taylor, J. Patterson, G. Recktemvala, and R. Martin, *J. Amer. Chem. Soc.*, **81,** 1068 (1959), and references therein.
[484] B. Stevens and J. Dubois, *J. Chem. Soc.*, **1962,** 2813.
[485] H. Backstrom and K. Sandros, *Acta Chem. Scand.*, **12,** 832 (1958).
[486] G. Hammond, P. Leermakers, and N. Turro, *J. Amer. Chem. Soc.*, **83,** 2395 (1961),
[487] G. Hammond, N. Turro, and P. Leermakers, *J. Phys. Chem.*, **66,** 1144 (1962).
[488] W. Kemula and A. Grabowska, *Nature*, **188,** 224 (1960), and references therein.
[489] E. Lippert and W. Leeder, *J. Phys. Chem.*, **66,** 2430 (1962).
[490] K. Inuzuka and R. Becker, *Nature*, **219,** 383 (1968).
[491] R. Hubbard, *J. Amer. Chem. Soc.*, **78,** 4662 (1956).
[492] L. Jurkowitz, I. Loeb, P. Brown, and G. Wald, *Nature*, **184,** 614 (1959).
[493] R. S. Becker, J. Kolc, R. Bost, H. Kietrich, P. Petrellis, and G. Griffin, *J. Amer. Chem. Soc.*, **90,** 3292 (1968).
[494] W. Gibbons and A. Trozzolo, *J. Amer. Chem. Soc.*, **88,** 172 (1966).
[495] A. Trozzolo, W. Yager, G. Griffen, H. Kristinnson, and I. Sarker, *J. Amer. Chem. Soc.*, **89,** 3357 (1967), and references therein.
[496] E. Fisher, *Fortschritte Chem. Forsch.*, **1,** 605 (1967).
[497] R. Exelby and R. Grinter, *Chem. Rev.*, **65,** 247 (1965).
[498] M. Cohen and S. Flavian, *J. Chem. Soc.*, 334 (1967), and references therein.
[499] M. Ottolenghi and D. McClure, *J. Chem. Phys.*, **46,** 4613 (1967); ibid, 4620 (1967).
[500] R. S. Becker and J. Roy, *J. Phys. Chem.*, **69,** 1435 (1965).
[501] R. S. Becker and J. Michl, *J. Amer. Chem. Soc.*, **88,** 5931 (1966).
[502] R. S. Becker and W. Richey, *J. Amer. Chem. Soc.*, **89,** 1298 (1967).
[503] J. Kolc and R. Becker, *J. Phys. Chem.*, **71,** 4045 (1967).
[504] R. S. Becker and J. Kolc, *J. Phys. Chem.*, **72,** 997 (1968).
[505] W. Richey and R. S. Becker, *J. Chem. Phys.*, **49,** 2092 (1968).
[506] A. Santiago and R. S. Becker, *J. Amer. Chem. Soc.*, **90,** 3654 (1968).

[507] R. S. Becker, E. Dolan, and D. Balke, *J. Chem. Phys.*, Jan. (1969).
[508] N. Tyer, J. Michl, and R. S. Becker, to be published.
[509] J. Michl and R. S. Becker, unpublished.
[510] G. Scheibe and F. Feichtmayr, *J. Phys. Chem.*, **66**, 2449 (1962).
[511] A. Weller, *Progress in Reaction Kinetics*, G. Porter, Ed., Pergamon, New York (1961).
[512] F. Metz, W. Servoss, and F. Welsh, *J. Phys. Chem.*, **66**, 2446 (1966).
[513] G. H. Dorion and K. O. Loeffer, U.S. patent 3,127,335 (1964).
[514] J. Hauser, D. Jerchel, and R. Kuhn, *Ber. deut Chem. Ges.*, **82**, 515 (1949).
[515] W. R. Brode, J. M. Gould, and G. M. Wyman, *J. Amer. Chem. Soc.*, **74**, 4641 (1952).
[516] R. B. Woodward and R. Hoffmann, *J. Amer. Chem. Soc.*, **87**, 2511 (1965), and references therein.
[517] G. Wettermark, *J. Phys. Chem.*, **66**, 2560 (1962).
[518] J. D. Margerum, L. J. Miller, E. Saito, M. S. Brown, H. S. Mosher, and R. Hardwick, *J. Phys. Chem.*, **66**, 2434 (1962).
[519] K. R. Huffman, M. Loy, and E. F. Ullman, *J. Amer. Chem. Soc.*, **87**, 5417 (1965), and references therein.
[520] A. Weller, *Z. Elektrochem.*, **60**, 1144 (1956).
[521] G. Gabor and E. Fischer, *J. Phys. Chem.*, **66**, 2478 (1962).
[522] P. De Mayo and S. T. Reid, *Quart. Rev.*, **15**, 393 (1961).
[523] K. A. Muszkat, D. Gegion, and E. Fischer, *Chem. Comm. No. 19*, 447 (1965).
[524] R. S. Becker and C. E. Earhart, Jr., Univ. of Houston, unpublished.
[525] H. Brockmann and R. Muhlmann, *Ber. deut Chem. Ges.*, **82**, 348 (1949).

Index